NUCLEAR RADIATION INTERACTIONS

NUCLEAR RADIATION INTERACTIONS

Sidney Yip
Massachusetts Institute of Technology, USA

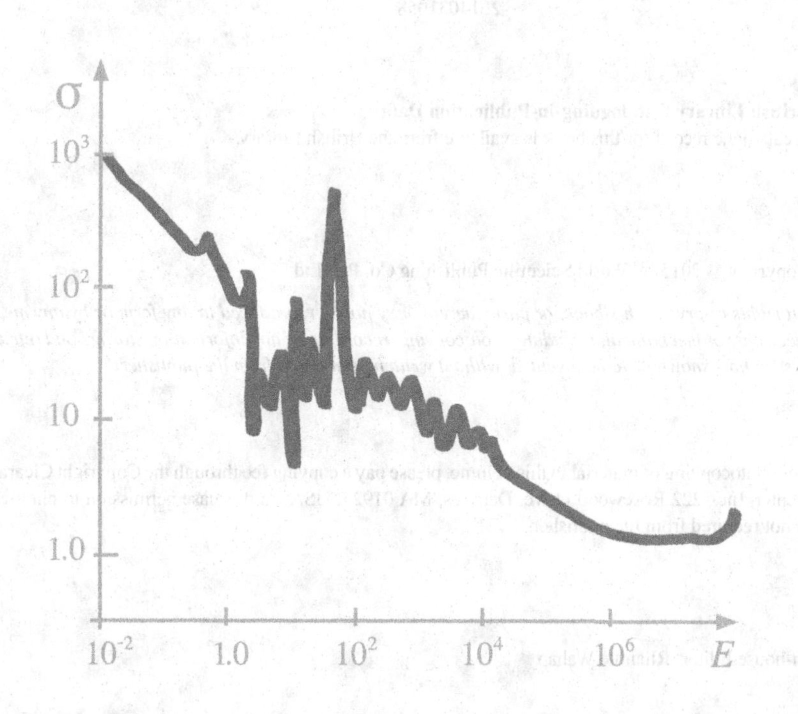

NEW JERSEY · LONDON · SINGAPORE · BEIJING · SHANGHAI · HONG KONG · TAIPEI · CHENNAI

Published by

World Scientific Publishing Co. Pte. Ltd.
5 Toh Tuck Link, Singapore 596224
USA office: 27 Warren Street, Suite 401-402, Hackensack, NJ 07601
UK office: 57 Shelton Street, Covent Garden, London WC2H 9HE

Library of Congress Cataloging-in-Publication Data
Yip, Sidney, author.
Nuclear radiation interactions / Sidney Yip (Massachusetts Institute of Technology, USA).
pages cm
Includes bibliographical references and index.
ISBN 978-9814368070 (alk. paper)
1. Matter--Effect of radiation on. 2. Radioactivity. 3. Nuclear physics. I. Title.
QC173.39.Y56 2014
539.7'2--dc23
2014031058

British Library Cataloguing-in-Publication Data
A catalogue record for this book is available from the British Library.

In-house Editor: Rhaimie Wahap

Typeset by Stallion Press
Email: enquiries@stallionpress.com

Printed in Singapore

To Nita, to the endless journey

Contents

Preface

This book treats the fundamentals of Nuclear Radiation Interactions, a topic considered foundational in the field of Nuclear Science and Engineering. It is based on lecture notes which I started writing in the early 1970s, while teaching a first-year graduate course in the Department of Nuclear Engineering (now Nuclear Science and Engineering) at MIT over a number of years. During this period, my motivation was to describe the basic concepts and models of interactions of nuclear radiation with matter in a compact and concise way. Although these topics were by no means neglected in the nuclear physics literature, a treatment suitable for nuclear engineering students with diverse backgrounds and careers goals was lacking in my opinion. In contrast to the traditional study of nuclear structure and associated properties, I felt the emphasis should be placed on the collisions and reactions that nuclear radiations undergo during interactions with their surroundings. I perceived the students to be aspiring nuclear scientists and engineers working with devices ranging from fission and fusion reactors to particle accelerators and nuclear detection systems. In these applications, the control and utilization of nuclear radiations present a scientific challenge as well as a technological opportunity. To design, operate and optimize nuclear systems for societal benefits, one needs to know how the radiations affect these devices and vice versa. It is through the manipulation of radiation interactions that new devices can be made safer, more powerful, durable, and economical. This notion has become the larger purpose for which the book attempts to provide a foundation. In other words, one should understand single-collision phenomena (unit processes) before going on to analyze and interpret the accumulated effects of many collisions which ultimately govern the system-level behavior.

In turning the former lecture notes into the present book, I have expanded on the previous materials, added a chapter on context, two more on nuclear reactions and neutron transport, and organized the entire work into a coherent and compact presentation. I believe this has allowed me to make clear the fundamental connections between nuclear interactions and neutron transport. I also feel it is worth while to ask the reader to keep in mind that even though the scientific understanding of nuclear radiation interactions is likely to remain timeless, their study in the years ahead should proceed as a contemporary endeavor with concern for balancing societal impact and benefits in an ever evolving world of science and technology.

The omission in this work of two topics, nuclear fusion and radiological sciences, deserves comment. My reason is not they are less important topics in the Nuclear Science and Engineering discipline. It is to include them would not serve my goal of a compact treatment focused on the fundamentals of neutron interactions. I believe readers of this book should be reasonably well prepared to access the relevant topical literature whenever an occasion arises.

I would like to acknowledge the pleasure of working with many former students and teaching assistants, particularly Michael Stawicki and Erik Johnson. I have received support of various kinds from colleagues and staff in the Nuclear Science and Engineering Department, Sow-Hsin Chen, Ben Forrget, Mujid Kazimi, Scott Kemp, Richard Lester, Rachel Morton, Michael Short, and Bilge Yildiz. I thank Bob Hsiung for artistic rendering of the cover figure. Rhaimie Wahap and Low Lerh Feng at World Scientific were very helpful in getting the work into book form.

When I started teaching this subject my father once mentioned I should write a book on it, a casual remark I related to my two brothers years later. The completion of this work, with unwavering support from my wife, is a personal closure for which I feel very fortunate and deeply grateful.

Sidney Yip
July 2014
Cambridge, MA

1

Context and Perspective

Nuclear Engineering is that branch of science and engineering concerned with the various beneficial uses of nuclear radiations. In this discipline, for which we will use the broader designation of *Nuclear Science and Engineering* (NSE), two areas of study are foundational. One deals with the basic understanding of individual interactions of nuclear radiations with the atoms in a materials medium. The other deals with the collective evolution of nuclear radiations in systems that are engineered for specific technological applications. These two areas are intimately connected, and it is the fundamental connection between the two that we feel is noteworthy in the present overview. We will use the term *Nuclear Interactions* to denote a collection of single reaction events, each being an interaction (collision) between a particular type of nuclear radiation (particle) with an atom (nucleus and electrons) in a specified medium, while the term *Radiation Transport* will denote the distribution of nuclear radiations after many sequences of collision-induced reactions and spatial migrations have occurred. With this distinction it should be clear that *Interactions* are the elementary (unit) processes that provide the essential information needed in the study of *Transport* phenomena at the system level.

This book is intended to be primarily a treatment of the fundamentals of *Nuclear Interactions* for students in (NSE). Additionally the connections between *Nuclear Interactions* and *Radiation Transport* are also addressed, albeit to a limited extent. These connections are important for the students' appreciation of the broader context of this work. To see this we briefly consider how the discipline of NSE has evolved during its approximately 60-year-old history (Sec. 1.1). In the first ~30 years, NSE experienced rapid growth as many countries embraced the peaceful uses of nuclear energy. A period of uncertainty then followed as the acceptance of nuclear fission power became a subject of intense public debate. Over the last decade interests in nuclear energy gradually returned until another incident occurred in 2011 (Sec. 1.2). Now the overall picture for nuclear science and technology appears likely to evolve further. Against this background, we give an outlook on the relevance of *Nuclear Interactions* and *Radiation Transport* in a world of increasing technological complexity (Sec. 1.3).

1.1 EVOLUTION OF *NUCLEAR SCIENCE AND ENGINEERING*

The historical development of NSE follows closely the development of nuclear fission power. As the use of this technology rises and falls worldwide, the discipline either thrived or struggled. Following World War II, the peaceful use of nuclear energy quickly gained attention in many countries. A number of nuclear engineering departments were established across the U.S., offering a multidisciplinary curriculum drawing heavily on nuclear physics (theoretical and experimental, particle transport), chemical engineering (fuel processing, isotope separation, chemical thermodynamics), and mechanical engineering (heat transfer and power engineering). *Nuclear Interactions* and *Radiation Transport* (equivalently nuclear reactor physics) emerged as two fundamental subjects in all the nuclear engineering curricula. This is the period of *Early Promise* extending from ~1950 to ~1980. The overlap between NSE and the discipline of nuclear physics was strong because of opportunities for research and applied technology.

With hindsight, it can be said the initial utilization of the emerging nuclear technology apparently did not adequately appreciate the full complexity of the technology, or the critical role of human error in reactor plant operation and control. Thus when an accident such as Three-Mile Island occurred (see Sec. 1.2), public opposition quickly gained momentum. Coupled to this fear of nuclear accident are concerns over safe disposals of the long-term waste. The public debates are factors that made a sophisticated technology more costly to introduce. In this period of *Lean Years*, the survival of nuclear engineering departments in the US became problematic for a number of universities. Correspondingly, there was little incentive for research in *Nucleat Interactions*, and efforts in developing more powerful computational methods in *Radiation Transport* were limited. It was during this period that NSE started to decouple from nuclear physics, as research opportunities moved to higher energies with less relevance to nuclear power. Around the turn of the last century interests in nuclear power started to increase as a result of two developments. One was the significant improvement in plant performance sustained over several years. The other, even more important, was the growing environmental concern of CO_2 emission. Additionally new demands for nuclear power have emerged in countries undergoing strong industrial growth. This period of *Partial Renaissance* continued for about a decade until the Fukushima incident (Sec. 1.2). The incident reminded the world of the vulnerabilities of nuclear power technology to unexpected environmental conditions, in this case a combination of earthquake and tsunami of unprecedented proportions.

1.2 HISTORICAL EVENTS

A survey of the historical events relevant to the study of *Nuclear Interactions* would entail a section that would be more lengthy and detailed than what we feel would be appropriate for this work. The decision was made at the outset to focus on only a few events to illustrate the broad impact that nuclear interactions can have in the evolution of scientific and technological enterprises in our society. All the chapters in the book, except for this and Chap. 2, deals with technical contents. Here we take the opportunity to connect with the history of nuclear science and engineering, in a highly selective (but not arbitrary) manner. In aiming to be brief and coherent, we find it necessary to leave many interesting details to further reading of the references cited.

We propose to draw a connection among just five events, the discovery of a new nuclear radiation (the particle neutron), the discovery of a new nuclear reaction (neutron fission), and a group of three incidents each involving a nuclear power plant accident. This is admittedly an unusual set of events being considered together in a book of this type. In the last section of the chapter we will suggest what implications the present discussion may have in the context of motivation and historical perspective. We hope to show the thematic foundation for this book may be characterized as the linking of science to systems to society in the study of nuclear phenomena.

We begin with the discovery of the neutron as a form of nuclear radiation in 1932 by James Chadwick. This will be followed by a second discovery, that of neutron-induced fission six years later by Otto Hahn and Fritz Strassmann. Both are bona fide major events in the history of science. Chadwick was awarded the Nobel Physics Prize in 1936 and Hahn the Nobel Chemistry Prize in 1944. Historically the period surrounding the fission discovery, when many other significant events also took place, is perhaps the most fascinating era of nuclear science development, and much of the events can be related to nuclear radiation interactions in one way or another [Rhodes, 1986]. We then skipped forward to recent times and the issue of nuclear safety facing our society. By nuclear safety we mean the safety of nuclear power reactor technology [Seghal, 2012] rather than anything pertaining to nuclear weapons technology and proliferation [Garwin and Charpak, 2002; Glasstone and Dolan, 1977; Satori, 1983; Murray, 2009].

Discovery of neutron

Among the nuclear particles which we will study, neutron is the most special and important. Neutron interactions are the fundamental processes underpinning nuclear science and technology. Without neutrons there would be no nuclear power as we know it, and the discipline of nuclear science and engineering very likely would not exist. While there is a great deal to be said about the neutron [Schofield 1982], for setting the context for this book we focus exclusively on the discovery of this nuclear particle. The scientific announcement of the discovery itself was anything but spectacular, certainly not by the current standards of how scientific breakthroughs are presented to the public. This contrast should not go unnoticed by the reader, keeping in mind how the neutron's discovery soon led to the discovery of nuclear fission, and from that event the development of nuclear technology in two major forms, energy production and weapons. In a one-page Letter to the Editor in the journal *Nature* James Chadwick presented a summary of experiments and analysis under the heading, "*Possible Existence of a Neutron*" [Chadwick 1932a]. The Letter was dated February 17, 1931. On May 10, 1932 he communicated to the Royal Society a full discussion of his findings in a paper now entitled "*The Existence of a Neutron*" [Chadwick 1932b]. With hindsight it is rather remarkable that a discovery with such great subsequent impact could be demonstrated in just a single page. A lesson for the students here is that nuclear science at that time was in its infancy, and much of what we now regard as common knowledge in fact was not known. If one follows Chadwick's arguments in deducing the existence of a particle with no charge and yet great ionizing power, one can see that he relied basically on the conservation laws and kinematic relations between energy and momentum. While Chadwick made use of the work of others to arrive at his own conclusions, he was, in effect, "forced" to break with traditional thinking to postulate the existence of the neutron. Both the Letter to the Editor and the follow-up paper are worthwhile reading for insights into early-day research (and incisive deductions) in nuclear radiation interactions, especially after the reader has studied Chap. 8.

Discovery of nuclear fission

Following the neutron discovery experiments using neutrons to bombard uranium soon began at various laboratories in Rome (E. Fermi), Paris (I. Curie and F. Joliot), and Berlin (O. Hahn, L. Meitner, later F. Strassmann) [Segrè, 1989]. The thinking was that the bombardment would lead to the absorption of neutrons, and therefore elements heavier than uranium would be produced. What was not expected was that the neutrons could actually cause the uranium nucleus to be unstable and undergo a "fission" reaction. The events leading up to the discovery of fission, and the process of discovery, now involving more than a single individual, are considerably more complicated compared to the neutron discovery. Moreover, with each fission event producing more than 2 neutrons on the average, a self-sustained chain reaction could be achieved to produce a large amount of energy. Because the fission discovery occurred during the war time in Europe, it soon became apparent that this process can have two applications, peaceful power generation and weapons.

The immediate events prior to the discovery could be summarized as follows. With the expectation of producing transuranic element 93 with chemical properties resembling rhenium (element 75), the Rome group reported finding products (actually fission products like $_{43}Tc$) which were interpreted to be element 93. Similar conclusions were made in Berlin and Paris. On the other hand, the possibility of fission had been pointed out as early as 1934 [Noddack, 1934]; however, no experiments were performed to verify the hypothesis. In 1937–1938 I. Curie and P. Savitch reported finding bombardment product chemically resembling lanthanum, but did not realize or prove that it was indeed $_{57}La^{141}$ [Curie and Savitch, 1938(a)]. In 1938, L. Meitner, an Austrian citizen, fled to Sweden (when Hitler annexed Austria) and continued work with nephew Otto Frisch in Copenhagen. In two papers, dated July 12, 1938 [Curie and Savitch, 1938(b)] and November 8, 1938 [Hahn *et al.*, 1938], the participating scientists still believed that neutron bombardment of uranium would lead to only transuranic products. The dramatic change, the realization that fission was being observed, was announced by Hahn and Strassmann, in a paper dated December 22, 1938 [Hahn and Strassmann, 1939]. In this publication four reactions were observed and reported. They were written out as:

"RaI"? → AcI → "Th"?
"RaII"? → AcII → "Th"?
"RaIII"? → AcIII → "Th"?
"RaIV"? → AcIV → "Th"?

The elements were identified as Ra and Th in quotation mark because the authors, while thinking they should be interpreted as Radium and Thorium, were no longer sure. In fact in the paper they argued to the contrary [Segrè, 1989]:

> "As chemists, in consequence of the experiments just described, we should change the schema given above and introduce the symbols Ba, La, Ce in place of Ra, Ac, Th. As 'nuclear chemists' working very close to the field of physics, we cannot yet bring ourselves to take such a drastic step, which goes against all previous experiences of nuclear physics."
>
> as quoted in Segrè (1989)

It is interesting that in those early days of nuclear research, the chemists seemed to defer to the physicists for the last word on fundamental scientific knowledge. In 1944, Otto Hahn was awarded the Nobel Prize in chemistry for the discovery of nuclear fission. One might ask whether the discovery was worthy of a Physics Prize. Also, to this date, discussions continue on whether Meitner also deserved recognition. She and Frisch came to the same conclusion regarding the possibility of fission, which they published in a paper received just two weeks later than the receipt date of the paper of Hahn and Strassmann [Meitner 1939]. Emilio Segrè, a collaborator of Fermi who himself was awarded the Nobel Prize in Physics (1959) for the discovery of antiproton, later observed [Segrè, 1989]:

> "The discovery of fission has an uncommonly complicated history; many errors beset it. Nature had, however, truly complicated the problem. One had to contend with the radioactivity of natural uranium and the presence of two long-lived isotopes – U^{235} and U^{238}. The heavier isotope, as is well-known, does not undergo fission when bombarded by slow neutrons. The lighter isotope, which makes up 0.7% of natural uranium, is responsible for all the slow-neutron fission. This is a tricky set-up. Above all, it seems to me that the human mind sees only what it expects."

As a follow-up to the above narrative, one can consider two major applications of the nuclear fission reaction (neutron interaction with uranium), nuclear weapons and power generation. On the former we make a few observations below and then refer the reader to the literature. On the latter we go a bit further in the next section.

In 1930's many scientists, including Rutherford, Milliken, Einstein, did not believe that one can get more energy out of nuclear reactions than what has to be put in. After fission was discovered at the end of 1938, Niels Bohr electrified the American scientific community when he reported Frisch's experiments in Copenhagen at the annual American Physical Society (March 1939). Within days, nuclear fission was widely accepted. Two questions were raised immediately, *Which fissioned,* U^{238} *or* U^{235}? *How many neutrons are emitted* (*self-sustaining chain reaction*)? The potentials for military applications were quickly recognized in several countries, the US, England, France, Germany, Soviet Union, Japan. Many European scientists, predominantly from Germany and Italy, but also other countries like Hungary and Austria, emigrated to the US. At the end of 1939 nuclear bomb research began in the US under the Manhattan Project. This led to the use of two nuclear weapons on Hiroshima (6 August 1945) and Negasaki (9 August 1945). The debate over nuclear proliferation remains a current challenge before the human society. For an absorbing and authoritative account of the historical events and the involvement of many scientists prominent in the field of nuclear physics and nuclear reactions, further reading of *Making of the Atomic Bomb* [Rhodes, 1986] is recommended.

> "In an enterprise such as the building of the atomic bomb the difference between ideas, hopes, suggestions and theoretical calculations, and solid numbers based on measurements, is paramount. All the committees, the politicking and the plans would have come to naught if a few unpredictable nuclear cross sections had been different from what they are by a factor of two."
>
> — Emilio Segre, as quoted in R. Rhodes (1986).

Nuclear plant accidents

The separation between neutron discovery and the discovery of its most unique interaction, fission, is a relatively short six years. For the next set of historical events we will discuss in the present context, nuclear power reactor incidents, the separation is 40 years, 1939 to 1979.

The demonstration of a critical pile, an assembly of uranium pellets and graphite blocks capable of self-sustained fission reactions, by a group of scientists led by E. Fermi at the University of Chicago in December 1942 marked the beginning of the nuclear power era [Glasstone and Polan, 1967]. The development of civilian light water reactors commenced with the construction of a pressurized water reactor (PWR) at Shippingport, PA in 1957 and the first commercial nuclear power plant Yankee-Rowe in Massachusetts in 1960. The notion of cheap nuclear energy was very popular during the 1960s, as a result a significant number of nuclear plants were put into service in the following decade. The number of plants currently operating in the U.S. stands at about 100, and the number of plants worldwide is about 500. A recent account of the nuclear power development from the standpoint of *first milestones* is a readily accessible way to follow the historical evolution of this aspect of nuclear technology [Marcus, 2010]. For a discussion of the safety concerns of these devices, see an even more recent assessment by the European Commission, with emphasis on the understanding and knowledge of of the complex physics of severe accidents, which involves principally the disciplines of probability theory, neutron physics, thermal hydraulics, high-temperature materials science, chemistry, and structural mechanics [Sehgal, 2012].

Three incidents in the development of nuclear power generation particularly stand out as part of the history of nuclear energy utilization worldwide. Like the scientific discoveries just discussed, they were totally unexpected. But unlike the discoveries which all can agree are beneficial events, these incidents were accidents which demonstrated the continuing need to safeguard the nuclear systems in use today and in the future. It is not our intention to go into any technical details of the safety of nuclear reactors [Seghal, 2012], rather we regard these incidents as significant historical events with implications for motivating the study of *Nuclear Interactions* (see Sec. 1.3).

TMI-2 (Three Mile Island – 2) is a nuclear power reactor in Harrisburg, PA which suffered on March 28, 1979 an accident in which the reactor core melted partially. This incident was caused by a combination of equipment (valve) failure and faulty operator actions. While there were no injuries or property damage and little radioactivity was released from the reactor containment, the event was a major psychological set-back in the public acceptance of nuclear power. The accident was initiated by a loss of feed-water to the steam generator (boiler) causing a dry-out condition. As the pressure in the reactor vessel increased, the reactor was shut down. A personnel-operated valve was then opened, but later when the pressure decreased, the valve did not close (malfunction). The emergency core cooling system was activated, injecting water into the vessel. Apparently a steam bubble formed and blocked further water injection, and the temperature reached a point where exothermic reaction between the steam and the Zircaloy cladding could take place. This chain of events led to the core melt at a speed and severity greater than expected.

On April 26, 1986 an accident occurred in one of the reactors in the nuclear power plant complex in Chernobyl, Ukraine, Soviet Union, which resulted in core melting, explosion and radioactivity release. The incident was set off during a special test conducted to determine the capacity of the turbine under essentially zero load. In contrast to TMI, where core melting was attributed to the lack of heat removal, the Chernobyl event took place on a much shorter time scale, the signature of a runaway chain reaction. It was estimated that sufficient reactivity was inserted into the core to cause the reactor power to rise to 100 times full power level in 4 seconds [Seghal, 2012]. The number of immediate deaths, 31, was considered small for such a disastrous event, while the long-term effect of the radioactivity exposure remains difficult to estimate.

On March 11, 2011 when the Fukushima nuclear power plant, situated on the coast 240 km Northeast of Tokyo, suffered a major earthquake (Richter scale, 9.0), followed an hour later by a huge tsunami of 14 m. The four reactors at the plant shut down properly when the earthquake struck. However, the emergency power (diesel engines) located at an elevation of 10 m was flooded by the tsunami, thus preventing the removal of decay heat from the scrammed reactors. The loss of cooling due to station blackout played a central role in the subsequent accident recovery and risk management. During this rather long period of time the reactor core became partially uncovered, allowing the clad temperature to reach a level sufficient to set off the exothermic reaction as in the TMI-2 scenario. The hydrogen produced leaked into the reactor building, eventually causing an explosion. Another aspect of the Fukushima incident, not encountered at TMI or Chenobyl, was the hazard presented by the spent-fuel pool, which had a significant inventory of radioactivity and required cooling to remove the decay heat.

1.3 RELEVANCE OF *NUCLEAR RADIATION INTERACTIONS*

We have collected five historical events to motivate what could be regarded as the "context" for the study of nuclear radiation interactions. The first two events are scientific discoveries; it is easy to understand why they would belong to being part of a "historical perspective". The other three events, while also unexpected, are technological in nature in that they illustrate the complexity of the nuclear systems being used by society. What is the connection? How can one relate "science" to "systems" to "society"? We point to a recent decision of the Department of Nuclear Science and Engineering at MIT to adopt the departmental educational triad of *science:systems:society.* [MIT Department of Nuclear Science and Engienering, "Engineering" 2010] We take this to be the motivation (context) for the study of *Nuclear Interactions* and its connection to *Radiation Transport*. More specifically, we see nuclear radiation interactions as part of the foundation of the discipline of nuclear science and engineering. The *Nuclear Interactions* concepts are fundamental in that they relate radioactivity and nuclear cross sections to nuclear enterprises. The *Nuclear Interactions-NSE* connection can be developed more explicitly by considering the study of nuclear interactions at the unit process level (cross sections) as the input to the study of radiation transport, which is concerned with the space-time distribution of radiations in nuclear systems for the benefit of society. If neutron discovery can be viewed as the beginning of nuclear fission technology, then one has a 'brief' 80-year history leading up to the present state of nuclear technology used in society. This is a rapid pace of technology development and utilization in the peaceful uses of nuclear energy.

There are implications here for the reader (student) even with such a brief look. Among them is the observation that radioactivity, here synonymous with nuclear radiation and reactions, is at the heart of the matter. The three incidents just described have a "common cause" — heat and stress distributions due to nuclear reactions in the reactor core drive the temperature and pressure levels in the core and the reactor vessel well beyond the limits of normal operation, as well as a "common consequence" — the release of radioactivity to the environment. It is an inescapable fact that radioactivity (the nuclear radiation and its various reactions) is necessary for the production of nuclear power (call this condition *a*). It is equally certain that radioactivity (the fission products and various decay radiation) must be kept away from the public (condition *b*). The fundamental challenge in the *NSE* discipline is the quest to satisfy *a* as efficiently as possible, while ensuring *b* is not violated. There is little doubt as history evolves society will learn how to manage this challenge better and better, including the valuable lessons provided by the historical events along the way.

The underlying principles upon which the *NSE* discipline is established must keep pace with the continuing learning process toward meeting the challenge of satisfying simultaneously conditions *a* and *b*. The basic concepts of *Nuclear Interactions*, the primary subject of this book, is part of these principles, foundational in serving as general (and usually simplified) starting points which then can be extended to more complicated (usually more realistic) situations.

In seeking to understand the implications of the increasingly more complex incidents, from TMI to Chernobyl to Fukushima, the overall goal for the future nuclear plants should be to optimize the reliability (safety) of these complex systems. The incidents were occasions when the distribution of radiation in space and time exceeded the design limits and subsequent response was less than effective from a preparedness standpoint. Although rare, these occasions did occur, and reactor incidents could happen again. In my opinion the only way to make this struggle between mankind (technological systems of our society) and nature more of a level playing field is to strive for more powerful science and technology, while maintaining a positive vision of the fruits of our knowledge and technological ingenuity [Domenici, 2004]. This requires not only safer and more efficient systems, but also an enlightened society capable of balancing risks and benefits in a realistic, global manner [Johnston, 2012].

REFERENCES

J. Chadwick, Possible existence of a neutron, *Nature* **129**, 312 (1932) [Chadwick, 1932a].

J. Chadwick, The existence of a neutron, *Proc. Roy. Soc.* **136A**, 692 (1932) [Chadwick, 1932b].

I. Curie and P. Savitch, *C. R. Acad. Sci. (Paris)* **206**, 906 and 1643 (1938).

I. Curie and P. Savitch, *J. Phys. Radium* **9**, 355 (1938).

P. Domenici, *A Brighter Tomorrow* (Rowman and Littlefield, Langham, 2004).

R. D. Evans, *The Atomic Nucleus* (McGraw Hill, New York, 1955).

E. Fermi, *Nuclear Physics*, Lecture Notes by J. Orear, A. H. Rosenfeld and R. A. Schluter (University of Chicago Press, 1949).

R. L. Garwin and G. Charpak, *Megawatts and Megatons: The Future of Nuclear Power and Nuclear Weapons* (University of Chicago Press, 2002).

S. Glasstone and P. J. Dolan, *The Effects of Nuclear Weapons,* third edition (US Government Printing Office: Washington, DC, 1977).

O. Hahn, L. Meitner, and F. Strassmann, *Naturwissenschaften* **26**, 475 (1938).

O. Hahn and F. Strassmann, *Naturwissenschaften* **27**, 11 (1939).

S. F. Johnston, *The Neutron's Children: Nuclear Engineers and the Shaping of Identity* (Oxford Univ. Press, 2012).

K. S. Krane, *Introductory Nuclear Physics* (Wiley, New York, 1987).

G. H. Marcus, *Nuclear Firsts: Milestones in the Road to Nuclear Power Development* (American Nuclear Society, La Grange Park, 2010).

MIT Department of Nuclear Science and Engineering, "*Science. Systems. Society. A New Strategy for the Department of Nuclear Science and Engineering*" (2010) http://web.mit.edu/nse/pdf/spotlights/2011/NSE_StrategicPlan_overview.pdf

L. Meitner and O. R. Frisch, Disintegration of uranium by neutrons: a new type of nuclear reaction, *Nature* **143**, 239 (1939).

R. L. Murray, *Nuclear Energy* (Butterworth-Heinemann, Amsterdam, 2009), Chap. 26.

I. Noddack, *Angew. Chem.* **47**, 653 (1934).

R. Rhodes, *The Making of the Atomic Bomb* (Simon & Schuster, New York, 1986).

L. Sartori, Effects of nuclear weapons, *Physics Today,* March 1983.

E. Segrè, *Nuclei and Particles* (W. A. Benjamin, 1965).

E. Segrè, The discovery of nuclear fission, *Physics Today* **42**, 38 (1989).

B. R. Sehgal, ed. *Nuclear Safety in Light Water Reactors* (Elsevier, Amsterdm, 2012).

P. Schofield, *The neutron and its applications, 1982: plenary and invited papers from the conference to mark the 50th anniversary of the discovery of the neutron held at Cambridge, 13–17 September 1982* (Institute of Physics, Bristol, 1983), Institute of Physics conference series, no. 64.

2

Organization

Nuclear Interactions is a very broad topic. Although there exist many forms of radiations, types of interactions, and different materials environments, fortunately, not all the combinations are equally relevant for our purposes. At the fundamental level this subject lies at the core of nuclear physics, itself a broad discipline. What makes our studies not so open-ended is that we also have in mind the practitocal use of *Nuclear Interactions* in the context of *Radiation Transport*. From the standpoint of nuclear science and technology, it is sufficient to focus on only three types of radiations, neutrons, gammas, and charged particles, each in a certain range of energies, as well a set of interactions that are most relevant to the discipline of Nuclear Science and Engineering. It is in this particular way of thinking that we strive to provide a treatment of the underlying principles of *Nuclear Interactions*.

This book is an attempt to discuss the fundamentals of *Nuclear Interactions* in a compact and concise manner. Aside from two opening chapters on context and organization respectively, the contents are grouped into three parts. Part (I) deals with traditional nuclear physics topics, mass and stability of nuclei, energy levels, radioactivity and nuclear decays; these basic concepts can be found in any text on nuclear physics. Part (II) deals with two-body collisions and nuclear reactions, and the concept of cross sections as studied through the interactions of specific nuclear radiation with matter. While these treatments are also available from existing text, they are generally not given the same emphasis as the topics in Part (I). In terms of the fundamental understanding and use of *Nuclear Interactions* it is the second part that is the most relevant. Through discussions of various types of cross sections, we not only demonstrate the scope and substance of *Nuclear Interactions*, but also provide a direct connection between the single-event collisions and reactions studied here with an introduction to *Radiation Transport*, the study of cumulative effects of radiation interactions in matter, as discussed in Part (III).

2.1 CHAPTER LAYOUT

The Table of Contents is a convenient summary of the contents of this book. To visualize how the chapters relate to each other in describing the fundamental concepts one may refer to the flow chart below.

The layout of the chapters is intended to convey a progression from two opening chapters, one on motivations and historical developments (Chap. 1), and the other on organization, bibliographic notes, and logistical details (Chap. 2). Technical discussions begin with basic nuclear properties and sources of nuclear data (Chap. 3), followed by considerations of binding energy (mass defect) leading to an empirical understanding of stability of nuclei (Chap. 4), and elementary quantum mechanical models describing nuclear energy levels (Chap. 5). The last topic treated in this first part of the book is the study of spontaneous nuclear disintegration and

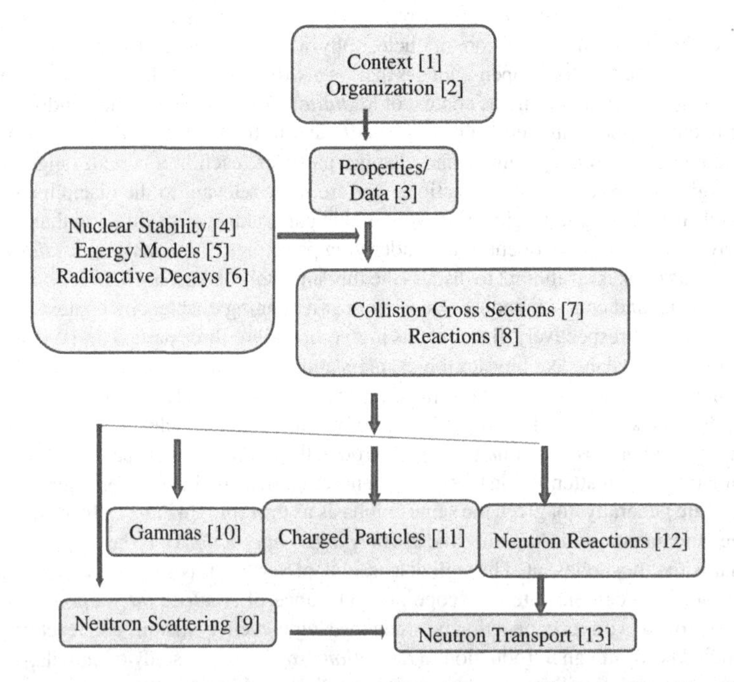

Fig. 2.1 Flow chart of the 13 Chapters in this book, with the chapter number in brackets. Flow of chapters collectively defining the subject matter of *Nuclear Radiation Interactions* follows the downward arrows. The group of Chaps. 3 through 6 makes up part (I) which is concerned with the nuclear physics fundamentals. Part (II) continues with cross section concepts for collisions, Chap. 7, and reactions, Chap. 8, followed by interactions of specific radiations with matter, chapters 9 through 12. Part (III) consists of Chap. 13 which treats the connections to neutron transport as the natural continuation of neutron scattering and neutron reactions.

decays (Chap. 6). This is the portion concerning the nuclear physics fundamentals discussed in all standard text. The flow then continues on to the second (main) part of the book pertaining to what we have referred to as the fundamentals of NRI (see Preface and Chap. 1).

We begin with the concept of collision cross sections and potential scattering as treated in elementary quantum mechanics (Chap. 7), and continue with nuclear reactions in terms of kinematics, cross section systematics, and basic elements of theoretical formulations (Chap. 8). Together these two chapters provide the background for the specific discussions to follow, in the form of individual chapters treating neutron scattering (Chap. 9), gamma attenuation (Chap. 10), charged particle penetration (Chap. 11), and neutron resonant reactions (Chap. 12). As mentioned in the Preface and discussed in Chap. 1, NRI is important in the discipline of Nuclear Science and Engineering for the physical understanding of the various cross sections needed for neutron or other radiation transport calculations in practical applications. The link between NRI and neutron transport is discussed in the final chapter (Chap. 13) to provide an appreciation of this essential connection. Whether this last chapter belongs to the same curriculum as the first twelve chapters or should be regarded as supplemental material is left to the discretion of the instructor and the class, or the interest of the reader.

2.2 BIBLIOGRAPHIC NOTES

Throughout the book the relation between each of the chapters and the external literature should be noted by the reader. As we strive for readability and brevity in our treatment, we also rely on the external resources for further discussions,

Author	Year	Title	Influence on the present book
Fermi	1949	*Nuclear Physics*	Overall emphasis, original source
Evans	1955	*The Atomic Nucleus*	Selected data, detailed discussions
Segrè	1965	*Nuclei and Particles*	Mainstream nuclear physics text
Meyerhof	1967	*Elements of Nuclear Physics*	Content selection, physical insight
Krane	1987	*Introductory Nuclear Physics*	Relatively recent text

including additional topics recommended to be studied. We have chosen five references from the existing literature as the peer group for comparison of topic selection and emphasis in two areas, nuclear physics fundamentals and nuclear radiation interactions.

Each member of the group is a well-known reference, and all have significant overlap with the topics discussed in the book. We briefly comment on the particular characteristics of each one along with indicating the specific overlap with our own treatments. The reader is encouraged to become familiar with this group of references through supplemental reading.

The lecture notes of E. Fermi were written up by three students in a course that he gave at the University of Chicago. The topics covered may be taken to reflect the personal views of the author, arguably the world's leading figure on the physics of neutrons and multiplying pile (see Chap. 1) at the time. (Fermi received the Nobel Physics Prize for Slow Neutron Physics in 1936.) The Lecture Notes contained 10 chapters in 242 pages under the headings of properties of nuclei, interaction of radiation with matter, alpha emission, beta decay, gamma decay, nuclear forces, mesons, nuclear reactions, neutrons, and cosmic rays. Coming only 10 years after the discovery of fission and just a few years after the realization of the first multiplying pile, the contents of the back to back chapters on nuclear reactions and neutrons are particularly revealing. Relative to the chapters in the present book, there is clear correspondence throughout. This reference is recommended for frequent consultation.

The text of Evans is an encyclopedic treatise on the physics of nuclei. It is an outgrowth of materials used over a number of years in a two-semester course offered in the physics department to graduate students at the Massachusetts Institute of Technology. Of the 26 chapters (960 pages) there are 8 chapters on radiation interactions, particularly charged particles and electromagnetic radiations. There is essentially no coverage of neutrons, in sharp contrast with our emphasis in the present book. Nonetheless, this is a valuable, supplemental reference.

The book of Segrè is an excellent example of a comprehensive nuclear physics text. It has broad coverage of topics of current interest at the time, consisting of three parts, tools, the nucleus, and subnuclear particles. In the first part are four chapters, passage of radiation through matter (13 sections), detection methods for nuclear reactions (13 sections), particle accelerators (9 sections), and radioactive decay (9 sections). In the second (the largest) part are seven chapters dealing with nuclear structure and systematics (12 sections), alpha emission (4 sections), gamma emission (8 sections), beta decay (11 sections), two-body systems and nuclear forces (9 sections), nuclear reactions (15 sections), and neutrons (18 sections). In the last part are three chapters on muons (5 sections), pions (8 sections), and strange particles (7 sections). The chapters are supplemented by nine appendices, mostly on aspects of the theoretical analysis of collision phenomena.

In contrast to the Segrè text, the book of Meyerhof, written at about the same time, is rather compact (260 pages compared to 723) and admirably concise. There are six chapters, basic nuclear concepts, nuclear structure, interactions of nuclear radiation with matter, radioactive decay, and nuclear forces. Even though the work is brief by comparison with the other text, it is outstanding in clarity and insight. For this reason we have drawn heavily from this source in several chapters in the present volume.

The book of Krane, the most recent of the group, is a significantly expanded version of a classic undergraduate text on nuclear physics [Halliday, 1955]. Its emphasis is on the breadth of the field of nuclear physics, with focus on experimental and phenomenological studies. The chapter outline is a comprehensive list of the fundamentals of nuclear physics, with 20 Chapters covering basic concepts, elements of quantum mechanics, nuclear properties, the force between nucleons, nuclear models, radioactive decay, nucler radiation detection, alpha decay, beta decay, gamma decay, nuclear reactions, neutron physics, nuclear fission, nuclear fusion, accelerators, nuclear spin and moments, meson physics, particle physics, nuclear astrophysics, and nuclear physics applications.

Comparatively speaking, the basic contents of all five text are similar — equilibrium properties and nuclear decays treated first before nuclear reactions and more specific applications. The following observations may be noted concerning the relevance of these references to the present book:

- The essential topics pertaining to either the first part, nuclear properties and structure, or the second part, nuclear reactions and interactions involving neutrons, gammas, and charged particles, have not changed over the years. These contents therefore may be regarded as fundamentals in the study of nuclear science and engineering.
- There is general agreement among the authors in the treatment of the main topics in each area of nuclear physics — properties, structure, decays, radioactivity, interactions. The same cannot be said about nuclear radiation interactions. For example, Fermi and Segrè have given much more emphasis to neutron scattering and diffusion, while these aspects were not considered by Evans, or given less attention by Meyerhof and Krane. This distinction should be appreciated by the present readers in using these references.
- Charged particle and electromagnetic (gamma) interactions were considered by all the text but not at the same level of emphasis. The special relevance of

these processes and phenomena to the field of *Nuclear Science and Engineering* should be appreciated by the reader.

- The natural relation between *Nuclear Interactions* and *Radiation Transport* as two fundamental parts of *NSE* appears not to be recognized by the literature just examined (see Preface and Chap. 1).

2.3 HOUSEKEEPING

This is a collection of notational details to augment the Table of Contents. We describe how the different parts of each chapter are organized and labeled for quick reference by a person who is not yet a familiar reader.

Equations, Figures, and Tables

Equations, figures, and tables are numbered consecutively in the order of their appearance in each chapter, with a prefix indicating the chapter. Thus Eq. (3.4), Fig. 3.4 and Table 3.4 denote the fourth equation, figure, and table respectively in Chap. 3. A few incidental equations, figures, and tables are not numbered if they are used only for a single occasion.

References

References cited in each chapter are listed at the end of that chapter with author and year enclosed by square brackets [Chadwick, 1932]. All cited references are also collected in the master bibliography at the end of the book. Titles of some (but not all) articles are given when they are of particular significance.

Homework and Quiz Problems

Problems suitable for homework exercise are listed at the end of the book grouped by chapters. They have no restrictions concerning the reference materials that can be used, provided the references are properly cited in the solutions to the problem. In addition, problems suitable for a quiz are also provided. These are closed-book questions in that no reference material is permitted in working out the solutions. We further distinguish two types of quiz questions, those dealing with the material discussed in the current chapter and those dealing with materials covered in several chapters. The two types of quiz problems will be designated by a single or double asterisk respectively. Problems can serve as reviews or summaries of the materials covered in the chapters. They also can serve as templates for the instructor to make up new problems with their own emphasis. Thus the collection of problems becomes part of the teaching materials as opposed to materials solely for exercise and testing.

REFERENCES

R. D. Evans, The *Atomic Nucleus* (McGraw Hill, New York, 1955).
E. Fermi, *Nuclear Physics*, Lecture Notes by J. Orear, A. H. Rosenfeld and R. A. Schluter (University of Chicago Press, 1949).
K. S. Krane, *Introductory Nuclear Physics* (Wiley, New York, 1987).
D. Halliday, *Introductory Nuclear Physics* (Wiley, New York, 1955).
W. E. Meyerhof, *Elements of Nuclear Physics* (McGraw Hill, New York, 1967).
E. Segrè, *Nuclei and Particles* (W. A. Benjamin, 1965).

PART 1

Nuclear Physics Background

3

Nuclear Properties and Data

We anticipate the readers of this book have very diverse backgrounds in the study of elementary nuclear physics, some will have no familiarity with nuclear physics while others have strong undergraduate preparation in physics or nuclear engineering. For a person in the former group it is possible to make up for the lack of previous experience by putting in extra effort in self-study, particularly in going through the nuclear physics topics in Part (I), beginning with this chapter. Here we discuss the basic properties of nuclear radiations that would provide orientation for the study of the chapters to follow. Because this information is available in all nuclear physics text, the readers are encouraged to get in the habit of consulting regularly one or more of the classic books discussed in Sec. 2.2.

There are several ways one can be introduced to the physical properties of nuclides. By regarding the nucleus as a composite particle made up of two constituents, protons and neutrons, one can think about what are the implications of its mass, size, charge, spin, electric and magnetic moments, etc. These are the basic properties discussed in Sec. 3.1. Another way is to take the existing nuclear data, such as the experimental results tabulated in Sec. 3.2, and think about the overall trends that these results display, or whether some of the properties can be fundamentally related to each other. Also one can use the data to validate theoretical predictions, or as input to numerical calculations. Still another way is to focus on the characteristic behavior of various neutron interaction cross sections, briefly previewed in Sec. 3.3. It will be seen in the later chapters that cross sections are the essential quantities in the study of *Nuclear Interactions* and *Radiation Transport*, albeit in different ways. Therefore, the earlier the reader can appreciate their complexity and significance, the better.

3.1 NUCLEI, RADIATIONS, AND INTERACTIONS

While all aspects of nuclei, radiations, and interactions can be said to belong to the field of nuclear physics, the intent of this book is to address a very specific and limited portion of this vast field of study. We will be interested only in those interactions involving a particular group of radiations (particles), namely those that are relevant to the discipline of nuclear science and engineering. Right at the outset of our discussion of basic nuclear properties, it will be appropriate to specify what are these radiations (particles) and their energy ranges of interest.

Throughout the book, we will deal primarily with three kinds of nuclear radiations (each alternatively can be considered as particles) — neutron, electron, and gamma radiations. Among these three, neutrons are of the greatest interest as far as the type of nuclear radiation interactions is concerned. The reason is that many of the most important nuclear systems in *Nuclear Science and* Engineering cannot perform their functions without neutrons. This means neutrons are important for both *Radiation Transport and Nuclear Interactions*. The neutron energies relevant to our discussions span a very wide range, from 10^{-3} eV to 10^6 eV. In contrast, electrons and gammas (high-energy photons) are of interest because they are emitted in nuclear process such as radioactive decay or in nuclear reactions. The relevant energies of these radiations are of the order of 10^6 eV. By limiting our attention only to these radiations our intention to give more in-depth discussions of their interactions with matter. As can be seen from the Table of Contents and mentioned in Chap. 2, we treat neutrons, gammas, and charged particles in separate chapters. Chapters 9, 12, and 13 are devoted to neutrons, and Chapters 10 and 11 to gammas and charged particles respectively.

Terminology and basic nuclear properties

To the lay person, the nucleus is the positively charged core of an atom, surrounded by a number of electrons moving in their orbits. For a neutral atom, the number of atomic electrons is equal to the charge of the nucleus. If the atom is ionized, the number of atomic electrons is less than the charge of the nucleus by the degree of ionization, one electron for singly ionized, two for doubly ionized, etc. Unless specified otherwise we consider the atoms in any material medium to be neutral.

The atomic nucleus itself is an assembly of nucleons, which are either neutrons or protons. The properties of a free-standing neutron or proton are not identical to a neutron or proton bound in the nucleus. The bound neutrons and protons that exist in the nucleus are called *nucleons*. The number of protons and neutrons are called respectively the *atomic number*, denoted as Z, and the *neutron number*, denoted as N. The sum of Z and N is called the *mass number*, denoted as A; it is the total number of nucleons. To specify a nucleus one has to give only two values to the three numbers, Z, A, and N, with the third given by $A = Z + N$. *Nuclides*, nuclei with particular specifications, are denoted by the symbol, ${}_{Z}X^{A}$. Here X is the symbol of the nucleus which is an atomic element. Known atomic elements are listed, for example, in the Periodic Table or the Chart of the Nuclides (see Table 3.4). They are denoted by the symbol X, which is either a single letter as in U for uranium, or two letters, as in Cu or Pu for copper or plutonium respectively. There is a one-to-one correspondence between Z and X, so specifying both is actually redundant (but could be helpful since not everyone remembers the entire list of atomic elements). For those nuclei which are encountered frequently in the book, Z is often not specified. For example, one would write just U^{235} since uranium is well known to have $Z=92$. With the number of neutrons being always $A-Z$, N is also often suppressed. When N is being shown, it appears in the lower right hand corner, as in ${}_{92}U^{235}_{143}$. There exist several uranium nuclides with different mass numbers, such as U^{233}, U^{235}, and U^{238}. Nuclides with the same Z but different A are called *isotopes*. By the same token, nuclides with the same A but different Z are called *isobars*, and nuclides with same N but different Z are called *isotones*. *Isomers* are nuclides with the same Z and A in different excited states. For a compilation of the nuclides that are known, see the *Table of Nuclides* (KAERI) for which a website is given in the references.

We are, in principle, interested in all the elements up to $Z=94$ (plutonium). There are about 20 more elements which are known, most with very short lifetimes; these are of interest mostly to nuclear physicists and chemists, but not so much to nuclear engineers. While each element can have several isotopes of significant abundance, not all the elements are of equal interest to us in this book. The number of nuclides we might encounter in our studies is probably no more than about 20.

A great deal is known about the properties of nuclides. In addition to the *Table of Nuclides*, the student can consult the *Table of Isotopes*, cited in the References (see also Sec. 3.2). It should be appreciated that the great interest in nuclear structure and reactions is not just for scientific knowledge alone; the fact that there are two applications that affects the welfare of our society — nuclear power and nuclear weapons — has everything to do with it. We begin this chapter with a review of the most basic physical attributes of nuclides, to provide motivation and a basis to introduce what we want to discuss throughout the book.

Nuclear mass

We adopt the unified scale where the mass of C^{12} is exactly 12. On this scale, one mass unit 1 mu ($C^{12}=12$) $=M(C^{12})/12=1.660420\times10^{-24}$ gm ($=931.478$ Mev), where $M(C^{12})$ is actual mass of the nuclide C^{12}. Studies of the atomic masses by mass spectrograph shows that a nuclide has a mass nearly equal to the mass number A times the proton mass. Three important rest mass values, in mass and energy units, to keep handy are given in Table 3.1 (see also Table 3.3).

The reason why we care about the mass is that it is *an indication of the stability* of the nuclide. One sees this directly from $E=Mc^2$, the higher the mass the higher

Table 3.1 Rest masses of electron, proton, and neutron.

	Mass Unit [M(C^{12})=12]	**MeV**
Electron	0.000548597	0.511006
Proton	1.0072766	938.256
Neutron	1.0086654	939.550

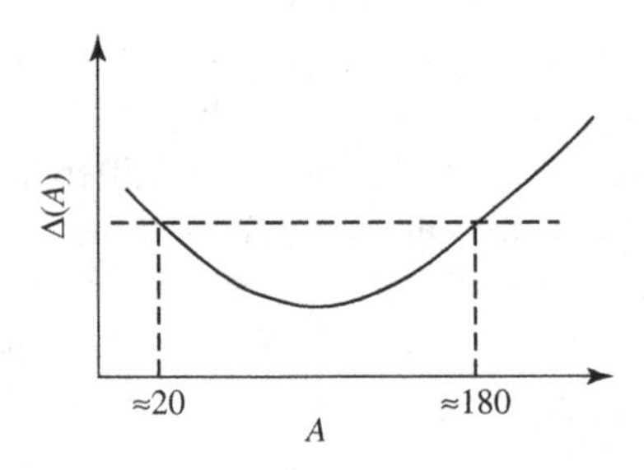

Fig. 3.1 Schematic variation of mass decrement $\Delta(A)$ showing that nuclides with mass numbers in the range ~(20–180) should be stable.

the energy and the less stable is the nuclide (think of the nuclide in an excited state). We will find that if a nuclide can lower its energy by undergoing disintegration, it will do so — this is the simple explanation of radioactivity. Notice the proton is lighter than the neutron, implying the former is more stable than the latter. Indeed, if the neutron is not bound in a nucleus (a free neutron) it will decay into a proton plus an electron (and additionally an antineutrino, for explanation see Chap. 6, beta decay) with a half-life of about 13 min.

Nuclear masses have been determined to quite high accuracy, precision of ~ 1 part in 10^8, by the methods of mass spectrography and energy measurements in nuclear reactions. Just using the mass data alone we can get an idea of the stability of nuclides. Let the mass defect be defined as the difference between the actual mass of a nuclide and its mass number, $\Delta = M - A$; it is also called the "mass decrement". If we plot Δ versus A, we get a curve sketched in Fig. 3.1. When $\Delta < 0$, it means that taking the individual nucleons when they are separated far from each other to make the nucleus gives a product whose mass is lighter than the sum of the components. This can only happen if energy is given off during the formation. In other words, to reach a final state (the product nuclide) with smaller mass than the initial state (collection of individual nucleons) one must take away some energy (mass). It also follows that the final state will be more stable than the initial state, since energy must be put back in if one wants to reverse the process to go from the nuclide to the individual nucleons. We therefore expect that $\Delta < 0$

means the nuclide is stable. Conversely, $\Delta > 0$ means the nuclide is unstable. Our sketch shows that very light elements ($A < 20$) and heavy elements ($A > 180$) are not stable, and that maximum stability occurs around A $\sim$ 50. We will return to discuss this behavior further when we consider the nuclear binding energy in Chap. 4 .

Nuclear size

According to Thomson's "electron" model of the nucleus ($\sim$1900), the size of a nucleus should be about an *Angstrom*, 10^{-8} cm. We now know this picture is wrong. The correct nuclear size was determined by Rutherford ($\sim$1911) in establishing his atomic nucleus hypothesis. The nuclear size was put at about 10^{-12} cm. The size of a nucleus is not a well-defined quantity because nuclei are not spherical when they contain many nucleons, see discussion of nuclear moments below. Also different experiments can give different results if the measurements involve different physical processes (phenomena), for example, scattering of electrons versus scattering of nuclear particles. Roughly speaking, we will take the nuclear radius to vary with the 1/3 power of the mass number, $R = r_o A^{1/3}$, with $r_o \sim 1.2 - 1.4 \times 10^{-13}$ cm. The lower value of the coefficient r_o comes from electron scattering which probes the charge distribution of the nucleus, while the higher value comes from nuclear scattering which probes the range of nuclear force. Since nuclear radii have magnitudes of the order of 10^{-13} cm, it is conventional to adopt a length unit called *Fermi* (F), $F \equiv 10^{-13}$ cm.

Because of particle-wave duality we associate a wavelength with the momentum of a particle. The corresponding wave is called the *deBroglie* wave. Before discussing the connection between a wave property, the wavelength, and a particle property, the momentum, let us first set down the relativistic kinematic relations between mass, momentum and energy of a particle with arbitrary velocity. Consider a particle with rest mass m_o moving with velocity v. There are two expressions we can write down for the total energy E of this particle. One is the sum of its kinetic energy E_{kin} and its rest mass energy, $E_o = m_o c^2$,

$$E_{tot} = E_{kin} + E_o = m(v)c^2 \tag{3.1}$$

The second equality introduces the relativistic mass $m(v)$ which depends on its velocity,

$$m(v) = \gamma\, m_o, \quad \gamma = \left(1 - \frac{v^2}{c^2}\right)^{-1/2} \tag{3.2}$$

where γ is called the *Einstein* factor. To understand (3.2) one should look into the Lorentz transformation and the special theory of relativity in any text. Equation (3.1) is a first-order relation for the total energy. Another way to express the total energy is a second-order relation

$$E^2 = c^2 p^2 + E_o^2 \tag{3.3}$$

where $p = m(v)v$ is the momentum of the particle. Equations (3.1)–(3.3) are the general relations between the total and kinetic energies, mass, and momentum. We next introduce the deBroglie wave by defining its wavelength λ in terms of the momentum of the corresponding particle,

$$\lambda = \frac{h}{p} \tag{3.4}$$

where h is the *Planck's* constant ($h/2\pi = \hbar = 1.055 \times 10^{-27}$ erg sec). Two limiting cases are worth noting.

Non-relativistic regime:

$$E_o \gg E_{kin}, \quad p = (2\, m_o E_{kin})^{1/2}, \quad \lambda = h/\sqrt{2 m_o E_{kin}} = \frac{h}{m_o v} \tag{3.5}$$

Extreme relativsitic regime:

$$E_{kin} \gg E_o, \quad p = \frac{E_{\text{kin}}}{c}, \quad \lambda = \frac{hc}{E} \tag{3.6}$$

Equation (3.6) applies as well to photons and neutrinos which have zero rest mass. The kinematical relations just discussed are general. In practice we can safely apply the non-relativistic expressions to neutrons, protons, and all nuclides because their rest mass energies are always much greater than any kinetic energies of these particles that we expect to encounter. This is not true for electrons, since we will be interested in electrons with energies in the MeV region and the rest mass of electron is 0.511 MeV. Thus, the two extreme regimes do not apply to electrons, and one should use (3.3) for the energy-momentum relation. Since photons have zero rest mass, they are always in the relativistic regime.

If a nucleus has a certain mass and size, then its density is known. For a spherical nucleus with radius $r = 2 \times 10^{-13}$ cm containing 10 nucleons, the nuclear density has a value of 3×10^{38} nucleons/cm^3. Comparing this to the typical densities of matter, 10^{19} and 10^{22} atoms/cm^3 for a dilute gas and a solid respectively, the nucleus is seen to be a very dense system. Another aspect of the nuclear size is the range of interactions. Because the nucleus is such a small object, the interactions among the nucleons must take place at very close range. Are these interactions attractive or repulsive? For the nucleus to be stable, the interactions must be mostly attractive. On the other hand, the interactions cannot be entirely attractive, otherwise, it will collapse unto itself. These considerations will come into play more quantitatively in Chap. 5 where we discuss simple models that allow us to calculate the energy levels of nuclides. In these models we will have to specify the interaction potential which varies with the distance of separation. For now we can just appreciate that nuclear interactions (or forces) are strong and occur over short distances. "Strong" and "short" in what sense? We can think of two other kinds of system for comparison, one is the electrons in an atom, and the other the atoms in a solid or liquid. Electrons interact with each other and with the nucleus through Coulomb forces which are long-ranged. The strength of the interaction may be measured by the binding energy of the electrons as they occupy various orbits surrounding the atom. Typically ionization energies vary linearly with Z, ranging from tens of eV to a few keV. Atoms in a solid interact through interatomic forces. For a solid at equilibrium at temperature T, the energies of thermal excitation may be characterized by k_BT, where k_B is the Boltzmann constant (see Table 3.3). At room temperature the energy is 0.025 eV. Thus nuclear interactions in the range of tens of MeV are about three to four orders of magnitude stronger than electronic interactions, which in turn are about three orders of magnitude stronger than interatomic interactions in condensed matter. As to the range of interaction, electrons and atoms interact at distances of the order of *Angstroms* Å, 10^{-8} *cm*, which are about five orders of magnitude greater than the range of nuclear interactions (see Chap. 5).

Nuclear charge

The charge of a nuclide $_ZX^A$ is positive and equal to Ze, where e is the magnitude of the electron charge, $e = 4.80298 \times 10^{-10}$ esu ($= 1.602189 \times 10^{-e19}$ Coulomb). We consider single atoms as exactly neutral, the electron-proton charge difference is $< 5 \times 10^{-19}e$, and the charge of a neutron is $< 2 \times 10^{-15}e$.

We can look to high-energy electron scattering experiments to get an idea of how nuclear density and charge density are distributed across the nucleus. Figure 3.2 shows two typical nucleon density distributions obtained by high-energy electron scattering. One can see two basic components in each distribution, a core region of constant density and a boundary region where the density decreases smoothly to zero. Notice the magnitude of the nuclear density is 10^{38} nucleons per cm^3, compared to the atomic density of solids and liquids which is in the range of 10^{24} nuclei per cm^3. What does this say about the packing of nucleons in a nucleus, or the average distance between nucleons versus the separation between nuclei? Indeed the nucleons are packed together much more closely than the nuclei in a solid. The

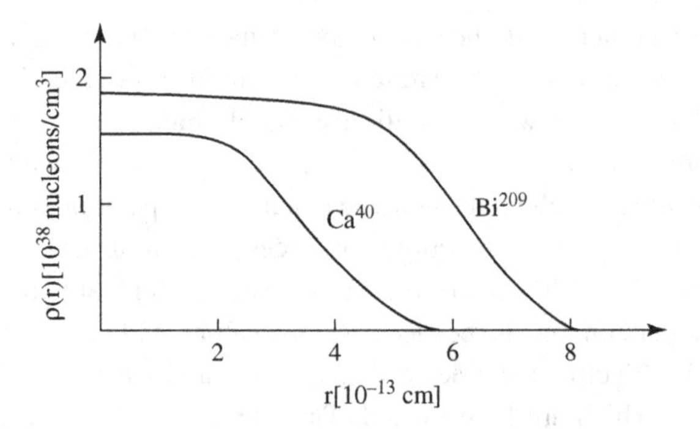

Fig. 3.2 Typical nucleon density distributions ρ(r) showing nuclei have a core of constant density and a rather diffuse boundary. See also Fig. 3.3.

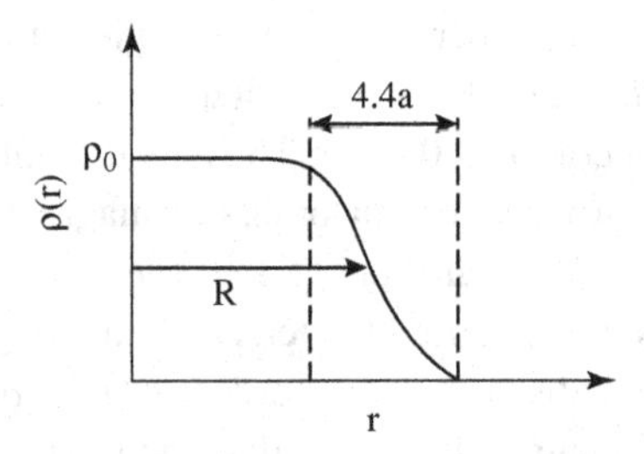

Fig. 3.3 Schematic of the nuclear density distribution, with R being a measure of the nuclear radius, and the width of the boundary region being given by 4.4a, see Eq. (3.7).

shape of the distributions shown in Fig. 3.2 can be fitted to the expression, called the Saxon distribution,

$$\rho(r) = \frac{\rho_o}{1 + \exp\left[\frac{(r-R)}{a}\right]} \tag{3.7}$$

where $\rho_o = 1.65 \times 10^{38}$ nucleons/cm^3, $R \sim 1.07\, A^{1/3}\, F$, and a ~0.55 F. A sketch of this distribution, given in Fig. 3.3, shows the core and boundary components of the distribution. Detailed studies based on high-energy electron scattering have also revealed that even the proton and the neutron have rather complicated structures. This is illustrated in Fig. 3.4. We note that mesons are unstable particles of mass between the electron and the proton: π-mesons (pions) play an important role in nuclear forces ($m_\pi \sim 270 m_e$), μ-mesons (muons) are important in cosmic-ray processes ($m_\mu \sim 207 m_e$).

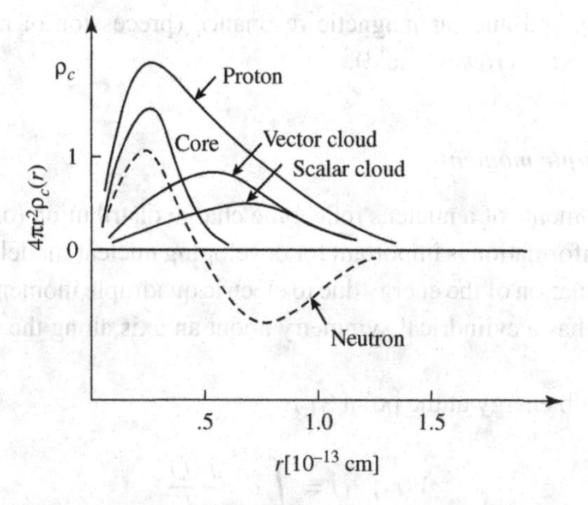

Fig. 3.4 Schematic of charge density distributions of the proton and the neutron showing how each can be decomposed into a core and two meson clouds, inner (vector) and outer (scalar). The core has a positive charge of $\sim 0.35e$ with probable radius $0.2\ F$. The vector cloud has a radius $0.85\ F$, with charge $0.5e$ and $-0.5e$ for the proton and the neutron respectively, whereas the scalar cloud has radius $1.4\ F$ and charge $0.15e$ for both proton and neutron [adapted from Marmier 1969, p. 18].

Nuclear spin and magnetic moment

Nuclear angular momentum is often known as nuclear spin $\hbar \underline{I}$. It is made up of two parts, the intrinsic spin of each nucleon and the orbital angular momenta of the nucleons. We call I the spin of the nucleus, which can take on integral or half-integral values. The following is usually accepted as facts. Neutron and proton both have spin 1/2 (in unit of $\hbar$). Nuclei with even mass number A have integer or zero spin, while nuclei of odd A have half-integer spin. Angular momenta are quantized.

Associated with the spin is a magnetic moment $\underline{\mu}_I$, which can take on any value because it is not quantized. The unit of magnetic moment is the magneton

$$\mu_n \equiv \frac{|e|\hbar}{2m_p c} = \frac{\mu_B}{1836.09} = 0.505 \times 10^{-23} \text{ ergs/gauss} \tag{3.8}$$

where μ_B is the Bohr *magneton*. The relation between the nuclear magnetic moment and the nuclear spin is

$$\underline{\mu}_I = \gamma \hbar \underline{I} \tag{3.9}$$

where γ here is the *gyromagnetic ratio* (no relation to the *Einstein* factor in special relativity). Experimentally, spin and magnetic moment are measured by hyperfine structure (splitting of atomic lines due to interaction between atomic and nuclear magnetic moments), deflections in molecular beam under a magnetic field

(Stern-Gerlach), and nuclear magnetic resonance (precession of nuclear spin in combined DC and microwave field).

Electric quadruple moment

The electric moments of a nucleus reflect the charge distribution (or shape) of the nucleus. This information is important for developing nuclear models. We consider a classical calculation of the energy due to electric quadruple moment. Suppose the nuclear charge has a cylindrical symmetry about an axis along the nuclear spin $\underline{I}$, see Fig. 3.5.

The Coulomb energy at the point S_1 is

$$V(r_1, \theta_1) = \int d^3r \frac{\rho(\underline{r})}{d} \tag{3.10}$$

where $\rho(\underline{r})$ is the charge density, and $d = |\underline{r}_1 - \underline{r}|$. We will expand this integral in a power series in $1/r_1$ by noting the expansion of 1/d in a Legendre polynomial series,

$$\frac{1}{d} = \frac{1}{r_1} \sum_{n=0}^{\infty} \left(\frac{r}{r_1} \right)^n P_n(\cos\theta) \tag{3.11}$$

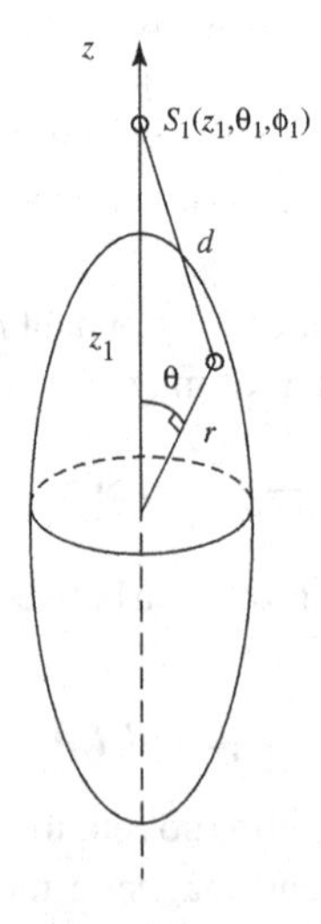

Fig. 3.5 Cylindrical coordinate system for calculating the Coulomb potential energy at a field point S_1 due to a charge distribution $\rho(\underline{r})$ on a spheroidal surface. The sketch is for $\underline{r}_1$ located along the z-axis.

Table 3.2 Spin and electric quadruple moments of nucleons and selected nuclei.

Nucleus	**I [$\hbar$]**	**Q [10^{-24} cm^2]**
n	1/2	0
p	1/2	0
H^2	1	0.00274
He^4	0	0
Li^6	1	−0.002
U^{233}	5/2	3.4
U^{235}	7/2	4
Pu^{241}	5/2	4.9

where $P_0(x) = 1$, $P_1(x) = x$, $P_2(x) = (3x^2 - 1)/2, \ldots$ Then (3.10) can be written as

$$V(r_1, \theta_1) = \frac{1}{r_1} \sum_{n=0}^{\infty} \frac{a_n}{r_1^n} \tag{3.12}$$

with

$$a_o = \int d^3r \rho(\underline{r}) = Ze \tag{3.13}$$

$$a_1 = \int d^3r z \rho(\underline{r}) = \text{electric dipole} \tag{3.14}$$

$$a_2 = \int d^3r \frac{1}{2}(3z^2 - r^2)\rho(\underline{r}) \equiv \frac{1}{2} eQ \tag{3.15}$$

The coefficients in the expansion for the energy, Eq. (3.12), are recognized to be the total charge, the dipole (here it is equal to zero), the quadruple, etc. In Eq. (3.15), Q is defined to be the quadrupole moment (in unit of 10^{-24} cm^2, or barns). Notice that if the charge distribution were spherically symmetric, $\langle x^2 \rangle = \langle y^2 \rangle = \langle z^2 \rangle = \langle r^2 \rangle / 3$, then $Q = 0$ We see also, $Q > 0$, if $3\langle z^2 \rangle > \langle r^2 \rangle$ and $Q < 0$, if $3\langle z^2 \rangle < \langle r^2 \rangle$. The corresponding shape of the nucleus in these two cases would be prolate or oblate spheroid, respectively. Some values of the spin and quadruple moments are given in Table 3.2.

3.2 PHYSICAL CONSTANTS

As in the study of any physical phenomena, one encounters quantities that specify the properties of the physical system in question which are known and do not vary.

They are the physical constants of the problem, for example, the mass of proton or electron, or the speed of light. For quick estimates and numerical calculations, it is useful to collect these constants in a convenient part of the book for easy reference. Here we give a list of the physical constants useful for the study of *Nuclear Interactions*. The entries are compiled from tables given in the text on nuclear [Meyerhof, 1967] and atomic [Livesey, 1966] physics, and from a nuclear data reference [Table of Isotopes 1978].

Among the physical constants that have appeared in various branches of physics [Cohen, 1995] one may select those regarded as universal constants along with quantities that are appropriate for the subject of this book, electromagnetic constants, atomic constants, basic properties of electron, proton, neutron, to go along with physicochemical constants. As more accurate measurements become available, some of the recommended values of the constants can undergo refinements [Cohen, 1995], generally at a level of accuracy that does not concern us. Regarding units and conversion factors, the reader should be aware of the *SI units* (International System of Units), founded on seven base units listed below [Nelson 1995]. By convention they are regarded as dimensionally independent, and all other units are *derived* units, formed by multiplying and dividing units within the system without numerical factors.

Among the physical constants listed in Table 3.3, perhaps the one used most frequently in this book is the reduced Planck constant $\hbar$. Its ubiquitous role lies in the relations expressing the energy E, scaler momentum p, and speed v of a free particle, $E = \hbar^2k^2/2m$, $p = \hbar k$, $v = \hbar k/m$, where k is the wavenumber ($k = 2\pi/\lambda$, λ being the de Broglie wavelength), and m is the particle mass. These relations apply to non-relativistic particles such as neutrons, protons, and nuclides in the energy range of our interest. For gamma particles (photons), we have $E = \hbar ck = \hbar\omega = h\nu$, where c is the speed of light, ω and ν are the circular and linear frequency respectively. We have already mentioned earlier that if a nucleus is regarded as a sphere, its radius would be given as $R \sim 1.07\,A^{1/3}\,F$ (see Fig. 3.3), where A is the mass number, and $F = 10^{-13}$ cm. So typical nuclear sizes are a few Fermi's. Notice that the classical radius of electron r_e is of the same order of magnitude (see Chap. 10).

We have also mentioned the importance of nuclear masses in its equivalence to energy which in turn is the decisive factor determining the stability of a nucleus (see Chap. 4 for discussion of stability and Chap. 5 for models describing energy levels). Because energy can be expressed in different ways, we give in Table 3.4 the common conversion factors. We will use the unit of electron-volt eV most frequently in this book. Compared to another common energy unit, erg, (still another unit Joule $= 10^7$ ergs), eV is a rather small energy unit. For nuclear interactions the

Table 3.3 Physical constants.

Symbol		Value	Unit (SI)	Unit (cgs)
$a_o = \hbar^2/m_e c^2$	Bohr radius	5.29177	10^{-11} m	10^{-9} cm
b	Barns	1.0	10^{-28} m^2	10^{-24} cm^2
$1/\alpha = \hbar c/e^2$	Fine structure constant	137.03604		
c	Speed of light	2.99792	10^8 m s^{-1}	10^{10} cm s^{-1}
e	Elementary charge	4.80324		10^{-10} esu
e/c		1.60218	10^{-19} C	10^{-20} emu
e/m_e	Charge-to-mass ratio (electron)	5.27276		10^{17} esu g^{-1}
F	Fermi	1.0	10^{-15} m	10^{-13} cm
h	Planck constant	6.62618	10^{-34} J s	10^{-27} erg s
$\hbar = h/2\pi$	Planck constant (reduced)	1.05459		
k_B	Boltzmann constant	1.38066	10^{-23} JK^{-1}	10^{-18} erg K^{-1}
$\lambda_C = h/m_e c$	Compton wavelength (electron)	2.42631	10^{-12} m	10^{-10} cm
m_e	Electron rest mass	5.48580	10^{-31} kg	10^{-28} g
m_p	Proton rest mass	1.00727	10^{-27} kg	10^{-24} g
m_n	Neutron rest mass	1.008665	10^{-27} kg	10^{-24} g
N_A	Avogardro number	6.02205	10^{23} mol^{-1}	10^{23} mol^{-1}
r_e	Classical electron radius	2.81793	10^{-15} m	10^{-13} cm

Table 3.4 Energy conversion factors.

1 eV	$= 1.6022 \times 10^{-12}$ erg	
	$= 8.0655 \times 10^3$ cm^{-1}	(wavenumber)
	$= 1.1605 \times 10^4$ deg K	
	$= 3.8293 \times 10^{-20}$ cal	
	$= 1.0735 \times 10^{-9}$ u	(unified mass)
m_p	$= 1.0073\,u$	$(m_p c^2 = 938.26$ MeV)
m_n	$= 1.0087\,u$	$(m_n c^2 = 939.55$ MeV)
m_H	$= 1.0078\,u$	$(m_H c^2 = 938.77$ MeV)
u	Mass unit (unified mass scale) $= 1.6605655\ 10^{-24}$ g	

energies involved are generally in the range of 10^6 eV (MeV). An important exception is neutron interaction which can involve neutron energies down to 10^{-2} eV. We call this low-energy range the thermal region because molecular motions at room temperature ($T = 300\ K$) have energies $k_B T$ of this magnitude (see Chap. 9). In Table 3.4 the unit of wavenumber cm^{-1} is convenient for discussion of spectroscopy, whereas the unit of calorie cal is commonly used in reference to chemical reaction energies.

3.3 NUCLEAR DATA

There is a very large body of information in the realm of nuclear data that is available to the scientific community. It has been said the study of nuclear properties — forces, structure, reactions — is one of the most intensive human endeavors known. It is certainly true in the 20th century judging from historical developments, first concerning scientific discoveries and followed soon after by technological applications, and the continuation of these activities worldwide into the 21st century with no sign of slowing down. For the students of Nuclear Science and Engineering a vast and growing amount of nuclear data has been generated using increasingly sophisticated experimental methods and facilities, organized using the latest data storage technologies, and presented in formats varying from the traditional printing of charts, tables, and compilations to electronic files with an interactive user interface. Fortunately there is no need to keep up with this explosion of information if one were content with only a basic understanding of the information that has become available. It is the opinion of the author that even though the amount of nuclear data will continue to increase in complexity and specificity, the underlying fundamentals have not changed significantly over the years. Of course the present assurance does not mean there will be no new discoveries ahead that are totally unanticipated.

In chemistry and physics an element is a substance composed of atoms, each having the same number of protons (atomic number Z) in its nucleus. Each element has a distinct name and a corresponding unique symbol which is either one or two letters (first is in cap). Table 3.5 is an alphabetical listing of 103 elements ($Z = 1$ to 103) and their atomic weights in mass unit $u = 1.6605655 \times 10^{-24}$ g (see Table 3.4). The values shown are for naturally occurring elements. For the heavy nuclides which are not naturally occurring, the values shown refer to the atomic number of the most stable isotope. Elements from atomic number 1 to 98 exist naturally, although some are found only in trace amounts, having been initially discovered by laboratory synthesis. Elements with Z from 99 to 118, with 113, 115, 117 and 118 yet to be confirmed, have only been synthesized in laboratories. The search for superheavy elements is an on-going area of research. Although interesting, it has no relevance to our studies.

The information in Table 3.5 may be considered to be the most basic of all nuclear data in that only the weights of the nuclides are given. We defer to Chap. 4 for a discussion of nuclear stability that one can derive from just the mass data alone.

Once we go beyond the mass of the elements there are a number of different properties one can consider, all may be regarded as basic. The amount of such data one can tabulate increases so quickly that one has to restrict the entries to

Table 3.5 Mean atomic weights of the elements (on the C^{12} scale). (Adapted from Livesey 1966.) Values in parentheses are the mass number A of the most stable isotope.

Element	Symbol	Atomic Weight	Atomic (Z) Number	Element	Symbol	Atomic Weight	Atomic Number
Actinium	Ac	(227)	89	Mendelevium	$M\gamma$	256	101
Aluminium	Al	26.98	13	Mercury	Hg	200.59	80
Americium	Am	243	95	Molybdenum	Mo	95.94	42
Antimony	Sb	121.75	51	Neodymium	Nd	144.24	60
Argon	A	39.95	18	Neon	Ne	20.18	10
Arsenic	As	74.92	33	Neptunium	Np	237	93
Astatine	At	210	85	Nickel	Ni	58.71	28
Barium	Ba	137.34	56	Niobium	Nb	92.91	41
Berkelium	Bk	247	97	Nitrogen	N	14.01	7
Beryllium	Be	9.01	4	(Nobelium)	No	254	102
Bismuth	Bi	208.98	83	Osmium	Os	190.2	76
Boron	B	10.81	5	Oxygen	O	16.00	8
Bromine	Br	79.91	35	Palladium	Pd	106.4	46
Cadmium	Cd	112.40	48	Phosphorus	P	30.97	15
Calcium	Ca	40.08	20	Platinum	Pt	195.09	78
Californium	Cl	251	98	Plutonium	Pu	244	94
Carbon	C	12.01	6	Polonium	Po	(210)	84
Cerium	Ce	140.12	58	Potassium	K	39.10	19
Cesium	Cs	132.91	55	Promethium	Pm	145	68
Chlorine	Cl	35.45	17	Protoaetinum	Pa	(231)	91
Chromium	Cr	52.00	24	Radium	Ra	(226)	88
Cobalt	Co	58.93	27	Radon	Rh	222	86
Copper	Cu	63.54	29	Rbenium	Re	186.2	75
Curium	Cm	247	96	Rhodium	Rh	102.91	45
Dysprosium	Dy	162.50	66	Rubidium	Rb	85.47	37
Einsteinium	Es	(252)	99	Ruthenium	Ru	101.07	44
Erblum	Er	167.25	68	Samarium	Sm	150.35	62
Europium	Eu	151.96	63	Scandium	Sc	44.96	21
Fermium	Fm	257	100	Selenium	Se	78.96	34
Fluorine	F	19.00	9	Silicon	Si	28.09	14
Francium	Fr	223	87	Silver	Ag	107.87	47
Gadolinium	Gd	157.25	64	Sodium	Na	22.99	11
Gallium	Ga	69.72	31	Strontium	Sr	87.62	38
Germanium	Ge	72.59	32	Sulfur	S	32.06	16
Gold	Au	195.97	79	Tantalum	Ta	180.95	73
Hafinium	Hf	178.49	72	Technesium	Tc	97	43
Helium	He	4.00	2	Tellurium	Te	127.60	52
Holmium	Ho	164.93	67	Terbium	Tb	158.92	65
Hydrogen	H	1.008	1	Thallium	Tl	204.37	81
Indium	In	114.82	49	Therium	Th	(232)	90
Iodine	I	126.90	53	Thullium	Tm	168.93	69
Indium	Ir	192.2	77	Tin	Sn	118.69	50
Iron	Fe	55.85	26	Titanium	Ti	47.90	22
Krypton	Kr	83.80	36	Tungsten	W	183.85	74
Lanthanum	La	138.91	57	Uranium	U	(238)	92
Lawrencium	Lw	257	103	Vanadium	V	50.94	23
Lead	Pb	203.19	82	Xenon	Xe	131.30	54
Lithium	Li	6.94	3	Yeterbium	Yb	173.04	70
Lutetium	Lu	174.97	71	Yurium	Y	88.91	39
Magnesium	Mg	24.31	12	Zinc	Zn	65.37	30
Mangacese	Mn	54.94	25	Zircoaium	Zr	91.23	40

Table 3.6 Selected nuclear stability data for light nuclides. (Adapted from Livesey 1966.)

Atomic Number	Nuclide	(Stable) % in Natural Element	(Unstable) Made of Decay	Half-Life	Q-value in MeV
0	n^1		e^-	12 min	0.783
1	H^1	99.985			
	$(D^3)\ H^2$	0.015			
	$(T^3)\ H^3$		e^-	12.26 yr	0.0181
2	He^3	0.00013			
	He^4	99.99987			
	He^5		$n + He^4$	$\sim 2\times10^{-22}$ sec	
	He^6		e^-	0.81 sec	3.51
	He^7		e^-	$\sim 5\times10^{-6}$ sec	$\sim$10
3	Li^5		$p + He^4$	$\sim 10^{-21}$ sec	
	Li^6	7.42			
	Li^7	92.58			
	Li^8		e^-	0.85 sec	16.0
	Li^9		e^-	0.17 sec	14.1
4	Be^6		$p + Li^5$	$\sim 4\times10^{-21}$ sec	
	Be^7		EC	53 days	0.86
	Be^8		$2He^4$	-3×10^{-18} sec	
	Be^9	100			
	Be^{10}		e^-	2.7×10^5 yr	0.56
	Be^{11}		e^-	13.6 sec	11.5
5	B^8		e^+	0.78 sec	18
	B^9		$p + Be^8$	3×10^{-10} sec	
	B^{10}	19.78			
	B^{11}	80.22			
	B^{12}		e^-	0.020 sec	13.4
	B^{12}		e^-	0.019 sec	13.4
6	C^{10}		e^+	19 sec	3.62
	C^{11}		e^+	20.5 min	1.95
	C^{12}	98.89			
	C^{13}	1.11			
	C^{14}		e^-	5730 yr	0.156
	C^{15}		e^-	2.25 sec	9.77
	C^{16}		e^-	0.74 sec	8.0

only a small (selected) portion of the information that is available. Table 3.6 is an example of this situation. It gives nuclear data regarding the stability of various isotopes of the elements, but here we show only the data for the very light elements, atomic number up to 6. Stability means the modes of decay and the half lives of each isotope (see Chap. 4 for detailed discussions on mass-based stability considerations, and Chap. 6 for radioactive decay and various decay modes). Comparison of Tables 3.5 and 3.6 illustrates the unavoidable tradeoff between including more data versus keeping the tabulation manageably compact. One can deduce from Table 3.6 a number of interesting facts, such as the neutron by itself is not a stable particle, it undergoes decay to a proton by emitting an electron of energy ~0.78 MeV, with a half-life of 12 min.

Recall again each element is an atom with a nucleus of definite number of protons. Associated with each element are several isotopes (nuclides with different neutron number N and therefore mass number A), such as the ones shown in Table 3.6 for the light elements. One can estimate roughly the total number of isotopes, known to be around 1000, by taking 100 Z values and 8 to 10 isotopes for each element. Compiling all the stability and decay information for these isotopes is an important task, and the resulting data are clearly beyond the scope of our study. Such a database actually exists as we will find below, see the *Nuclide Chart* and the *Table of Isotopes* in the following discussions. But there are also Z atomic electrons associated with each element, assuming all the atoms are neutral. What can we say about their role in determining physical properties of the elements?

Before the survey of nuclear data progresses to more properties and greater details, let us take the opportunity to examine the well-known *Periodic Table*, which is a way of organizing all the known elements into an arrangement that brings out a periodicity in their chemical behavior. Although detailed considerations of the periodic table are more in the realm of atomic physics than nuclear interactions, it is nevertheless worthwhile to comment briefly on how nuclear and atomic data have an intrinsic overlap (or analogy). One finds a similar underlying trend in building up from light to heavy nuclides, or from light to heavy atoms. In the former we add protons and appropriate number of neutrons to keep the system (nucleus) stable, while in the latter we add electrons to keep the system (atom) neutral. One might ask how do the additional protons and neutrons go into the nucleus, or how do the additional electrons go into the atom? Basically this question is about the *mechanism* of formation of a nucleus or an atom. Such issues are commonly addressed by formulating a model description of the interactions among the constituents of the system (nucleons or electrons). Nuclear models that attempt to calculate the energy levels of a nucleus are discussed in Chap. 5. Roughly speaking, one finds both the nucleons and the electrons go into states with definite angular momenta (orbital plus spin), and these states have a shell structure such that when one shell is filled (closed), the additional particles must go into the next shell. The significance of the periodic table is therefore elements belonging to the same "group" (column in Table 3.7) tend to display the same chemical behavior.

Table 3.7 Periodic classification of the elements (based on atomic number Z) in the standard 18-column form. Symbols for elements with Z = 104 to 118 are not shown. (Adapted from http://en.wikipedia.org/wiki/Periodic_table.)

Group→ ↓Period	1	2	3	4	5	6	7	8	9	10	11	12	13	14	15	16	17	18
1	1 H																	2 He
2	3 Li	4 Be											5 B	6 C	7 N	8 O	9 F	10 Ne
3	11 Na	12 Mg											13 Al	14 Si	15 P	16 S	17 Cl	18 Ar
4	19 K	20 Ca	21 Sc	22 Ti	23 V	24 Cr	25 Mn	26 Fe	27 Co	28 Ni	29 Cu	30 Zn	31 Ga	32 Ge	33 As	34 Se	35 Br	36 Kr
5	37 Rb	38 Sr	39 Y	40 Zr	41 Nb	42 Mo	43 Tc	44 Ru	45 Rh	46 Pd	47 Ag	48 Cd	49 In	50 Sn	51 Sb	52 Te	53 I	54 Xe
6	55 Cs	56 Ba	*	72 Hf	73 Ta	74 W	75 Re	76 Os	77 Ir	78 Pt	79 Au	80 Hg	81 Tl	82 Pb	83 Bi	84 Po	85 At	86 Rn
7	87 Fr	88 Ra	**	104 Rf	105 Db	106 Sg	107 Bh	108 Hs	109 Mt	110 Ds	111 Rg	112 Cn	113 Uut	114 Fl	115 Uup	116 Lv	117 Uus	118 Uuo
		*	57 La	58 Ce	59 Pr	60 Nd	61 Pm	62 Sm	63 Eu	64 Gd	65 Tb	66 Dy	67 Ho	68 Er	69 Tm	70 Yb	71 Lu	
		**	89 Ac	90 Th	91 Pa	92 U	93 Np	94 Pu	95 Am	96 Cm	97 Bk	98 Cf	99 Es	100 Fm	101 Md	102 No	103 Lr	

Since this is a recurring, or periodic, occurrence, as one moves to heavier atoms, it provides the motivation for organizing the Table to reflect the electronic structure of atoms.

Table 3.7 shows a tabular arrangment that emphasizes the electronic structure of the elements. Its interest lies not so much in the nuclear properties as in classifying the chemical behavior of the elements. The table is based on the ordering of known nuclides in a consecutive sequence according to the atomic number Z, $Z = 1$ to 118. However, the table does not have a regular shape in the horizontal or vertical directions. The particular arrangements of the square blocks of elements, each with a unique symbol and Z value, is such that the Z-ordering sequence is broken up at several characteristic Z values between (1, 118). A discussion as to why the *Periodic Table* has the structure seen in Table 3.7 can be an instructive exercise.

Suppose one tries to arrange the 103 blocks of elements from Table 3.5 into a matrix with 18 columns and 7 rows, labeling each column as a "*group*" and each row a "*period*". The periodic table, shown in Table 3.7, therefore spans 18 groups and 7 periods. As the 103 blocks of elements go into a 7×18 rectangular matrix, clearly the matrix will not be full. One sees there are three gaps, the first gap (period 1) extends from group 2 to group 17, the second and third gaps (periods 2 and 3) extend from groups 3 to 12. Moreover, there are additional blocks of elements to be inserted into periods 6 and 7. Specifically a row of 10 elements are to be inserted in period 6, right after Ba (Z = 56). This extra row is labeled "Lanthanides"; it consists of La (Z = 57) ... Lu (Z = 72). Similarly, there is another extra row of element, labeled "Actinides", that should be inserted into period 7, after Ra (Z = 88). This row consists of Ac (Z = 89) ... Lw (Z = 103). The reasons for the gaps and the insertions have to do with electron configurations in different shells, and the number of electronic states that can be accommodated in each shell. The configurations are labeled by the orbital and spin angular momenta of the state (see Chap. 5 for explanation of angular momentum labels and the number of states that can be accommodated, see also any standard atomic physics text for explanation of the Periodic Table). One can show the gaps at periods 1 through 3 are the consequences of occupation of the states with angular momentum quantum number $\ell = 3$ (d shell), in period 4, starting with Ti (Z = 22) and ending with Zn (Z = 30), in period 5, stating with Zr (Z = 40) and ending with Cd (Z = 48), and in period 6, starting with Hf (Z = 72) and ending with Hg (Z = 80). The d-shell electrons give the corresponding elements special chemical properties. The 10 elements in periods 4, 5 and 6 form what is called the *First* (3*d*), *Second* (4*d*), *and Third* (5*d*) *Transition Series* respectively. Similarly the extra rows of 15 elements to be inserted, starting with La (Z = 57) and ending with Lu (Z = 72) in period 6, and starting with Ac (Z = 88) and ending with Lw (Z = 103) in period 7, appear because of the occupation of electronic configuration states with orbital angular momentum quantum number $\ell = 4$ (f shell). These two rows of elements are known as the *Lathanides* and *Actinides* respectively. They correspond to the occupation of the so-called $4f$ and $5f$ shells, in contrast to the 3*d* to 5*d* shells for the *Transition Series*. In Table 3.5 the heaviest element indicated is Lawrencium at Z = 103. In Table 3.7 heavier elements are indicated for $Z = 104$ to $Z = 118$. We need not go into any further details about these heavy elements because they have no relevance to any of our discussions in this book.

There is no doubt that knowing what kind of nuclear data information is available and how they have been organized will better guide us in the study of NRI in the chapters to follow. We will next survey briefly the various efforts at organizing a very large nuclear data base some of which are still currently on-going.

Nuclide chart

Among the body of nuclear data available to the community, the Chart of Nuclides [Knolls Atomic Power Laboratory, 2009] remains one of the most valuable and convenient sources of information for students in the field of nuclear science and engineering. The concept, originated in 1935 [Fea, 1935], was to graphically represent the stable nuclides and the transitions they can undergo to neighboring nuclides by various means of radioactive decay. In contrast to the Periodic Table which shows the grouping of chemical elements and therefore the chemical behavior of the elements, the Chart of Nuclides, of which several versions are available [Karlsruhe Nuclide Chart, etc.], focuses on the *nuclear* behavior of the elements.

Imagine a two-dimensional plot of all the known nuclides. Let the X (horizontal) axis denote the number of neutrons N and the Y (vertical) axis the number of protons Z, with the atomic numbers Z range from 1 (hydrogen) to 102 (nobelium). For the decay modes one has the following: alpha decay, beta decay, proton emission, neutron emission, positron emission or electron capture, and spontaneous fission. Each entry, in the form of a square box, in this chart is a unique nuclide. The basic information shown in the box consists of symbol and mass number of the element, atomic weight (mass, Carbon-12 scale), percent natural abundance, spin and parity, if available, thermal neutron absorption or activation cross section. Additionally, the various nuclear transitions which the isotope in question can undergo and the associated transition energy are given.

The chart of nuclides is a highly compact (and useful) form of nuclear data made available to the public by various issuing institutions. Although all the data displayed originally come from measurements and evaluations reported in the literature, the precise source is not available to the user. In cases where information on the actual source and the uncertainties involved are needed (not anticipated for most readers of this book), one can consult a more comprehensive compilation of nuclear data, the *Table of Isotopes*, to be discussed next.

Web Links on Chart of Nuclides

- *http://knollslab.com/nuclides.html* (Chart of Nuclides — Knolls Atomic Power Laboraory)
- *http://atom.kaeri.re.kr* (Chart of Nuclides — Korea)
- *http://www.nucleonica.com/wiki/index.php?title=Category%3AKNC* (Karlsruhe nuclide chart)
- *http://www-nds.iaea.org/livechart* (The live chart of nuclides — IAEA)
- *http://www.nndc.bnl.gov/chart* (Interactive chart of nuclides — Brookhaven National Laboratory)

Table of Isotopes

The Table of Isotopes is perhaps the most authoritative general reference compilation of basic properties for nuclei of all mass numbers A. First published in 1940, the 7th edition in 1978 treats more than 2600 known isotopes, stable and radioactive [Lederer, 1978]. For each mass number the table shows, in annotated energy-level diagram, the associated isotopes and the decay relationships between them, following a mass-chain decay scheme. Reported experimental uncertainties are given, all data are referenced, along with extensive references to additional data. The properties listed include (compare with those given by the Chart of Nuclides): isotopic mass, natural abundance, nuclear spin, neutron capture and fission cross sections, decay modes and branching, decay energies, half-life, certainty and means of identification, means of production, energies and intensities of all radiations, angular and polarization correlations of radiations, half-lives of excited states, and level schemes for each nucleus as determined from radioactive decay and from nuclear reactions.

The 8th edition, published in 1999, is freely available on line,
http://www.wiley-vch.de/books/info/0-471-35633-6/toi99/doc_info/cover99.pdf

To explore the information on the internet, simply Google *table of isotopes*.

Nuclear data compilation and evaluation

Given the importance of having accurate knowledge of cross sections and other data in nuclear calculations, it is not surprising that there is an entire industry of nuclear data technology that has been long established and continues to be active in maintaining and improving the database. While detailed information may have changed over the years, the role of nuclear data in nuclear system analysis and design remains as vital as ever. Compilation and evaluation of nuclear data are carried out at many national laboratories and research centers worldwide. In the US, early efforts of this activity began at the National Neutron Cross Section Center at the Brookhaven National Laboratory. Perhaps the most important contribution of this Center is a library of evaluated nuclear data, known as the *Evaluated Nuclear Data File* (***ENDF***), which was developed for the storage and retrieval of nuclear data needed for neutronic and photonic calculations. Initially there were two libraries. *ENDF/A* was a collection of useful evaluated data sets. ***ENDF/B*** contained the reference data sets recommended by the *Cross Section Evaluation Working Group*. At any given time there was only evaluated data set for a particular material. Most users were concerned only with the *B* version. Evaluation of nuclear data involved the assignment of the most credible value after consideration of all the pertinent information. The evaluation was supported by documentation giving a description of how the value was determined and an estimate of its uncertainty. In 1975, the information contained in *ENDF* consisted of resonance parameters, cross section tables, angular distributions, energy distributions, double differential data in angle and energy, scattering law data, and fission parameters. Four primary centers were charged with responsibilities for collecting and disseminating nuclear data information to the world-wide community. The Neutron Cross Section Center, Brookhaven National Laboratory, served the US and Canada. The Neutron Data Compilation Center, Saclay, France, served Western Europe and Japan. The Nuclear Data Center, Obninsk, USSR, served the Soviet Union. The *Nuclear Data Section, IAEA*, served essentially the rest of the world.

As an illustration of the *ENDF/B* library, the *ENDF/B-II* file contained 3 basic types of data stored on 13 magnetic tapes.

- Scattering law data for 12 moderator materials. For example, $S(\alpha, \beta)$ for a series of temperatures, 10 temperatures between 296°K and 2000°K in the case of graphite.
- Neutron cross sections for 78 fissile, fertile, structural, and other materials, each being an element or an isotope — total and any significant partial cross sections, reactions producing outgoing neutrons, angular and energy distributions, radioactive decay chain data, fission product yield, $\nu(E)$.
- Photon interaction cross sections for 87 elements from $Z = 1$ to 94 — photon cross sections, angular and energy distributions of secondary photons.

Over the years a number of active compilation groups have issued library files: KEDAK from Karlsruhe, Germany, JAERI (Japan), UKNDL from England, and various libraries from US national laboratories such as Oak Ridge, Los Alamos, and Lawrence Livermore.

As just indicated the *ENDF* evaluations are among the most useful:

http://www.nndc.bnl.gov/exfor/endf00.jsp

The nndc website (*http://www.nndc.bnl.gov/*) on which ENDF is a useful collection of tools to visualize and query continuous energy data.

Another popular set of evaluations is JEFF:

http://www.oecd-nea.org/dbdata/jeff/

Moreover, there are evaluations from Japan (JENDL), China, Russia, ..., see the handbook of nuclear engineering for more details.

http://link.springer.com/referencework/10.1007/978-0-387-98149-9/page/1

3.4 NEUTRON CROSS SECTION BEHAVIOR

Among the nuclear data we will want to understand in detail, the most important and arguably most interesting are the neutron cross sections and their variations with energy. This behavior is a central part of the studies to be discussed in Chaps. 8, 9, and 12. To give the reader an orientation we preview here a few examples of the characteristic variation of neutron cross section with energy only from a phenomenological standpoint. Compilations of neutron cross section curves provide direct access to actual data on the energy variation of neutron cross sections. One of the first, and still instructive, reference of this kind is the report BNL-325, Neutron Cross Sections [Hughes 1955], and subsequent editions [McLane 1988], produced by the Nuclear Data Center at the Brookhaven National Laboratory. We will return in the three later chapters to discuss further the physical meaning of these variations.

We begin with the simple cross section behavior of elastic scattering. The variation of the elastic scattering cross section with energy of hydrogen is shown in Fig. 3.6. It is seen the cross section is a constant over the energy range starting around 0.1 eV and extending up 100 eV and beyond. Below 0.1 eV the cross section increases with decreasing energy. Throughout the entire range the variation is smooth.

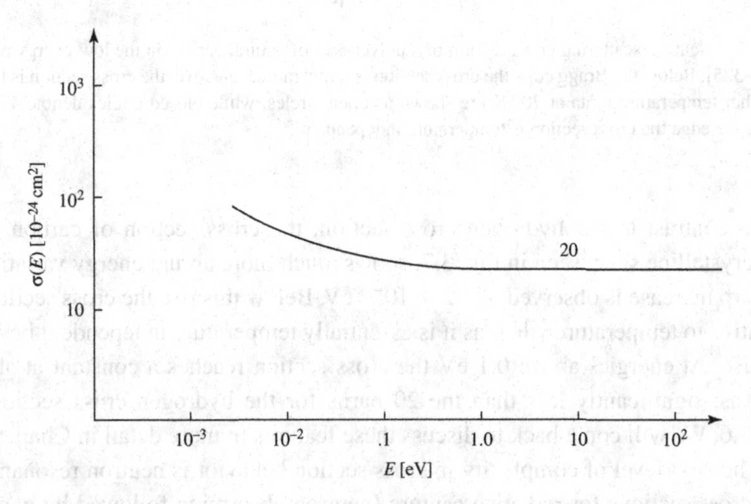

Fig. 3.6 Neutron scattering cross section of hydrogen in the low-energy region as measured in H_2O. The cross section is practically a constant at 20.4 barns starting at about 0.1 eV and remains so up to 1 keV and beyond [BNL-325].

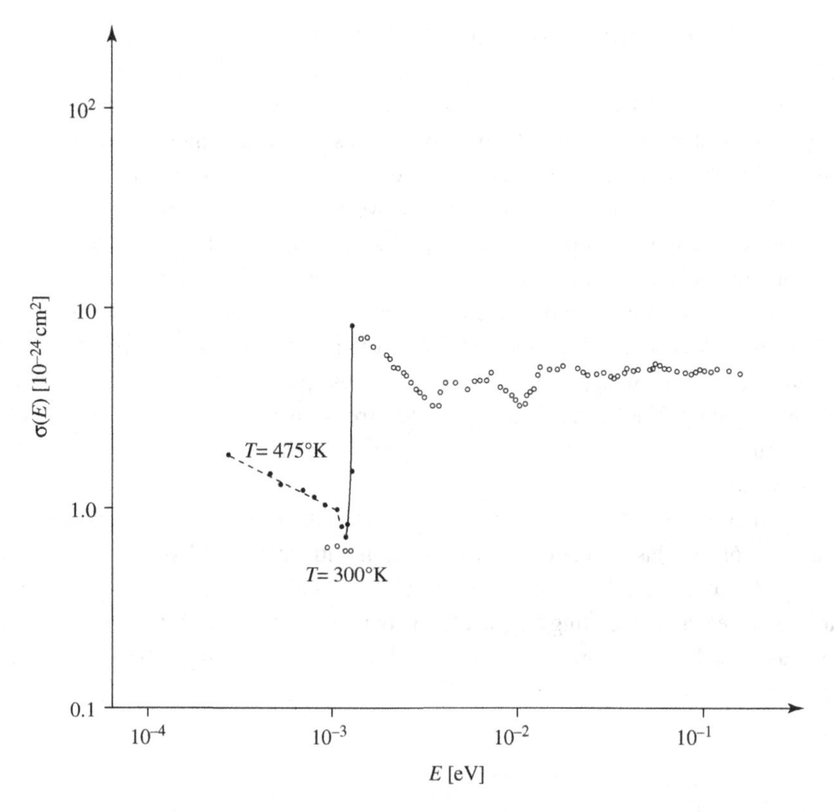

Fig. 3.7 Neutron scattering cross section of a polycrystal of natural carbon in the low-energy region [BNL-325]. Below the Bragg edge the cross section is temperature sensitive, the cross section is larger at higher temperature (data at 700 K are shown as open circles, while closed circles denote 475 K). Above the edge the cross section is temperature independent.

In contrast to the hydrogen cross section, the cross section of carbon in a polycrystalline state, seen in Fig. 3.7, shows much more abrupt energy variations. A sharp increase is observed at $\sim 2 \times 10^{-3}$ eV. Below this rise the cross section is sensitive to temperature whereas it is essentially temperature independent beyond the rise. At energies above 0.1 eV the cross section reaches a constant at about 5 barns, significantly less than the 20 barns for the hydrogen cross section in Fig. 3.6. We will come back to discuss these features in more detail in Chap. 9.

The next level of complexity in cross section behavior is neutron resonances. The cross sections for radiative capture (neutron absorption followed by gamma emission) and resonant elastic scattering have characteristics that can be explained by relatively simple models (see Chap. 12). An example of each is given below.

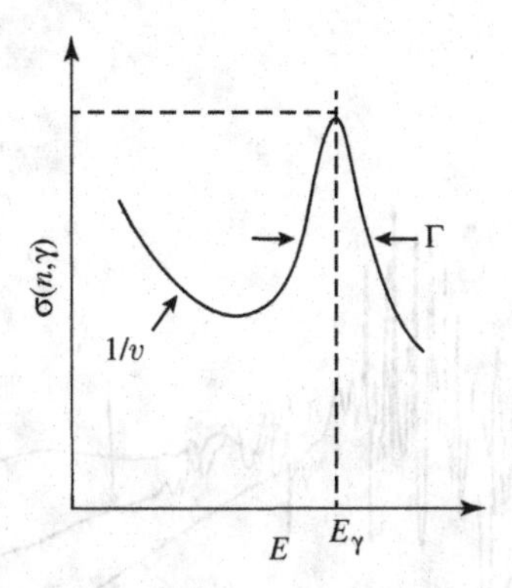

Fig. 3.8 Schematic energy variation of a neutron capture resonance, showing a characteristic $1/v$ behavior below the resonance and a full width at half maximum Γ. The cross section peaks at the resonance energy E_γ.

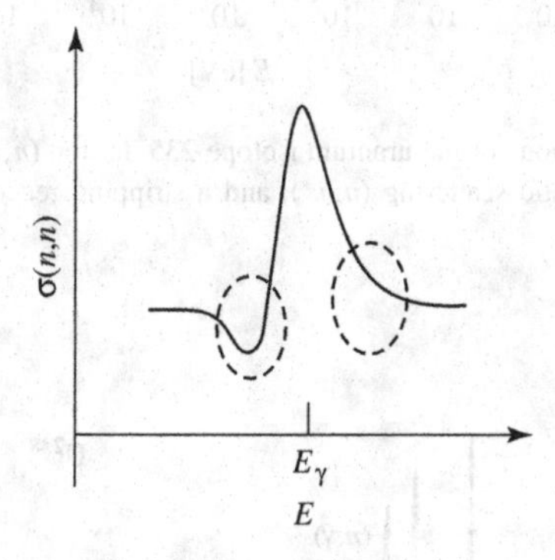

Fig. 3.9 Schematic energy variation of neutron elastic scattering showing interference effects between resonant and potential scattering. Regions of destructive and constructive interference at energies below and above E_γ respectively are indicated by the dashed lines.

The most complex behavior we will encounter pertains to neutron cross sections for a heavy nuclide. In the presence of many nucleons many resonances are possible to give rise to a rich spectrum of peaks and valleys in the energy variations. In the case of uranium, the possibility of fission is an additional complication. We compare the cross sections of U^{235} and U^{238}, bearing in mind the former is known to undergo fission reaction at low neutron energies, while the latter can undergo fission only when the incident neutron has a sufficiently high energy. The neutron cross section behavior, shown in Figs. 3.10 and 3.11, are more complicated than those we will mostly encounter in this book. In fact they are so complex that it is possible to discuss them only very qualitatively.

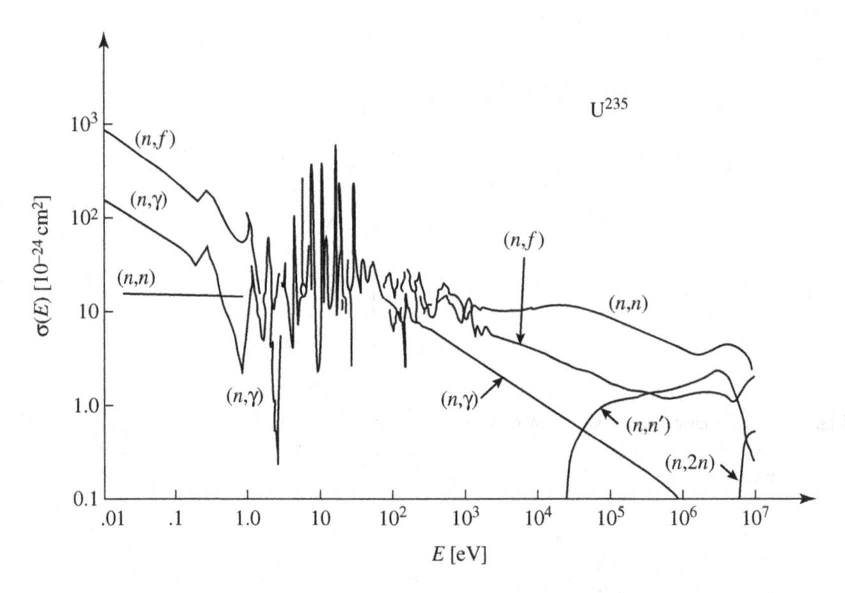

Fig. 3.10 Various cross sections of the uranium isotope 235, fission (n, f), radiative capture (n, γ), elastic scattering (n,n), inelastic scattering (n, n'), and a stripping reaction $(n, 2n)$. [Adapted from El-Wakil, 1962.]

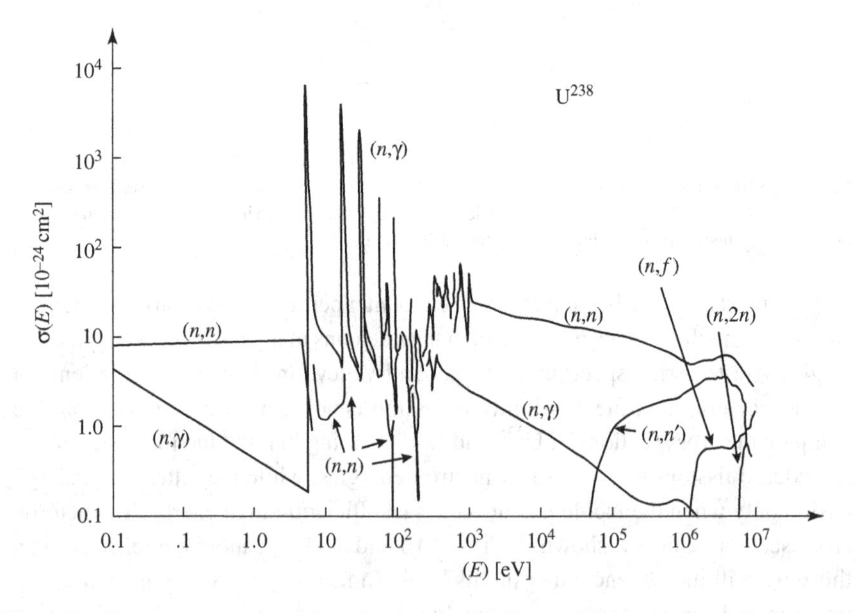

Fig. 3.11 Neutron cross sections of uranium isotope 238 for comparison with Fig. 3.10. [Adapted from El-Wakil, 1962.]

The overall characteristic features in each of the three types of examples discussed here should be noted for later references. These include energy-independent behavior for potential scattering, $1/v$ variation for capture and fission below the resonances, and threshold behavior for inelastic scattering. The student should also pay attention to the magnitudes of the cross sections and the wide energy range covered. In the thermal region, around 0.025 eV, the energy dependence is rather simple, all monotonic variations. Starting at about 1 eV and extending up to about 1 keV, there are many sharp resonances. Above 10 keV the cross sections return to smooth behavior. In the context of using uranium as the fuel for a nuclear reactor, we can see the reason for building thermal reactors — to take advantage of the large fission cross sections for neutrons at thermal energies. Since neutrons from fission are emitted in the MeV range, one also has the problem of slowing down the neutrons past the resonance region where there is a high probability of capture in order to reach the thermal energy range to sustain the chain reaction. Thus, the optimum design of a nuclear system, whether it is a nuclear reactor, a particle accelerator, or anything else, often comes down to a matter of nuclear properties of specific materials, the cross sections.

REFERENCES

E. R. Cohen and B. N. Taylor, The fundamental physical constants, *Physics Today* (August 1995), p. BG9.

M. M. El-Wakil, *Nuclear Power Engineering* (McGraw-Hill, New York, 1962).

R. D. Evans, *The Atomic Nucleus* (McGraw-Hill, New York, 1955).

G. Fea, *Il Nuovo Cimento* **2**, 368 (1935).

E. Fermi, *Nuclear Physics*, Lecture Notes compiled by J. Orear, A. H. Rosenfeld, and R. A. Schluter (University Chicago Press, 1949), revised edition.

D. J. Hughes and J. A. Harvey, Neutron Cross Sections, BNL-325 (Brookhaven National Laboratory, National Nuclear Data Center, 1955), vol. 2, Curves.

K. S. Krane, *Introductory Nuclear Physics* (Wiley, New York, 1987).

C. M. Lederer *et al.*, ed. *Table of Isotopes* (Wiley, New York, 1978).

D. L. Livesey, *Atomic and Nuclear Physics* (Blaisdell, Waltham, 1966).

P. Marmier and E. Sheldon, *Physics of Nuclei and Particles* (Acadmic Press, New York, 1969), vol. I, pp. 64–95.

V. McLane, C. L. Dunford, P. F. Rose, Neutron Cross Sections, 4th ed. (Academic Press, New York, 1988), vol. 2 Neutron Cross Section Curves.

W. E. Meyerhof, *Elements of Nuclear Physics* (McGraw Hill, New York, 1967).

R. A. Nelson, *Physics Today* (August 1995), p. BG15.

S. Pearlstein, Evaluated Nuclear Data Files, in *Advances in Nuclear Science and Technology* (Academic Press, New York, 1975), vol. 8.

E. Segrè, *Nuclei and Particles* (W. A. Benjamin, 1965).

4

Stability of Nuclei

The concept of an atomic structure may be traced back to the speculations by the Ionian philosophers in the fifth century B.C. that all matter can be regarded as assemblies of discrete particles (atoms). Modern atomic physics began with the discovery of X rays by Rontgen in 1895, radioactivity by Becquerel in 1896, and electron by Thomson in1897. An authoritative account of the historical development of the concept of *the Atomic Nucleus* can found in the Introduction of the text by Evans, *The Atomic Nucleus* [Evans, 1955]. Since we know the nucleus is an assembly of protons and neutrons, a natural question is how does the stability of a nuclide vary with changes in the number of the nucleons. Stability should depend on the interactions among the nucleons, so the issue of nuclear stability has to do with understanding the mass, or equivalently the energy, of the nuclides. We might expect the total energy is not the most relevant quantity in the discussion if by stability we mean removing or adding one or more nucleons to the assembly rather than breaking up the nuclide into the individual nucleons. Indeed in this chapter we are interested in the relative stability of nuclides with respect to their near neighbors in the sense of isotopes and isobars.

Our knowledge of nuclear mass, as noted in the previous chapter, is obtained from a large body of experimental data. Now we begin to derive physical understanding of these results through the concept of binding energy, which is an energy difference, effectively the mass decrement sketched in Fig. 3.1. Our goal is to represent the binding energy in terms of contributions from the different interactions between the nucleons, so that one can start to build nuclear models that can account for the experimental data, or make predictions in the absence of measurements.

4.1 MASS AND STABILITY

The stability of nuclei is an appropriate starting point for the study of nuclear radiation interactions in that unstable nuclei undergo transitions that can lead to the emission of particles and/or electromagnetic radiation (gammas). If the transition is spontaneous, it is called a radioactive decay, the topic that will be treated in Chap. 6.

If the transition is induced by the bombardment of particles or radiation, then it is called a nuclear reaction, the topic that will be studied in Chap. 8, concerning neutron elastic scattering, and in Chap. 12, concerning neutron reactions.

The mass of a nucleus is the decisive factor governing its stability, as we have already indicated. Knowing the mass of a particular nucleus in question and those of the neighboring nuclei, one can tell whether or not the nucleus is stable. On the other hand, one should anticipate the relation between mass and stability can be quite complicated. Increasing the mass of a stable nucleus by adding a nucleon can make the resulting nucleus unstable, but this is not always true. Starting with the simplest nucleus, the proton, suppose we add one neutron after another. This would generate the series,

$$\mathrm{H} + n \rightarrow \mathrm{H}^2\ (stable) + n \rightarrow \mathrm{H}^3\ (unstable) \xrightarrow{\beta^-} \mathrm{He}^3\ (stable) + n \rightarrow \mathrm{He}^4\ (stable)$$

Because He^4 is a double-magic nucleus, it is particularly stable (see the discussion of the nuclear shell model in Chap. 5). If we continue to add a nucleon we would find exprimentally the resulting nuclei are unstable,

$$\mathrm{He}^4 + n \rightarrow \mathrm{He}^5 \rightarrow \mathrm{He}^4, \quad \text{with } t_{1/2} \sim 3 \times 10^{-21}\mathrm{s}$$
$$\mathrm{He}^4 + H \rightarrow \mathrm{Li}^5 \rightarrow \mathrm{He}^4, \quad \text{with } t_{1/2} \sim 10^{-22}\mathrm{s}$$

One may ask: With He^4 so stable how is it possible to build up the heavier elements starting with neutrons and protons? This question was considered in the study of the origin of elements [Penzias, 1979]. The answer is that the following reactions can occur

$$\mathrm{He}^4 + \mathrm{He}^4 \rightarrow \mathrm{Be}^8$$
$$\mathrm{He}^4 + \mathrm{Be}^8 \rightarrow \mathrm{C}^{12}\ (stable)$$

Although Be^8 is unstable, its lifetime of $\sim 3 \times 10^{-16}$ s is long enough to enable the next reaction to proceed (in a statistical sense). Once C^{12} is formed, it can react with another He^4 to give O^{16}. In this way, the heavy elements can be formed.

4.2 BINDING ENERGY

Instead of the mass of a nucleus one can use the binding energy to express the same information. The binding energy concept is useful for discussing the calculation of nuclear masses and of energy released or absorbed in nuclear reactions. We define the binding energy of a nucleus with mass $M(A, Z)$ as

$$B(A, Z) \equiv [ZM_H + NM_n - M(A, Z)]c^2 \tag{4.1}$$

where M_H is the hydrogen mass and $M(A, Z)$ is the atomic mass. Strictly speaking one should subtract out the binding energy of the electrons; however, the error in not doing so is quite small, so we will ignore it. According to Eq. (4.1), the nuclear binding energy $B(A, Z)$ is the difference between the mass of the constituent nucleons, when they are far separated from each other, and the mass of the nucleus. Therefore, one can interpret $B(A, Z)$ as the *work required* to separate the nucleus into the individual nucleons (separated far from each other), or equivalently, as the *energy released* during the assembly of the nucleus from the constituents.

Taking the actual data on nuclear mass for various A and Z, one can calculate $B(A, Z)$ and plot the results in the form shown in Fig. 4.1.

The most striking feature of the B/A curve is the approximate constancy at ~8 MeV per nucleon, except for the very light nuclei. It is instructive to see what this behavior implies. If the binding energy of a pair of nucleons is a constant, say C, then for a nucleus with A nucleons, in which there are $A(A-1)/2$ distinct pairs of nucleons, the B/A would be $\sim C(A-1)/2$. Since this is not what one sees in Fig. 4.1, one can surmise that a given nucleon is not bound equally to all the other nucleons; in other words, nuclear forces, being short-ranged, extend over only a few neighbors. The constancy of B/A implies a saturation effect in

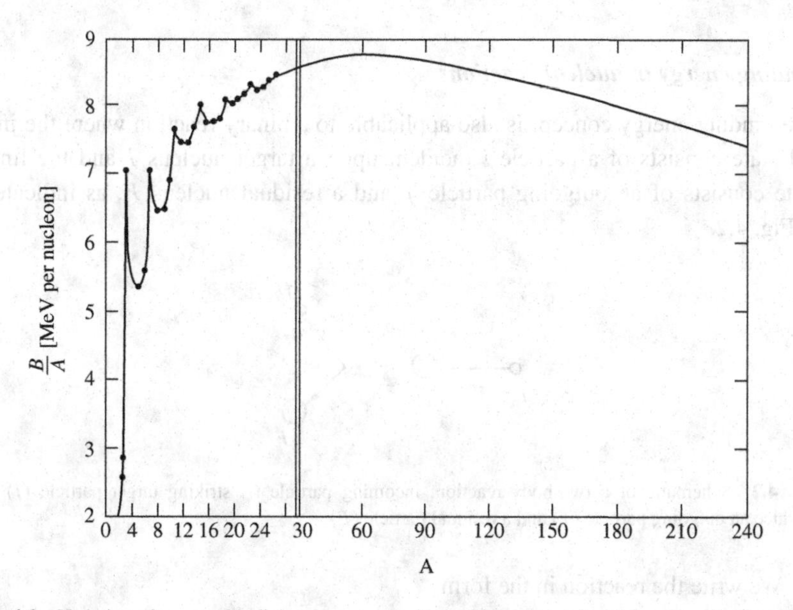

Fig. 4.1 Variation of average binding energy per nucleon with mass number A for naturally occurring nuclides (and ^{8}Be). Note scale change at $A = 30$. [Adapted from Evans, 1955, p. 299.]

nuclear forces, the interaction energy of a nucleon does not increase any further once it has acquired a certain number of neighbors. This number seems to be about 4 or 5.

One can understand the initial rapid increase of B/A for the very light nuclei as the result of the competition between volume effects, which make B increase with A like A, and surface effects, which make B decrease (in the sense of a correction) with A like $A^{2/3}$. The latter should be less important as A becomes large, hence B/A increases (see the discussion of the semi-empiricial mass formula in the next section). At the other end of the curve, the gradual decrease of B/A for $A > 100$ can be understood as the effect of Coulomb repulsion which becomes more important as the number of protons in the nucleus increases.

As a quick application of the B/A curve we make a rough estimate of the energies release in fission and fusion reactions. Suppose we have symmetric fission of a nucleus with $A{\sim}240$ producing two fragments, each $A/2$. The reaction gives a final state with B/A of about 8.5 MeV, which is about 1 MeV greater than the B/A of the initial state. Thus the energy released per fission reaction is about 240 MeV. (A more accurate estimate gives ~200 MeV.) For fusion reaction we take $H^2 + H^2 \rightarrow He^4$. The B/A values of H^2 and He^4 are 1.1 and 7.1 MeV/nucleon respectively. The gain in B/A is 6 MeV/nucleon, so the energy released per fusion event is ~24 MeV.

Binding energy in nuclear reactions

The binding energy concept is also applicable to a binary reaction where the initial state consists of a particle i incident upon a target nucleus I and the final state consists of an outgoing particle f and a residual nucleus F, as indicated in Fig. 4.2,

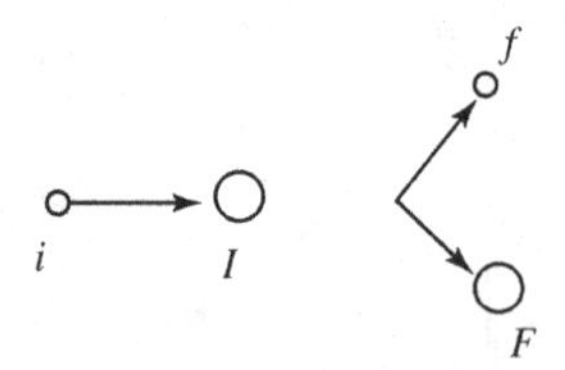

Fig. 4.2 Schematic of a two-body reaction, incoming particle (i) striking target particle (I) to produce an outgoing particle (f) and a residual particle (F).

We write the reaction in the form

$$i + I \rightarrow f + F + Q \tag{4.2}$$

where Q is an energy called the "Q-value of the reaction". Corresponding to (4.2), we have the definition

$$Q \equiv [(M_i + M_I) - (M_f + M_F)]c^2 \tag{4.3}$$

where the masses are understood to be atomic masses. Every nuclear reaction has a characteristic Q-value; the reaction is called exothermic (endothermic) for $Q > 0 (< 0)$ where energy is given off (absorbed). Thus an endothermic reaction cannot take place unless additional energy, called the threshold, is supplied. Often the source of this energy is the kinetic energy of the incident particle. One can also express Q in terms of the kinetic energies of the reactants and products of the reaction by invoking the conservation of *total* energy, which must hold for any reaction,

$$T_i + M_i c^2 + T_I + M_I c^2 \to= T_f + M_f c^2 + T_F + M_F c^2 \tag{4.4}$$

Combining this with Eq. (4.3) gives

$$Q = T_f + T_F - (T_i + T_I) \tag{4.5}$$

Usually T_I is negligible compared to T_i because the target nucleus follows a Maxwellian distribution at the temperature of the target sample (typically room temperature), while for nuclear reactions the incident particle can have a kinetic energy upto the MeV range. Exceptions are thermal neutron reactions, see Chaps. 9, 12, and 13, or any exothermic reaction in general. Since the rest masses can be expressed in terms of binding energies, another expression for Q is

$$Q = B(f) + B(F) - B(i) - B(I) \tag{4.6}$$

As an example, the nuclear medicine technique called *boron neutron capture therapy (BNCT)* is based on the reaction

$$_5B^{10} + n \to {}_3Li^7 + He^4 + Q \tag{4.7}$$

In this case, $Q = B(Li) + B(\alpha) - B(B) = 39.245 + 28.296 - 64.750 = 2.791$ MeV. The reaction is exothermic, therefore it can be induced by a thermal neturon. In practice, the simplest way of calculating Q values is to use the rest masses of the reactants and products. For many reactions of interest Q values are in the range 1–5 MeV. An important exception is fission, where $Q \sim 170 - 210$ MeV, depending on what one considers to be the fission products. Notice that if one defines Q in terms of the kinetic energies, as in Eq. (4.5), it may appear that the value of Q would depend on whether one is in the laboratory or the center-of-mass coordinate system. This is illusory because the equivalent definition, Eq. (4.3), is clearly independent of whatever coordinate system is adopted.

Separation energy

Recall the definition of binding energy, Eq. (4.1), involves an initial state where all the nucleons are separated far from each other. One can define another binding energy where the initial state is one where only one nucleon is separated off. The energy required to separate particle a from a nucleus is called the *separation energy* S_a. It is also the energy released, or energy available for reaction, when particle a is captured. The concept of a separation energy is quite general. We will apply it typically to a neutron, proton, deuteron, or α-particle. The energy balance takes the form

$$S_a = [M_a(A', Z') + M(A - A', Z - Z') - M(A, Z)]c^2 \tag{4.8}$$

where particle a is treated as a "nucleus" with atomic number Z' and mass number A'. For a neutron,

$$\begin{aligned} S_n &= [M_n + M(A - 1, Z) - M(A, Z)]c^2 \\ &= B(A, Z) - B(A - 1, Z) \end{aligned} \tag{4.9}$$

S_n is sometimes called the binding energy of the *last neutron*.

Clearly S_n will vary from one nucleus to another. In the range of A where B/A is roughly constant we can estimate from the B/A curve that $S_n \sim S_p \sim 8$ MeV. This is a rough figure, for heavy nuclei S_n is more like 5–6 MeV. It turns out that when a nucleus $M(A - 1, Z)$ absorbs a neutron, there is ~1 MeV (or more, can be up to 4 MeV) difference in the value of S_n depending on whether the particular neutron absorbed is an even or odd neutron (see Fig. 4.3). This difference is the reason that U^{235} can undergo fission with thermal neutrons, whereas U^{238} can fission only with fast neutrons ($E > 1$ MeV), as we have noted in comparing the cross section behavior shown in Figs. 3.10 and 3.11.

Generally speaking the following systematic behavior is observed in neutron and proton separation energies,

$$\begin{aligned} &S_n\ (\text{even } N) > S_n(\text{odd } N) \quad \text{for a given } Z \\ &S_p\ (\text{even } Z) > S_p(\text{odd } Z) \quad \text{for a given } N \end{aligned}$$

This effect is attributed to the pairing property of nuclear forces — the existence of extra binding between a pair of identical nucleons in the same state which have total angular momenta pointing in opposite directions (see Sec. 4.3). This is also the reason for the exceptional stability of the α-particle. Because of pairing the even-even (even Z, even N) nuclei are more stable than the even-odd and odd-even nuclei, which in turn are more stable than the odd-odd nuclei.

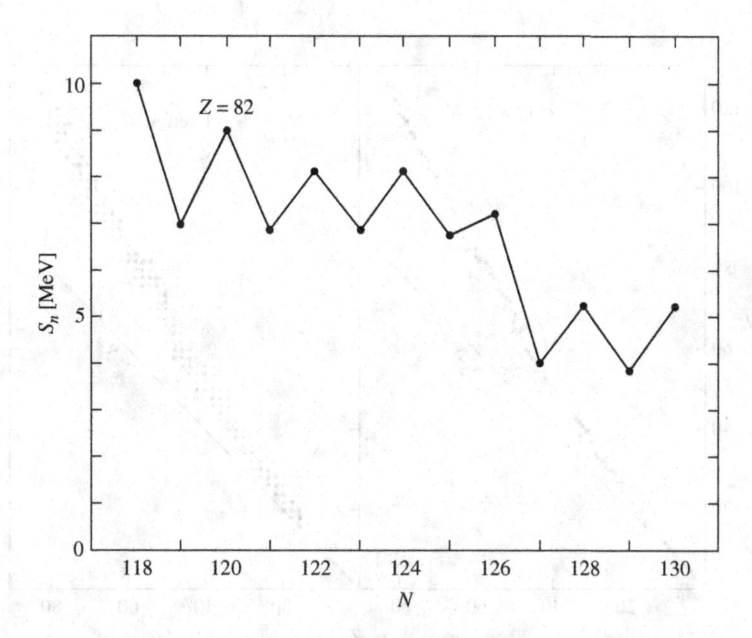

Fig. 4.3 Variation of the *neutron separation* energies of lead isotopes with neutron number of the absorbing nucleus showing the dependence on the odd or even nature of the neutron being separated. [Adapted from Meyerhof, 1967.]

Abundance systematics of stable nuclides

One can construct a stability chart by plotting the neutron number N versus the atomic number Z for all the stable nuclides. The results, shown in Fig. 4.4, show that $N \sim Z$ for low A, but $N > Z$ at high A.

Again, one can readily understand that in heavy nuclei the Coulomb repulsion will favor a neutron-proton distribution with more neutrons than protons. It is a little more involved to explain why there should be an equal distribution for the light nuclides (see the following discussion on the semi-empirical mass formula). We will simply note that to have more neutrons than protons means that the nucleus has to be in a higher energy state, and is therefore less stable. This symmetry effect is most pronounced at low A and becomes less important at high A. In connection with Fig. 4.4 we note:

(i) In the case of odd A, only one stable isobar exists, except $A = 113, 123$.
(ii) In the case of even A, only even-even nuclides exist, except $A = 2, 6, 10, 14$.

Still another way to summarize the trend of stable nuclides is shown in Table 4.1 (from Meyerhof, 1967)

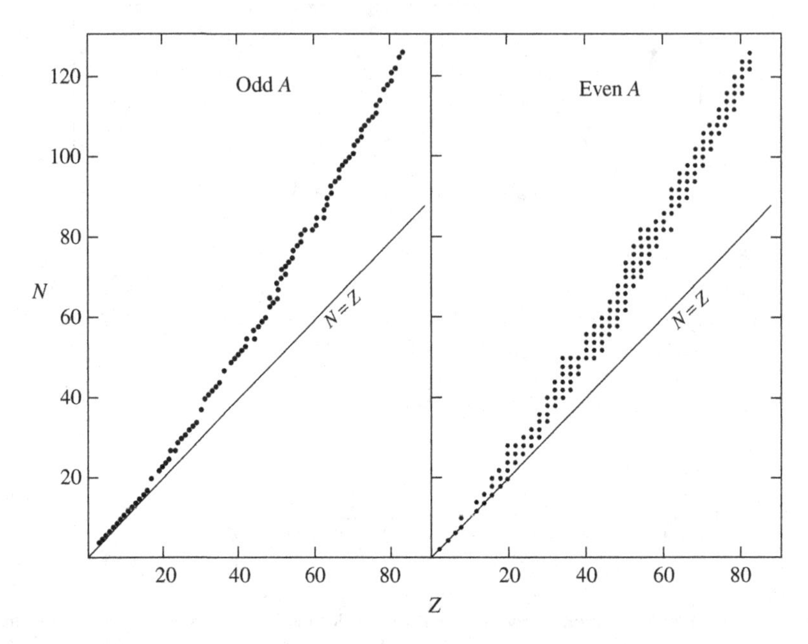

Fig. 4.4 Neutron and proton numbers of stable nuclides which are odd (left) and even isobars (right) showing the increasing deviation from the $N = Z$ symmetry which holds for the very light nuclides. [Adapted from Meyerhof, 1967.]

Table 4.1 Systematics of stability trends in nuclei.

A	*Z*	*N*	Type	Alterative designation	Number of stable + long-lived nuclides	Degree of stability	Usual number of stable isotopes per element
Even	Even	Even	e-e	Even mass, even *N*	$166 + 11 = 177$	Very pronounced	Several (2 and 3)
Odd	Even	Odd	e-o	Odd mass, odd *N*	$55 + 3 = 58$	Fair	1
Odd	Odd	Even	o-e	Odd mass, even *N*	$51 + 3 = 54$	Fair	1
Even	Odd	Odd	o-o	Even mass, odd *N*	$6 + 4 = 10$	Low	0
					$278 + 21 = 299$		

4.3 MASS PARABOLAS

The binding energy curve we have discussed is an overall representation of how the stability of nuclides varies across the entire range of mass number A. The curve shown in Fig. 4.1 is based on experimental data on atomic masses. One way to analyze this curve is to decompose the binding energy into various contributions from the interactions among the nucleons. An empirical formula for the binding energy consisting of contributions representing volume, surface, Coulomb and other effects was first proposed by von Weizsäcker in 1935. Such a formula is useful because it not only allows one to calculate the mass of a nucleus, thereby eliminating the need for a table of mass data, but also it leads to qualitative understanding of the essential features of nuclear binding. More detailed theories exist [Bruecker, 1961], but they are beyond the scope of our study.

The empirical mass formula we consider here was derived on the basis of the liquid crop model of the nucleus. The essential assumptions are the nucleus is composed of incompressible matter, thus $R{\sim}A^{1/3}$, the nuclear force is the same among neutrons and protons (excluding Coulomb interactions), and the nuclear force saturates (meaning it is very short ranged).

The empirical mass formula is usually given in terms of the binding energy,

$$B(A, Z) = a_v A - a_s A^{2/3} - a_c \frac{Z(Z-1)}{A^{1/3}} - a_a \frac{(N-Z)^2}{A} + \delta \qquad (4.10)$$

The coefficients a are to be determined (by fitting the mass data), where the subscripts v, s, c, and a referring to volume, surface, Coulomb, and asymmetry respectively. The last term in Eq. (4.10) represents the pairing effects,

$$\begin{aligned} \delta &= a_p/\sqrt{A} && \text{even-even nuclei} \\ &= 0 && \text{even-odd, odd-even nuclei} \\ &= -a_p/\sqrt{A} && \text{odd-odd nuclei} \end{aligned}$$

where coefficient a_p is also a fitting parameter. A set of values for the five coefficients in Eq. (4.10) is [Meyerhof, 1967]:

a_v	a_s	a_c	a_a	a_p	
16	18	0.72	23.5	11	MeV

Since the fitting to experimental data is not perfect one can expect to find several slightly different sets of coefficients in the literature. The average accuracy of Eq. (4.10) is about 2 MeV except where strong shell effects are present. One can also add a term, $\sim$1 to 2 MeV, to Eq. (4.10) to represent the shell effects of extra binding for nuclei with closed shells of neutrons or protons.

A simple way to interpret Eq. (4.10) is to regard the first term as a first approximation where the binding energy is taken to be proportional to the volume of the nucleus, or the mass number A. This assumes every nucleon is like every other nucleon which by itself would be an over-simplification. The remaining terms then can be regarded as corrections to this first approximation. Notice the terms representing surface, Coulomb and asymmetry all enter with a negative sign, each one subtracting from the volume term. It is not surprising the surface term should vary with $A^{2/3}$, or R^2. The Coulomb term is also quite self-evident, considering that $Z(Z-1)/2$ is the number of pairs that can be formed from Z protons, and the $1/A^{1/3}$ factor comes from the $1/R$ variation with distance of separation. The asymmetry term in Eq. (4.10) is less obvious, so we digress to derive it [Meyerhof, 1967].

What we wish to estimate is the energy difference between an actual nucleus, where $N > Z$, and an ideal nucleus, where $N = Z = A/2$. This is then the energy to transform a symmetric nucleus, in the sense of $N = Z$, to an asymmetric one, $N > Z$. For fixed N and Z, the number of protons that we need to transform into neutrons is ν, with $N = (A/2) + \nu$ and $Z = (A/2) - \nu$. Thus, $\nu = (N - Z)/2$. Now consider a set of energy levels for the neutrons and another set for the protons, each one filled to a certain level. To transform ν protons into neutrons the protons in question have to go into unoccupied energy levels above the last neutron. This means the amount of energy involved is ν (the number of nucleons that have to be transformed) times $\nu\Delta$ (energy change for each nucleon to be transformed) $= \nu^2\Delta = (N - Z)^2\Delta/4$, where Δ is the spacing between energy levels (assume to be the same for all the levels). To estimate Δ, we note that $\Delta \sim E_F/A$, where E_F is the *Fermi energy* (see Chap. 5) which is known to be independent of A. Thus $\Delta \sim 1/A$, and we have the expression for the asymmetry term in Eq. (4.10).

The magnitudes of the various contributions to the binding energy curve are depicted in Fig. 4.5. The initial rise of B/A with A is seen to be due to the decreasing importance of the surface contribution as A increases. The Coulomb repulsion effect grows in importance with A, causing a maximum in B/A at $A \sim 60$, and a subsequent decrease of B/A at larger A. Except for the extreme ends of the mass number range the semi-empirical mass formula generally can give binding energies accurate to within 1% of the experimental values (Evans, p. 382). This means that atomic masses can be calculated correctly to about 1 part in 10^4. However, there are conspicuous discrepancies in the neighborhood of magic nuclei. Attempts have been made to take into account the nuclear shell effects by generalizing the mass

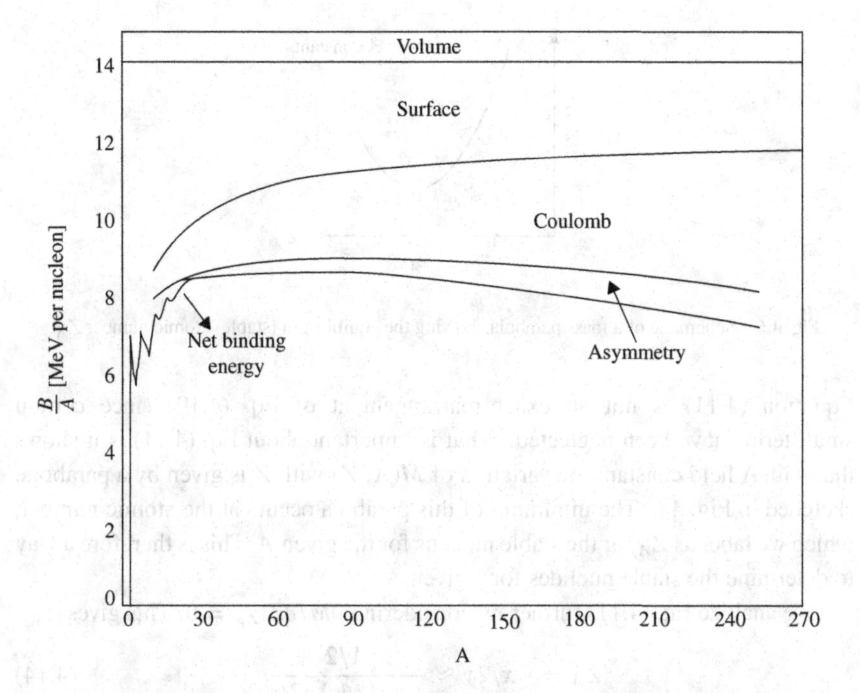

Fig. 4.5 Relative contributions to the binding energy per nucleon showing the importance of the various terms in the semi-empirical Weizsäcker formula. [Adapted from Evans, 1955.]

formula. In addition to what we have already mentioned, one can consider another term representing nuclear deformation [Marmier 1969, pp. 39].

The mass formula can be used to determine the constant r_o in the expression for the nuclear radius, $R = r_o A^{1/3}$. The radius appears in the coefficients a_v and a_s. In this way one obtains $r_o = 1.24 \times 10^{-13}$ cm.

Mass parabolas and stability line

The mass formula also can be rearranged to give an expression for the mass $M(A, Z)$ of a nuclide,

$$M(A, Z)c^2 \cong A[M_n c^2 - a_v + a_a + a_s/A^{1/3}] + xZ + yZ^2 - \delta \qquad (4.11)$$

where

$$x = -4a_a - (M_n - M_H)\,c^2 \cong -4a_a \qquad (4.12)$$

$$y = \frac{4a_a}{A} + \frac{a_c}{A^{1/3}} \qquad (4.13)$$

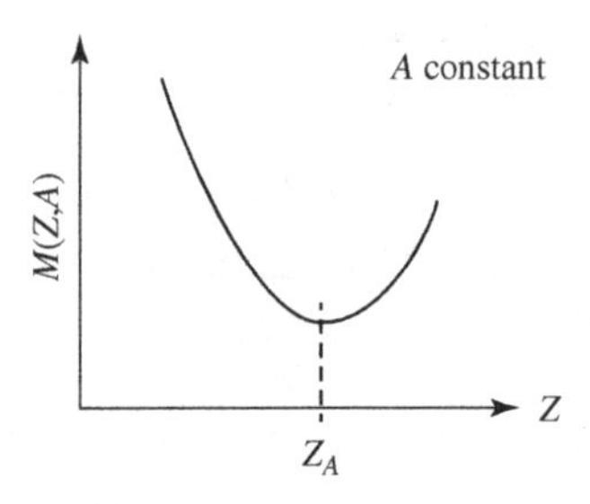

Fig. 4.6 Schematic of a mass parabola showing the equilibrium (stable) atomic number Z_A.

Equation (4.11) is not an exact rearrangement of Eq. (4.10) since certain small terms have been neglected. What is important about Eq. (4.11) is it shows that with A held constant the variation of $M(A, Z)$ with Z is given by a parabola, sketched in Fig. 4.6. The minimum of this parabola occurs at the atomic number, which we label as Z_A, of the stable nucleus for the given A. This is therefore a way to determine the stable nuclides for a given A.

We analyze Eq. (4.11) further by considering $\partial M/\partial Z|_{Z_A} = 0$. This gives

$$Z_A = -x/2y \approx \frac{A/2}{1 + \frac{1}{4}\left(\frac{a_c}{a_a}\right)A^{2/3}} \tag{4.14}$$

Notice that if we had considered only the volume, surface and Coulomb terms in $B(A, Z)$, then we would have found instead of Eq. (4.14) the expression

$$Z_A \approx \frac{(M_n - M_H)c^2 A^{1/3}}{2a_c} \sim 0.9A^{1/3} \tag{4.15}$$

This is a very different result because for a stable nucleus with $Z_A = 20$ the corresponding mass number given by Eq. (4.15) would be ~9000, which is clearly unrealistic. Fitting Eq. (4.14) to the experimental data gives $a_c/4a_a = 0.0078$, or $a_a \sim 20 - 23$ MeV. We see therefore the deviation of the stability line from $N = Z = A/2$ is the result of Coulomb effects, which favor $Z_A < A/2$, becoming relatively more important than the asymmetry effects, which favor $Z_A = A/2$.

We can ask what happens when a nuclide is unstable because it is proton-rich. The answer is that a nucleus with too many protons for stability can emit a *positron* (positive electron e^+ or β^+) and thus convert a proton into a neutron. In this process a *neutrino* (ν) is also emitted. An example of a positron decay is

$${}_9F^{16} \rightarrow {}_8O^{16} + \beta^+ + \nu \tag{4.16}$$

By the same token if a nucleus has too many neutrons, then it can emit an electron (e^- or β^-) and an *antineutrino* $\bar{\nu}$, converting a neutron into a proton. An electron decay for the isobar $A = 16$ is

$$_7N^{16} \rightarrow {_8O^{16}} + \beta^- + \bar{\nu} \tag{4.17}$$

A competing process with positron decay is *electron capture* (EC), where an inner shell atomic electron is captured by the nucleus so the nuclear charge is reduced by 1. (Notice orbital electrons can spend a fraction of their time inside the nucleus.) The atom as a whole would remain neutral but it is left in an excited state because a vacancy has been created in one of its inner shells.

As far as atomic mass balance is concerned, the requirement for each process to be energetically allowed is:

$$M(A, Z+1 > M(A, Z) + 2m_e \quad \beta^+ - decay \tag{4.18}$$

$$M(A, Z) > M(A, Z+1) \quad \beta^- - decay \tag{4.19}$$

$$M(A, Z+1) > M(A, Z) \quad \text{EC} \tag{4.20}$$

where $M(A, Z)$ now denotes atomic mass. Notice that *EC* is a less stringent condition for the nucleus to decrease its atomic number. If the energy difference between initial and final states is less than twice the electron rest mass (1.02 MeV), the transition can take place via *EC* while it would be energetically forbidden via positron decay. The reason for the appearance of the electron rest mass in Eq. (4.18) may be explained by looking at an energy balance in terms of nuclear mass $M'(A, Z)$, which is related to the atomic mass by $M(A, Z) = Zm_e + M'(A, Z)$, if we ignore the binding energy of the electrons in the atom. For β^+*-decay* the energy balance is

$$M'(A, Z) = M'(A, Z-1) + m_e + \nu \tag{4.21}$$

which we can rewrite as

$$Zm_e + M'(A, Z) = (Z-1)m_e + M'(A, Z-1) + 2m_e + \nu \tag{4.22}$$

The LHS is just $M(A, Z)$ while the RHS is at least $M(A, Z-1) + 2m_e$ with the neutrino having a variable energy. Thus one obtains Eq. (4.18). Another way to look at this condition is that in addition to the positron emitted, the daughter nuclide also has to eject an electron (from an outer shell) in order to preserve charge neutrality.

Having discussed how a nucleus can change its atomic number Z while preserving its mass number A, we can predict what transitions will occur as an unstable nuclide moves along the mass parabola toward the point of stability. Since the pairing term δ vanishes for odd-A isobars, there will be only a single mass parabola in this case in contrast to two mass parabolas for the even-A isobars. One might then expect that when A is odd there can be only one stable

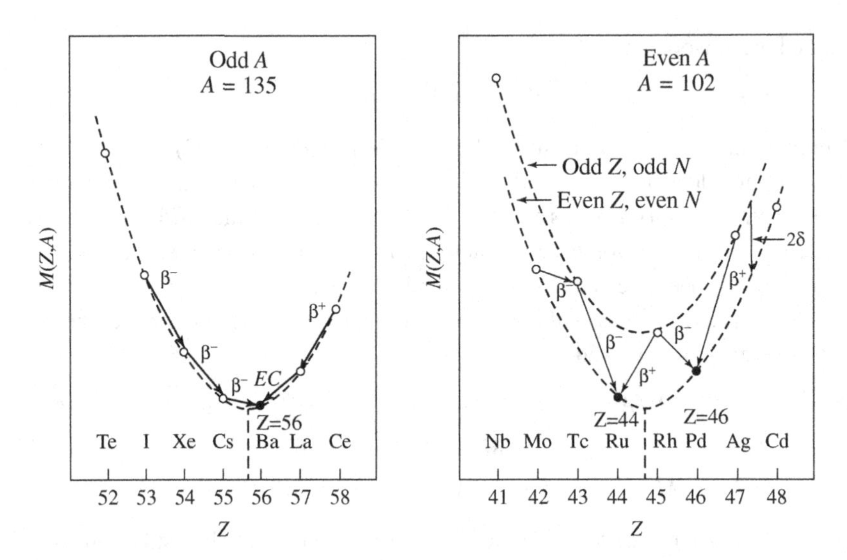

Fig. 4.7 Mass parabolas for an odd and an even isobars. Stable and radioactive nuclides are denoted by closed and open circles respectively. [Adapted from Meyerhof, 1967.]

isobar. This is generally true with two exceptions, at $A = 113$ and 123. In these two cases the discrepancies arise from small mass differences which cause one of the isobars in each case to have exceptionally long half-life. In the case of even A there can be stable even-even isobars (three is the largest number found). Since the odd-odd isobars lie on the upper mass parabola, one would expect there should be no stable odd-odd nuclides. Yet there are several exceptions, H^2, Li^6, B^{10} and N^{14}. One explanation is that there are rapid variations of the binding energy for the very light nuclides due to nuclear structure effects that are not taken into account in the semi-empirical mass formula. For certain odd-odd nuclides both conditions for β^+ and β^- decays are satisfied, and indeed both decays do occur in the same nucleus. Examples of odd- and even-A mass parabolas are shown in Fig. 4.7.

REFERENCES

K. A. Brueckner *et al.*, Properties of finite nuclei, *Physical Review* **121**, 255 (1961).

R. D. Evans, *The Atomic Nucleus* (McGraw Hill, New York, 1955).

E. Fermi, *Nuclear Physics*, Lecture Notes by J. Orear, A. H. Rosenfeld and R. A. Schluter (University of Chicago Press, 1949).

K. S. Krane, *Introductory Nuclear Physics* (Wiley, New York, 1987).

P. Marmier and E. Sheldon, *Physics of Nuclei and Particles* (Acadmic Press, New York, 1969), vol. I, pp. 39.

W. E. Meyerhof, *Elements of Nuclear Physics* (McGraw Hill, New York, 1967).

A. Penzias, The origin of elements, *Reviews of Modern Physics* **51**, 425 (1979).

E. Segrè, *Nuclei and Particles* (W. A. Benjamin, 1965).

5

Energy-Level Models

In this chapter, we study the energy levels of a particle confined in a potential well using elementary quantum mechanics. For readers either unfamiliar with quantum mechanics or in need of a review, this is a good opportunity to build a foundation that will prove useful throughout the studies in this book. The basic problem we face is to formulate a method to calculate the energy levels that can exist in a potential well of a given depth and width. As a start we consider the problem in one dimension. Then we extend the calculation to a three-dimensional well, thereby introducing the concept of orbital angular momentum. Beyond this extension we will also consider the existence of a nuclear spin which can couple with the orbital angular momentum to give rise to a complex interaction which is needed to explain a number of observed nuclear properties. Moreover, the basic formalism of quantum mechanical description of a particle in a potential carries over later to the problem of nuclear interaction, as we will see in Chaps. 7 and 8.

From the standpoint of nuclear physics background we have two objectives in mind. First is to give the reader an understanding of how discrete (quantized) energy levels depend on a given confining potential. For some students this may be a simple review of bound-state calculations. Secondly we wish extend to more realistic situations — three dimensional systems and spin-orbit coupling — in order to understand well-known experimental observations such as the magic numbers, and to predict spin and parity of nuclides in their ground states.

5.1 PARTICLE IN A SQUARE WELL

We will solve the Schrödinger wave equation by considering the simplest problem in quantum mechanics, a particle in a potential well [Schiff, 1955; Liboff, 1980]. The student will see from this calculation the problem is treated by dividing the system into two regions, the *interior* where the particle feels the potential, and the *exterior* where the particle is a free particle (zero potential). The solutions to the wave equation have to be different in these two regions to reflect the binding of the particle, namely, the wave function is oscillatory in the interior region and exponentially decaying (non-oscillatory) in the exterior region. Matching these two solutions at the boundary where the potential goes from a finite value (interior) to zero (exterior) gives a condition on the wavenumber (or wavelength), which turns out to be the condition of *quantization*. The meaning of quantization is that solutions exist only if the wavenumbers take on certain discrete values, which then translate into discrete energy levels for the particle. For a given potential well of certain depth and width, only a discrete set of wave functions can exist in the potential well. These wave functions are the eigenfunctions of the Hamiltonian (energy) operator, with corresponding energy levels as the eigenvalues. Finding the wavefunctions and the spectrum of eigenvalues is what we mean by solving the Schrödinger wave equation for the particle in a potential well. Changing the shape of the potential means a different set of eigenfunctions and the eigenvalues. The procedure to find them, however, is the same.

For a one-dimensional system the time-independent wave equation is [Schiff, 1955; Liboff, 1980]

$$-\frac{\hbar^2}{2m}\frac{d^2\psi(x)}{dx^2} + V(x)\psi(x) = E\psi(x) \tag{5.1}$$

We will use this equation to investigate the bound-states of a particle in a square well potential of depth V_o and width L. The physical meaning of Eq. (5.1) is the statement of energy conservation, the total energy E, a negative and constant quantity in the present problem of bound-state calculations, is the sum of *kinetic* and *potential* energies. Since Eq. (5.1) holds at every point in space, the fact that the potential energy $V(x)$ varies in space means the kinetic energy of the particle also will vary in space. For a square well potential, $V(x)$ has the form

$$\begin{aligned} V(x) &= -V_o \qquad -L/2 \leq x \geq L/2 \\ &= 0 \qquad \text{elsewhere} \end{aligned} \tag{5.2}$$

as shown in Fig. 5.1. Taking advantage of the piecewise constant behavior of the potential, we divide the configuration space (our entire system) into an *interior* region, where the potential is constant and negative, and an *exterior* region where the potential is zero. For the interior region the wave equation can be put into the standard form of a second-order differential equation with constant coefficient,

$$\frac{d^2\psi(x)}{dx^2} + k^2\psi(x) = 0 \quad |x| \leq L/2. \tag{5.3}$$

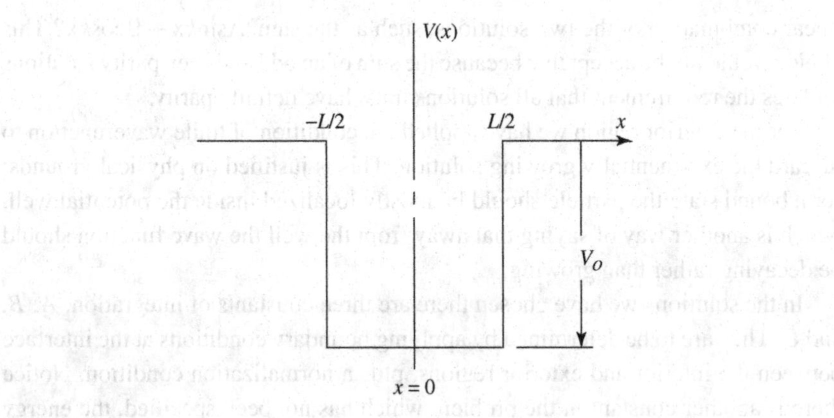

Fig. 5.1 The square well potential centered at the origin with depth V_o and width L.

In Eq. (5.3), we purposely introduced the wavenumber k, with $k^2 = 2m(E+V_o)/\hbar^2$ always positive, so that k is real. We will be doing this consistently in writing out the wave equation to be solved. In other words, the wavenumber we introduce is always real, whereas the sign of the second term in the wave equation can be positive, as in Eq. (5.3), or negative, as in Eq. (5.4) below. For k^2 to be positive we understand the solutions where absolute value of $E > V_o$ will be excluded from our considerations.

For the exterior region, the wave equation similarly can be put into the form

$$\frac{d^2\psi(x)}{dx^2} - \kappa^2\psi(x) = 0 \quad |x| \geq L/2 \tag{5.4}$$

where $\kappa^2 = -2mE/\hbar^2$. To obtain the solutions of physical interest to Eqs. (5.3) and (5.4), we keep in mind the solutions should have certain symmetry properties, in this case they should have definite parity, or inversion symmetry (see the discussion at the end of Sec. 5.2). This means that when $x \to -x$, $\psi(x)$ must be either invariant or it must change sign. The reason for this requirement is that the Hamiltonian H is symmetric under inversion (the potential is symmetric given our choice of the coordinate system (see Fig. 5.1)). Thus we take for our solutions

$$\begin{aligned} \psi(x) &= A\sin kx \quad && |x| \leq L/2 \\ &= Be^{-\kappa x} && x > L/2 \\ &= Ce^{\kappa x} && x < -L/2. \end{aligned} \tag{5.5}$$

We have used the condition of *definite parity* in choosing the interior solution to be a sine function, an odd function. The choice of a solution with odd parity is arbitrary because an even-parity solution, *coskx*, would be equally acceptable. What about a linear combination of the two solutions, such as the sum, $A\sin kx + B\cos kx$? This choice would not be acceptable because the sum of an odd and even parity solutions violates the requirement that all solutions must have definite parity.

For the exterior region we have applied the condition of finite wavefunction to discard the exponentially growing solution. This is justified on physical grounds; for a bound state the particle should be mostly localized inside the potential well, which is another way of saying that away from the well the wave function should be decaying rather than growing.

In the solutions we have chosen there are three constants of integration, A, B, and C. They are to be determined by applying boundary conditions at the interface between the interior and exterior regions, plus a normalization condition. Notice there is another constant in the problem which has not been specified, the energy eigenvalue E. All we have said thus far is E should be negative. We have already utilized the boundary condition at infinity and the inversion symmetry condition of definite parity. The conditions which we have not yet applied are the continuity conditions at the interface between the interior and exterior regions. At $x_o = \pm L/2$, the boundary conditions are

$$\psi_{\text{int}}(x_o) = \psi_{ext}(x_o) \tag{5.6}$$

$$\left.\frac{d\psi_{\text{int}}(x)}{dx}\right|_{x_o} = \left.\frac{d\psi_{ext}(x)}{dx}\right|_{x_o} \tag{5.7}$$

with subscripts *int* and *ext* denoting the interior and exterior solutions respectively.

The four conditions at the interface do not allow us to determine the four constants because our system of equations is homogeneous. As in situations of this kind, the proportionality constant is fixed by the normalization condition. We therefore obtain $C = -B$, $B = A\sin(kL/2)\exp(\kappa L/2)$, and

$$\cot(kL/2) = -\kappa/k \tag{5.8}$$

with the constant A determined by normalization. Eq. (5.8) is the most important result of this calculation; it is sometimes called a *dispersion relation*. It is a relation which determines the allowed values of E, a quantity that appears in both k and κ. These are then the discrete (quantized) energy levels which the particle can have in the particular potential well given, a square well of width L and depth V_o. Equation (5.8) is the consequence of choosing odd-parity solutions for the interior wave. For the even-parity solutions, $\psi_{\text{int}}(x) = A'\cos kx$, the corresponding dispersion relation is

$$\tan(kL/2) = \kappa/k \tag{5.9}$$

Since both solutions are equally acceptable, one has two distinct sets of energy levels, given implicitly by Eqs. (5.8) and (5.9). To see these levels more explicitly, further analysis is necessary.

We consider a graphical analysis of Eqs. (5.8) and (5.9). We first put the two equations into dimensionless form,

$$\xi \cot \xi = -\eta \quad (odd\text{-}parity) \tag{5.10}$$

$$\xi \tan \xi = \eta \quad (even\text{-}parity) \tag{5.11}$$

where $\xi = kL/2$, $\eta = \kappa L/2$. Then we notice that

$$\xi^2 + \eta^2 = 2mL^2|V_o|/4\hbar^2 \equiv \Lambda \tag{5.12}$$

is a constant for fixed values of V_o and L. In Fig. 5.2, we plot the left- and right-hand sides of Eqs. (5.10) and (5.11), and obtain from their intersections the allowed energy levels. The graphical method thus reveals the following features. There exists a minimum value of Λ below which no odd-parity solutions are allowed. On the other hand, there is always at least one even-parity solution. The first even-parity energy level occurs at $\xi < \pi/2$, whereas the first odd-parity level occurs at $\pi/2 < \xi < \pi$. Thus, the even- and odd-parity levels alternate in magnitudes, with the lowest level being even in parity. We should also note that the solutions depend on the potential function parameters only through the variable Λ, or the combination of V_oL^2, so that the effect of any change in well depth can be compensated by a change in the square of the well width.

At this point it can be noted we anticipate that for a particle in a potential well in three dimensions (next section), the cosine solution to the wave function has to

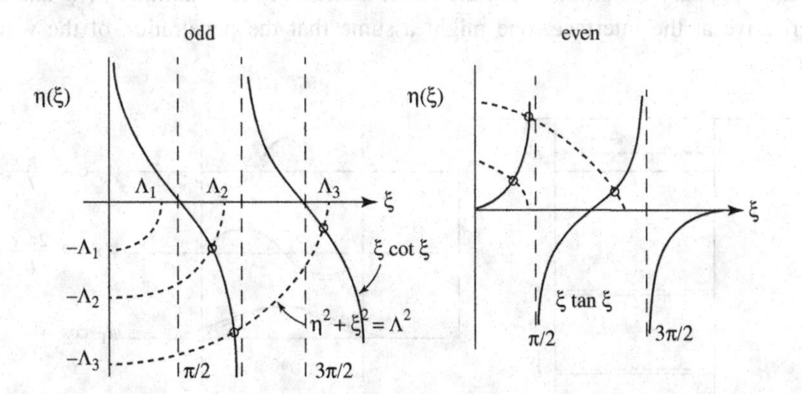

Fig. 5.2 Schematic sketch of graphical solutions of Eqs. (5.10) and (5.11) showing there would be no odd-parity solutions if Λ is not large enough (the potential is not deep enough or not wide enough), while there is at least one even-parity solution no matter what values are the well depth and width.

be discarded because of the condition of regularity (wave function must be finite) at the origin. This means that there will be a minimum value of Λ or $V_o L^2$ below which no bound states can exist. This is a feature of problems in three dimensions which does not pertain to problems in one dimension.

We now summarize our results for the allowed energy levels of a particle in a square well potential and the corresponding wave functions.

$$\psi_{\text{int}}(x) = A \sin kx \quad \text{or} \quad A' \cos kx \quad |x| < L/2 \tag{5.13}$$

$$\begin{aligned} \psi_{ext}(x) &= Be^{-\kappa x} & x > L/2 \\ &= Ce^{\kappa x} & x < -L/2 \end{aligned} \tag{5.14}$$

where the energy levels are

$$E = -|V_o| + \frac{\hbar^2 k^2}{2m} = -\frac{\hbar^2 \kappa^2}{2m} \tag{5.15}$$

The constants B and C are determined from the continuity conditions at the interface, while A and A' are to be fixed by the normalization condition. The discrete values of the bound-state energies, k or κ, are obtained from Eqs. (5.8) and (5.9). In Fig. 5.3, we show a sketch of the three lowest-level solutions, the ground state with even parity, the first excited state with odd parity, and the second excited state with even parity. Notice that the number of excited states that one can have depends on the value of V_o because our solution is valid only for negative E. This means that for a potential of a given depth, the particle can be bound only in a finite number of states.

To obtain more explicit results it is worthwhile to consider an approximation to the boundary condition at the interface. Instead of the continuity of ψ and its derivative at the interface, one might assume that the penetration of the wave

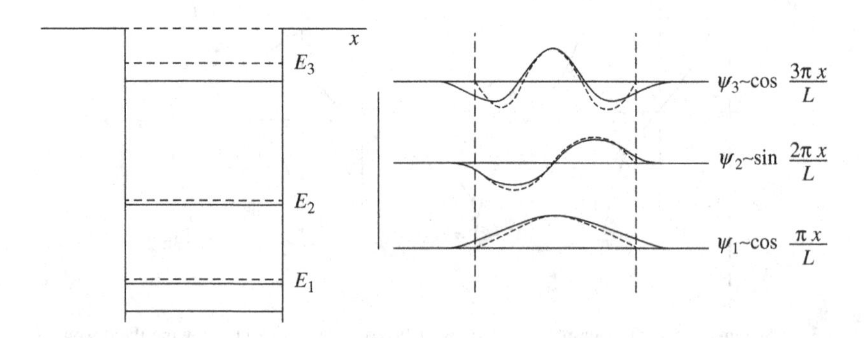

Fig. 5.3 Ground-state and first two excited-state solutions [adapted from Cohen, 1971, p. 16]. Approximate solutions given by the condition of vanishing wavefunction at the potential boundary are indicated by the dashed lines.

function into the exterior region can be neglected, and therefore require that ψ vanishes at $x = \pm L/2$. Applying this condition to Eq. (5.13) gives $kL = n\pi$, where n is any integer, or equivalently,

$$E_n = -|V_o| + \frac{n^2\pi^2\hbar^2}{2mL^2}, \quad n = 1, 2, \ldots \tag{5.16}$$

This shows explicitly how the energy eigenvalue E_n varies with the level index n, which is the quantum number for the one-dimensional problem under consideration. The corresponding wave functions under this approximation are:

$$\begin{aligned} \psi_n(x) &= A_n \cos(n\pi x/L), \quad n = 1, 3, \ldots \\ &= A'_n \sin(n\pi x/L) \quad n = 2, 4, \ldots \end{aligned} \tag{5.17}$$

The first three solutions in this approximate calculation are also shown in Fig. 5.3. We see that requiring the wave function to vanish at the interface has the effect of confining the particle in a potential well of width L with infinitely steep walls (the infinite well potential or limit of $V_o \to \infty$). It is therefore to be expected the problem becomes independent of V_o and there is no limit on the number of excited states. Clearly, the approximate solutions become the more useful the greater is the well depth, and the error is always an overestimate of the energy levels as a result of squeezing of the wave function. Physically this makes the wave have a shorter period or a larger wavenumber.

5.2 BOUND STATES IN THREE DIMENSIONS

We will next extend the bound-state calculation to three-dimensional systems. The problem we want to solve is the same as before, except that we wish to determine the bound-state energy levels and corresponding wave functions for a particle in a three-dimensional *spherical* well potential. Even though this is a three-dimensional potential, we can take advantage of its spherical symmetry in angular space and reduce the calculation to an equation still involving only one variable, the radial distance between the particle position and the origin. In other words, the spherical potential is still a function of one variable,

$$\begin{aligned} V(r) &= -V_o \quad r < r_o \\ &= 0 \qquad \text{otherwise} \end{aligned} \tag{5.18}$$

Here r is the radial position of the particle relative to the origin. Any potential that is a function only of r, the magnitude of the position vector $\underline{r}$ and not the position vector itself, is called a central-force potential. As we will see, this form of the potential makes the solution of the Schrödinger wave equation particularly simple. For a system where the potential or interaction energy has no angular dependence, one can reformulate the problem by factorizing the wave function into a component that involves only the radial coordinate and another component that involves only the angular coordinates. The wave equation is then reduced to a system of one-dimensional equations coupled through indices on the wave function one of which is a one-dimensional equation describing the radial component of the wave function. As to the justification for using a central-force potential for our discussion, this will depend on which properties of the nucleus we wish to study.

We again begin with the time-independent wave equation [Schiff, 1955; Liboff, 1980]

$$\left[-\frac{\hbar^2}{2m}\nabla^2 + V(r)\right]\psi(\underline{r}) = E\psi(\underline{r}) \tag{5.19}$$

Since the potential function has spherical symmetry, it is natural for us to carry out the analysis in the spherical coordinate system rather than the Cartesian system. A position vector $\underline{r}$ then is specified by the radial coordinate r and two angular coordinates, θ and φ, the polar and azimuthal angles respectively, see Fig. 5.4. In this coordinate system the *Laplacian operator* ∇^2 is of the form

$$\nabla^2 = D_r^2 + \frac{1}{r^2}\left[\frac{-L^2}{\hbar^2}\right] \tag{5.20}$$

where D_r^2 is an operator involving only the radial coordinate,

$$D_r^2 = \frac{1}{r^2}\frac{\partial}{\partial r}\left[r^2\frac{\partial}{\partial r}\right] \tag{5.21}$$

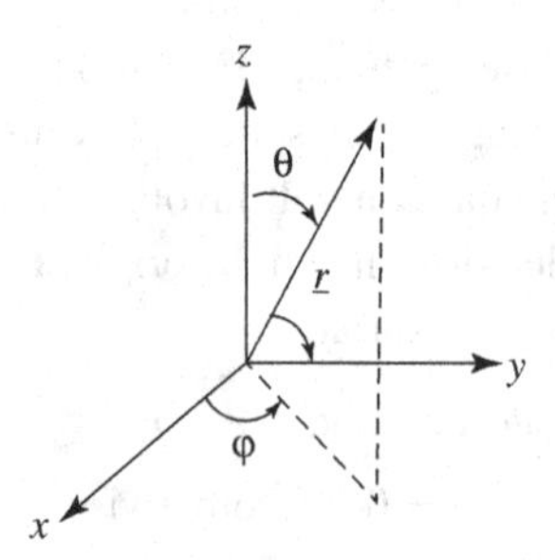

Fig. 5.4 The spherical coordinate system. A point in space is located by the vector $\underline{r}$, which can be resolved into a radial coordinate r, and polar and azimuthal angles θ and φ, respectively.

and the operator L^2 involves only the angular coordinates,

$$-\frac{L^2}{\hbar^2} = \frac{1}{\sin\theta}\frac{\partial}{\partial\theta}\left[\sin\theta\frac{\partial}{\partial\theta}\right] + \frac{1}{\sin^2\theta}\frac{\partial^2}{\partial\varphi^2} \tag{5.22}$$

In terms of these operators the wave equation Eq. (5.19) becomes

$$\left[-\frac{\hbar^2}{2m}D_r^2 + \frac{L^2}{2mr^2} + V(r)\right]\psi(r\theta\varphi) = E\psi(r\theta\varphi) \tag{5.23}$$

For any potential $V(r)$ the angular variation of ψ is always determined by the operator $L^2/2mr^2$. Therefore one can study the operator L^2 separately and then use its properties to simplify the solution of Eq. (5.23). This needs to be done only once, since the angular variation is independent of whatever form one takes for $V(r)$. It turns out that L^2 is very well known (it is the square of $\underline{L}$ which is the angular momentum operator); it is the operator that describes the angular motion of a free particle in three-dimensional space.

We first summarize the basic properties of L^2 before discussing any physical interpretation. It can be shown that the eigenfunction of L^2 are the spherical harmonics functions, $Y_\ell^m(\theta, \varphi)$,

$$L^2 Y_\ell^m(\theta,\varphi) = \hbar^2\ell(\ell+1)Y_\ell^m(\theta,\varphi) \tag{5.24}$$

where

$$Y_\ell^m(\theta,\varphi) = \left[\frac{2\ell+1}{4\pi}\frac{(\ell-|m|)!}{(\ell+|m|)!}\right]^{1/2} P_\ell^m(\cos\theta)e^{im\varphi} \tag{5.25}$$

and

$$P_\ell^m(\mu) = \frac{(1-\mu^2)^{m/2}}{2^\ell \ell!}\frac{d^{\ell+m}}{d\mu^{\ell+m}}(\mu^2-1)^\ell \tag{5.26}$$

with $\mu = \cos\theta$. The function $P_\ell^m(\mu)$ is called the *associated Legendre polynomials*, which are in turn expressible in terms of *Legendre polynomials* $P_\ell(\mu)$,

$$P_\ell^m(\mu) = (1-\mu)^{|m|/2}\frac{d^{|m|}}{d\mu^{|m|}}P_\ell(\mu) \tag{5.27}$$

with $P_o(x) = 1$, $P_1(x) = x$, $P_2(x) = (3x^2 - 1)/2$, $P_3(x) = (5x^3 - 3x)/2$, etc. Special functions like Y_ℓ^m and P_ℓ^m are quite extensively discussed in standard texts [Schiff, 1955, p. 70] and reference books on mathematical functions [Morse, 1953, p. 1264]. For our purposes it is sufficient to regard them as well known and tabulated quantities like *sines* and *cosines*, and whenever the need arises we will invoke their special properties as given in the mathematical handbooks.

It is clear from Eq. (5.24) that $Y_\ell^m(\theta, \varphi)$ is an eigenfunction of L^2 with corresponding eigenvalue $\ell(\ell+1)\hbar^2$. Since the angular momentum of the particle, like its energy, is quantized, the index ℓ can take on only positive integral values or zero,

$$\ell = 0, 1, 2, 3, \ldots$$

Similarly, the index m can have integral values from $-\ell$ to ℓ,

$$m = -\ell, -\ell+1, \ldots, -1, 0, 1, \ldots, \ell-1, \ell$$

For a given ℓ, there can be $2\ell+1$ values of m. The significance of m can be seen from the property of L_z, the projection of the orbital angular momentum vector $\underline{\boldsymbol{L}}$ along a certain direction in space (in the absence of any external field, this choice is up to the observer). Following convention we will choose this direction to be along the *z-axis* of our coordinate system, in which case the operator L_z has the representation, $L_z = -i\hbar\partial/\partial\varphi$, and its eignefunctions are also $Y_\ell^m(\theta, \varphi)$, with eigenvalues $m\hbar$. The indices ℓ and m are called orbital and magnetic quantum numbers. Since the angular space is two-dimensional (two degrees of freedom), it is to be expected that two quantum numbers will emerge from our analysis. By the same token we should expect three quantum numbers in our description of three-dimensional systems. We should regard the particle as existing in various states which are specified by a unique set of quantum numbers, each one is associated with a certain orbital angular momentum which has a definite magnitude and orientation with respect to our chosen direction along the z-axis. The particular angular momentum state is described by the function $Y_\ell^m(\theta, \varphi)$ with ℓ known as the *orbital angular momentum quantum number*, and m the *magnetic quantum number*. It is useful to keep in mind that $Y_\ell^m(\theta, \varphi)$ is actually a rather simple function for low order indices. For example, the first four spherical harmonics are:

$$Y_0^0 = 1/\sqrt{4\pi}, \qquad Y_1^{-1} = \sqrt{3/8\pi}e^{-i\varphi}\sin\theta,$$

$$Y_1^0 = \sqrt{3/4\pi}\cos\theta, \quad Y_1^1 = \sqrt{3/8\pi}e^{i\varphi}\sin\theta$$

Two other properties of the spherical harmonics are worth mentioning. First is that $\{Y_\ell^m(\theta, \varphi)\}$, with $\ell = 0, 1, 2, \ldots$ and $-\ell \leq m \leq \ell$, is a *complete set* of functions in the space of $0 \leq \theta \leq \pi$ and $0 \leq \varphi \leq 2\pi$ in the sense that any arbitrary function of θ and φ can be represented by an expansion in these functions.

Another property is orthonormality,

$$\int_0^{\pi} \sin\theta d\theta \int_0^{2\pi} d\varphi Y_\ell^{m*}(\theta,\varphi) Y_{\ell'}^{m'}(\theta,\varphi) = \delta_{\ell\ell'}\delta_{mm'} \tag{5.28}$$

where $\delta_{\ell\ell'}$ denotes the *Kronecker delta function*; it is unity when the two subscripts are equal, otherwise the function is zero.

Returning to the wave equation Eq. (5.23) we look for a solution as an expansion of the wave function in spherical harmonics series,

$$\psi(r\theta\varphi) = \sum_{\ell,m} R_\ell(r) Y_\ell^m(\theta,\varphi) \tag{5.29}$$

Because of Eq. (5.24) the L^2 operator in Eq. (5.23) can be replaced by the factor $\ell(\ell+1)\hbar^2$. In view of Eq. (5.28), we can eliminate the angular part of the problem by multiplying the wave equation by the complex conjugate of a spherical harmonic and integrating over all solid angles (recall an element of solid angle is $\sin\theta d\theta d\varphi$), obtaining

$$\left[-\frac{\hbar^2}{2m}D_r^2 + \frac{\ell(\ell+1)\hbar^2}{2mr^2} + V(r)\right] R_\ell(r) = ER_\ell(r) \tag{5.30}$$

This is an equation in one variable, the radial coordinate r, although we are treating a three-dimensional problem. We can make this equation look like a one-dimensional problem by transforming the dependent variable R_ℓ. Define the radial function

$$u_\ell(r) = rR_\ell(r) \tag{5.31}$$

Inserting this into (5.30) we get

$$-\frac{\hbar^2}{2m}\frac{d^2u_\ell(r)}{dr^2} + \left[\frac{\ell(\ell+1)\hbar^2}{2mr^2} + V(r)\right] u_\ell(r) = Eu_\ell(r) \tag{5.32}$$

We will call Eq. (5.32) *the radial wave equation.* It is the basic starting point of three-dimensional problems involving a particle interacting with a central potential field.

We observe Eq. (5.32) is actually a system of uncoupled equations, one for each fixed value of the orbital angular momentum quantum number ℓ. With reference to the wave equation in one dimension, the extra term involving $\ell(\ell+1)$ in Eq. (5.32) represents the contribution to the potential field due to the centrifugal motion of the particle. The $1/r^2$ dependence makes the effect particularly important near the origin; in other words, centrifugal motion gives rise to a barrier which tends to keep the particle away from the origin. This effect is of course absent in the case of $\ell = 0$, a state of zero orbital angular momentum, as one would expect. The first few ℓ states usually are the only ones of interest in most of our discussions (because they tend to have the lowest energies); they are given special spectroscopic designations,

$$\text{notation}: \quad \mathbf{s, p, d, f, g, h, \ldots}$$
$$\ell = \quad \mathbf{0, 1, 2, 3, 4, 5, \ldots}$$

where the first four letters stand for "sharp", "principal", "diffuse", and "fundamental" respectively. After f the letters are assigned in alphabetical order, as in $h, i, j, \ldots$ The wave function describing the state of orbital angular momentum ℓ is often called the ℓth partial wave,

$$\psi_\ell(r\theta\varphi) = R_\ell(r)Y_\ell^m(\theta\varphi) \tag{5.33}$$

Notice that in the case of s-wave the wave function is spherically symmetric since Y_0^0 is independent of θ and φ.

Interpretation of orbital angular momentum

In classical mechanics, the angular momentum of a particle in motion is defined as the vector product, $\underline{L} = \underline{r} \times \underline{p}$, where $\underline{r}$ is the particle position and $\underline{p}$ its linear momentum. $\underline{L}$ is directed along the axis of rotation (right-hand rule), as shown in Fig. 5.5.

$\underline{L}$ is called an axial or pseudo-vector in contrast to $\underline{r}$ and $\underline{p}$, which are polar vectors. Under inversion, $\underline{r} \to -\underline{r}$, and $\underline{p} \to -\underline{p}$, but $\underline{L} \to \underline{L}$. Quantum mechanically, L^2 is an operator with eigenvalues and eigenfunctions given in Eq. (5.24). Thus the magnitude of L is $\hbar\sqrt{\ell(\ell+1)}$, with $\ell = 0, 1, 2, \ldots$ being the orbital angular momentum quantum number. We can specify the magnitude and one Cartesian component (usually called the *z-component*) of $\underline{L}$ by specifying ℓ and m, an example is shown in Fig. 5.6(a). What about the x- and y-components? They are undetermined, in that they cannot be observed simultaneously with the observation of L^2 and L_z. From Fig. 5.5 we see the angular momentum is related to the linear

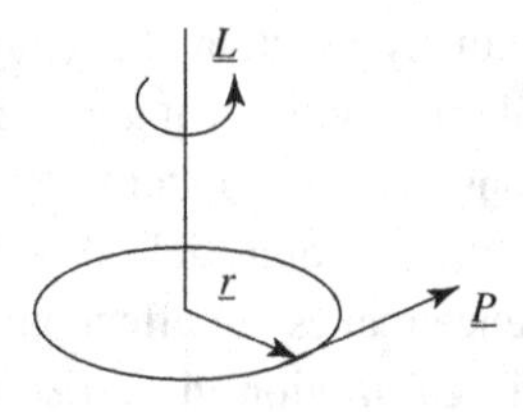

Fig. 5.5 Angular momentum of a particle at position $\underline{r}$ moving with linear momentum $\underline{p}$ (classical definition).

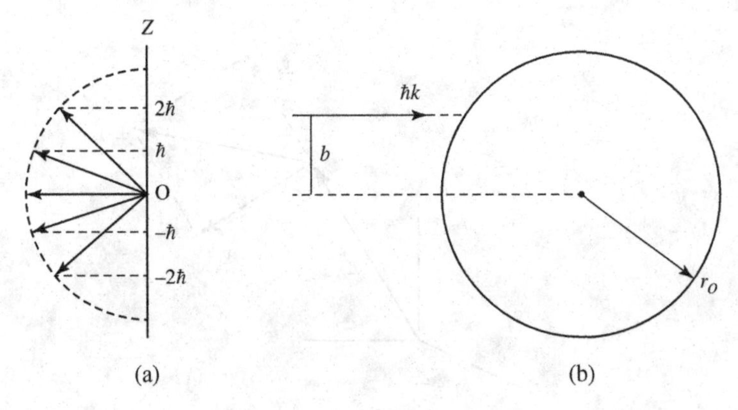

Fig. 5.6 (a) The $\ell(\ell+1) = 5$ projections along the z-axis of an orbital angular momentum with $\ell = 2$. Magnitude of L is $\sqrt{6}\hbar$. (b) A particle of momentum $\hbar k$ is incident upon a sphere of interaction at impact parameter b. The extent of the interaction region is r_o.

momentum $p = \hbar k$ by $\hbar\ell = \hbar kr$, where r is the moment arm. Now imagine a particle with liner momentum $\hbar k$ incident upon a sphere of interaction the extent of which is specified by a distance parameter r_o (later r_o becomes the range of the interaction potential $V(r)$). Suppose also the direction of incidence is offset from the center of the interaction sphere by a distance b, which in classical mechanics is known as the impact parameter, as shown in Fig. 5.6(b). For this simple physical set-up we can take the angular momentum to be $\hbar\ell = \hbar kb$. Then we can say that for $b < r_o$, the incident particle will enter into the region of interaction so an interaction is likely, whereas for $b > r_o$, the particle will pass by the sphere of interaction so interaction will be unlikely. Although quite simple and qualitative, this argument is often used to justify an important feature of the method of partial wave analysis in the calculation of cross sections. For a given incoming energy (or momentum) and a fixed range of interaction, there is a maximum value of the angular momentum $\hbar\ell^*$ beyond which the probability of interaction will be very small. The angular momentum cutoff ℓ^* is determined by the product kr_o, $\ell^* \sim kr_o$. We have seen that for nuclear interactions $r_o \sim 1.4\ A^{1/3}\ F$, which is a very small distance (nuclear forces are very short-ranged) on the spatial scale of atoms, typically *angstroms* Å (10^{-8} cm). This means that in neutron interactions involving neutron energies of order 1 keV or less, the value of kr_o will be appreciably less than unity, in which case only the $\ell = 0$ (s-wave) interaction is important.

Another useful interpretation of the orbital angular momentum is to look at the energy conservation equation in terms of radial and tangential motions. By this we

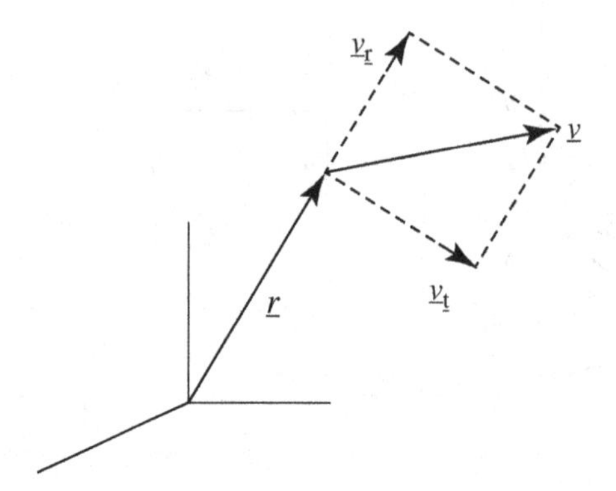

Fig. 5.7 Decomposing the velocity vector $\underline{v}$ of a particle at the position $\underline{r}$ into radial and tangential components, $\underline{v}_r$ and $\underline{v}_t$.

mean the total energy can be written as:

$$E = \frac{1}{2}m(v_r^2 + v_t^2) + V = \frac{1}{2}mv_r^2 + \frac{L^2}{2mr^2} + V \tag{5.34}$$

where the decomposition into radial and tangential velocities is depicted in Fig. 5.7. Equation (5.34) can be compared with the radial wave equation Eq. (5.32).

Thus far we have confined our discussions of the wave equation to its solutions in spherical coordinates. There are situations where it will be more appropriate to work in another coordinate system. As a simple example of a bound-state problem, we can consider the system of a free particle contained in a cubical box of dimension L along each side. In this case it is clearly more convenient to write the wave equation in Cartesian coordinates,

$$-\frac{\hbar^2}{2m}\left[\frac{\partial^2}{\partial x^2} + \frac{\partial^2}{\partial y^2} + \frac{\partial^2}{\partial z^2}\right]\psi(xyz) = E\psi(xyz) \tag{5.35}$$

$0 < x, y, z < L$. The boundary conditions are $\psi = 0$ whenever x, y, or z is 0 or L. Since both the equation and the boundary conditions are separable in the three coordinates, the solution is of the product form,

$$\begin{aligned}\psi(xyz) &= \psi_{n_x}(x)\psi_{n_y}(y)\psi_{n_z}(z) \\ &= (2/L)^{3/2}\sin(n_x\pi x/L)\sin(n_y\pi y/L)\sin(n_z\pi z/L)\end{aligned} \tag{5.36}$$

where n_x, n_y, n_z are positive integers (excluding zero), and the energy becomes a sum of three contributions,

$$E_{n_x n_y n_z} = E_{n_x} + E_{n_y} + E_{n_z}$$
$$= \frac{(\hbar\pi)^2}{2mL^2}[n_x^2 + n_y^2 + n_z^2]. \tag{5.37}$$

We see that the wave functions and corresponding energy levels are specified by the set of three quantum numbers (n_x, n_y, n_z). While each state of the system is described by a unique set of quantum numbers, there can be more than one state at a particular energy level. Whenever this happens, the level is said to be *degenerate*. For example, (112), (121), and (211) are three different states, but they are all at the same energy, so the level at $6(\hbar\pi)^2/2mL^2$ is triply degenerate. The concept of degeneracy is useful in our later discussion of the nuclear shell model where one has to determine how many nucleons can be put into a certain energy level. In Fig. 5.8, we show the energy level diagram for a particle in a cubical box. Another way to display the information is through a table, such as Table 5.1.

The energy unit is seen to be $\Delta E = (\hbar\pi)^2/2mL^2$. We can use this expression to estimate the magnitude of the energy levels for electrons in an atom, for which $m = 9.1 \times 10^{-28}$ gm and $L \sim 3 \times 10^{-8}$ cm, and for nucleons in a nucleus, for which $m = 1.6 \times 10^{-24}$ gm and $L \sim 5\,F$. The energies come out to be ~30 eV and 6 MeV respectively, values which are typical in atomic and nuclear physics.

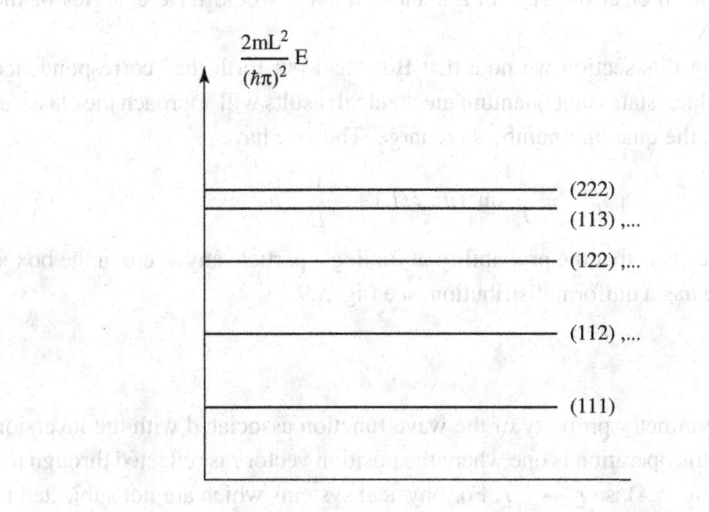

Fig. 5.8 Bound states of a particle in a cubical box of width L.

Table 5.1 The first few energy levels of a particle in a cubical box which correspond to Fig. 5.8.

n_x	n_y	n_z	$\frac{2mL^2}{(\hbar\pi)^2}E$	**Degeneracy**
1	1	1	3	1
1	1	2	6	3
1	2	1		
2	1	1		
1	2	2,...	9	3
1	1	3,...	11	3
2	2	2	12	1

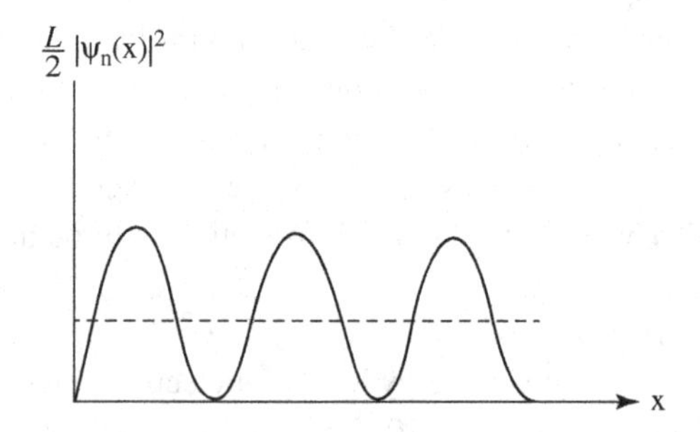

Fig. 5.9 The behavior of $\sin^2 nx$ in the limit of large n.

Notice that if an electron were in a nucleus, then it would have energies of the order 10^{10} eV!

In closing this section we note that Bohr had put forth the "correspondence principle" which states that quantum mechanical results will approach the classical results when the quantum numbers are large. Thus we have

$$|\psi_n|^2 = \frac{2}{L}\sin^2(n\pi x/L) \to \frac{1}{L} \quad n \to \infty \tag{5.38}$$

What this means is that the probability of finding a particle anywhere in the box is *1/L,* i.e., one has a uniform distribution, see Fig. 5.9.

Parity

Parity is a symmetry property of the wave function associated with the inversion operation. This operation is one where the position vector $\underline{r}$ is reflected through the origin (see Fig. 5.4), so $\underline{r} \to -\underline{r}$. For physical systems which are not subjected to an external vector field, we expect these systems will remain the same under an inversion operation, or the Hamiltonian is invariant under inversion. If $\psi(\underline{r})$ is a solution to the wave equation, then applying the inversion operation we get

$$H\psi(-\underline{r}) = E\psi(-\underline{r}) \tag{5.39}$$

which shows that $\psi(-\underline{r})$ is also a solution. A general solution is therefore obtained by adding or subtracting the two solutions,

$$H[\psi(\underline{r}) \pm \psi(-\underline{r})] = E[\psi(\underline{r}) \pm \psi(-\underline{r})]. \tag{5.40}$$

Since the function $\psi_+(\underline{r}) = \psi(\underline{r}) + \psi(-\underline{r})$ is manifestly invariant under inversion, it is said to have *positive parity*, or its parity, denoted by the symbol π, is $+1$. Similarly, $\psi_-(\underline{r}) = \psi(\underline{r}) - \psi(-\underline{r})$ changes sign under inversion, so it has *negative parity*, or $\pi = -1$. The significance of Eq. (5.40) is that a physical solution of our quantum mechanical description should have definite parity; this is the condition we have previously imposed on our solutions in solving the wave equation in one dimension, Sec. 5.1. Notice there are functions which do not have definite parity, for example, $A \sin kx + B \cos kx$. This is the reason that we take either the sine function or the cosine function for the interior solution in the one-dimensional problem above. In general, one can accept a solution as a linear combination of individual solutions all having the same parity. A linear combination of solutions with different parities has no definite parity, and is therefore unacceptable.

In spherical coordinates, the inversion operation of changing $\underline{r}$ to $-\underline{r}$ is equivalent to changing the polar angle θ to $\pi-\theta$, and the azimuthal angle φ to $\varphi+\pi$. The effect of the transformation on the spherical harmonic function $Y_\ell^m(\theta, \varphi) \sim e^{im\varphi} P_\ell^m(\theta)$ is

$$e^{im\varphi} \rightarrow e^{im\varphi} e^{im\pi} = (-1)^m e^{im\varphi}$$

$$P_\ell^m(\theta) \rightarrow (-1)^{\ell-m} P_\ell^m(\theta)$$

so the parity of $Y_\ell^m(\theta, \varphi)$ is $(-1)^\ell$. In other words, the parity of a state with a definite orbital angular momentum is even if ℓ is even, and odd if ℓ is odd. All eigenfunctions of the Hamiltonian with a spherically symmetric potential are therefore either even or odd in parity.

5.3 NUCLEAR SHELL MODEL

There are similarities between the electronic structure of atoms and nuclear structure. Atomic electrons are arranged in orbits (energy states) subject to the laws of quantum mechanics. (Recall the discussions of the Periodic Table and the Chart of Nuclides in Chap. 3.) The distribution of electrons in these states follows the Pauli exclusion principle. Atomic electrons can be excited up to normally unoccupied states, or they can be removed completely from the atom. From such phenomena one can deduce the structure of atoms. In nuclei there are two groups of like particles, protons and neutrons. Each group is separately distributed over certain energy states subject also to the Pauli exclusion principle. Nuclei have excited states, and nucleons can be added to or removed from a nucleus.

Electrons and nucleons have intrinsic angular momenta called intrinsic spins. The total angular momentum of a system of interacting particles reflects the details of the forces between particles. For example, from the coupling of electron angular momentum in atoms we infer an interaction between the spin and the orbital motion of an electron in the field of the nucleus (the spin-orbit coupling). In nuclei, there is also a coupling between the orbital motion of a nucleus and its intrinsic spin (but of different origin). In addition, nuclear forces between two nucleons depend strongly on the relative orientation of their spins.

The structure of nuclei is arguably more complex than that of atoms. In an atom the nucleus provides a common center of attraction for all the electrons and inter-electronic forces generally play a small role. The predominant force (Coulomb) is well understood. Nuclei, on the other hand, have no center of attraction; the nucleons are held together by their mutual interactions which are much more complicated than Coulomb interactions.

All atomic electrons are alike, whereas there are two kinds of nucleons. This allows a richer variety of structures. Notice that there are ~100 types of atoms, but more than 1000 different nuclides (see Table 3.5 and Sec. 3.3). Neither atomic nor nuclear structures can be understood without quantum mechanics.

Experimental basis

There exists considerable experimental evidence pointing to the shell-like structure of nuclei, each nucleus being an assembly of nucleons. Each shell can be filled with a given number of nucleons of each kind. These numbers are called magic numbers; they are **2**, **8**, **20**, **28**, **50**, **82**, and **126**. Nuclei with magic number of neutrons or protons, or both, are found to be particularly stable, as can be observed from the following data [Meyerhof 1967]. Figure 5.10 shows the abundance of stable isotones (same N) is particularly large for nuclei with magic neutron numbers. Figure 5.11 shows that the neutron separation energy S_n is particularly low for nuclei with one more neutron than the magic numbers, where (see Chap. 3)

$$S_n = [M(A-1, Z) + M_n - M(A, Z)]c^2 \quad (5.41)$$

This means that nuclei with magic neutron numbers are more tightly bound.

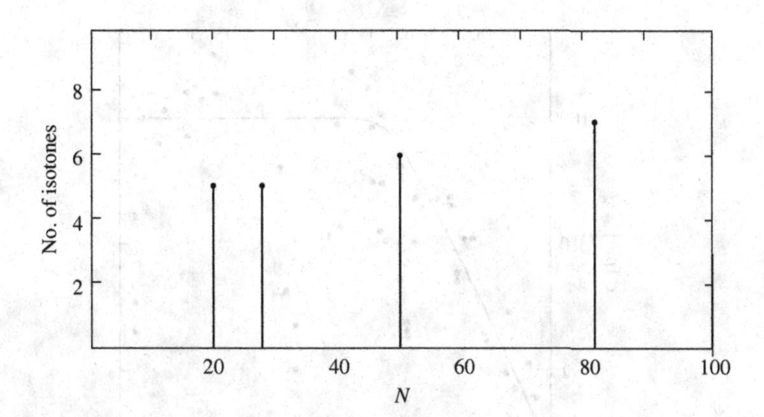

Fig. 5.10 Histogram of stable isotones showing nuclides with neutron numbers 20, 28, 50, and 82 are more abundant than those with non-magic neutron numbers for which the isotone numbers are mostly 1, 3, or 4. [Adapted from Meyerhof, 1967.]

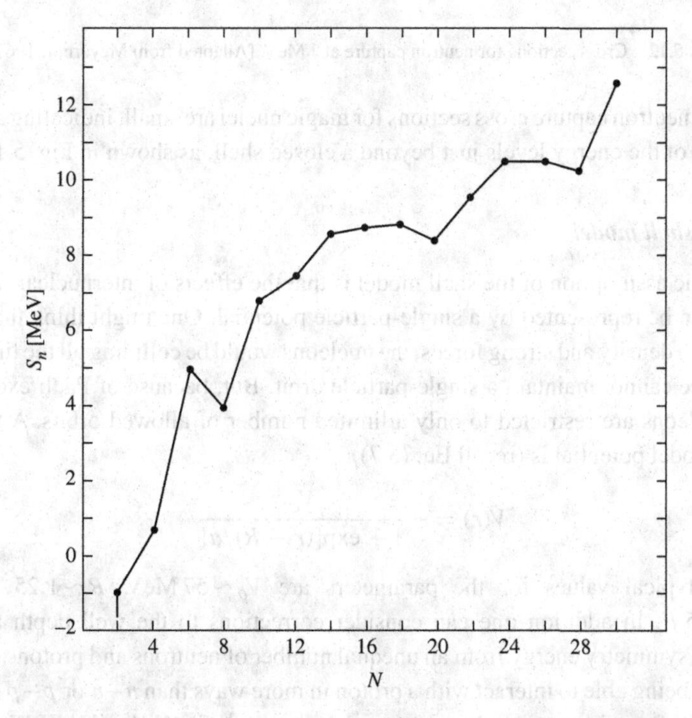

Fig. 5.11 Variation of neutron separation energy with neutron number of the final nucleus $M(A, Z)$. [Adapted from Meyerhof, 1967.]

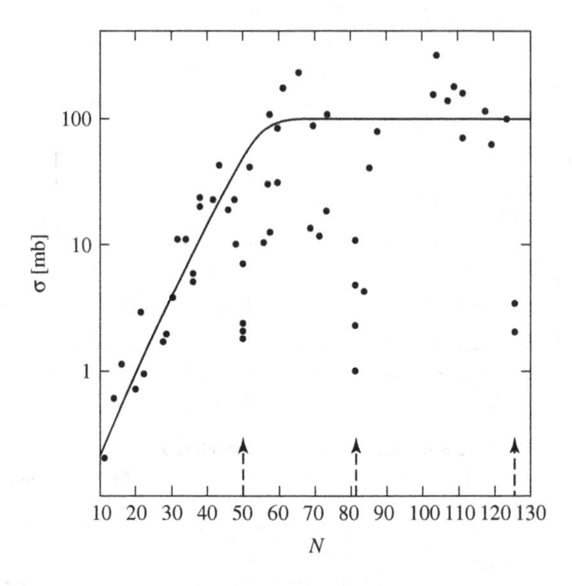

Fig. 5.12 Cross sections for neutron capture at 1 MeV. [Adapted from Meyerhof, 1967.]

The neutron capture cross sections for magic nuclei are small, indicating a wider spacing of the energy levels just beyond a closed shell, as shown in Fig. 5.12.

Simple shell model

The basic assumption of the shell model is that the effects of internuclear interactions can be represented by a single-particle potential. One might think that with very high density and strong forces, the nucleons would be colliding all the time and therefore cannot maintain a single-particle orbit. But, because of Pauli exclusion the nucleons are restricted to only a limited number of allowed orbits. A typical shell-model potential is (recall Eq. (3.7))

$$V(r) = -\frac{V_o}{1 + \exp[(r - R)/a]} \tag{5.42}$$

where typical values for the parameters are $V_o \sim 57\,\text{MeV}$, $R \sim 1.25\,\text{A}^{1/3}\,F$, a $\sim$0.65 F. In addition one can consider corrections to the well depth arising from (i) symmetry energy from an unequal number of neutrons and protons, with a neutron being able to interact with a proton in more ways than $n-n$ or $p-p$ (therefore $n-p$ force is stronger than $n-n$ and $p-p$), and (ii) Coulomb repulsion. For a given spherically symmetric potential $V(r)$, one can examine the bound-state energy levels that can be calculated from radial wave equation for a particular

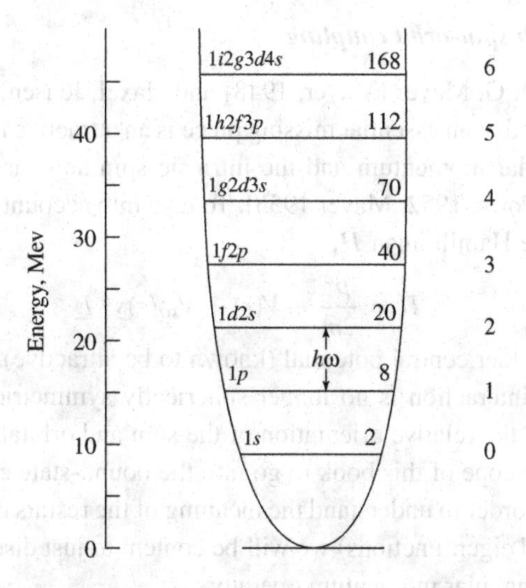

Fig. 5.13 Energy levels of nucleons in a parabolic potential well. In the spectroscopic notation (n, ℓ), n refers to the number of times the orbital angular momentum state ℓ has appeared. Also shown at certain levels are the cumulative number of nucleons that can be put into all the levels up to the indicated level. Oscillator quantum numbers are given in the right-hand column. [Adapted from Meyerhof, 1967.]

orbital angular momentum ℓ,

$$-\frac{\hbar}{2m}\frac{d^2 u_\ell}{dr^2} + \left[\frac{\ell(\ell+1)\hbar^2}{2mr^2} + V(r)\right] u_\ell(r) = E u_\ell(r) \tag{5.43}$$

Figure 5.13 shows the energy levels of the nucleons for a harmonic oscillator potential, $V(r) = m\omega^2 r^2/2$. In this case, one has the expression

$$E_\nu = \hbar\omega(\nu + 3/2)(n_x + n_y + n_z + 3/2) \tag{5.44}$$

where $\nu = 0, 1, 2, \ldots$, are quantum numbers. One should notice the degeneracy in the oscillator energy levels. The quantum number ν can be divided into *radial* quantum number $n(1, 2, \ldots)$ and *orbital* angular moments quantum numbers $\ell(0, 1, \ldots)$ as shown in Fig. 5.13. One can see from these results that a central force potential is able to account for the first three magic numbers, 2, 8, 20, but not the remaining four, 28, 50, 82, 126. This situation does not change when more rounded potential forms are used. The implication is that something very fundamental about the single-particle interaction picture is still missing in the description of nuclear energy levels.

Shell model with spin-orbit coupling

It remains for M. G. Mayer [Mayer, 1948] and Haxel, Jensen, and Suess [Haxel, 1949] to suggest that an essential missing piece is an attractive interaction between the orbital angular momentum and the intrinsic spin angular momentum of the nucleon [Duckworth, 1952; Mayer 1955]. To take into account this interaction we add a term to the Hamiltonian H,

$$H = \frac{p^2}{2m} + V(r) + V_{so}(r)\underline{s} \cdot \underline{L} \tag{5.45}$$

where V_{so} is another central potential (known to be attractive). This modification means that the interaction is no longer spherically symmetric; the Hamiltonian now depends on the relative orientation of the spin and orbital angular momenta. It is beyond the scope of this book to go into the bound-state calculations for this Hamiltonian. In order to understand the meaning of the results of such calculations (eigenvalues and eigenfunctions) we will be content to just discuss the concept of addition of two angular momentum operators.

The presence of the spin-orbit coupling term in Eq. (5.45) means that we will have a different set of eigenfunctions and eigenvalues for the new description. What are these new quantities relative to the eigenfunctions and eigenvalues we had found for the problem without the spin-orbit coupling interaction? We first observe that in labeling the energy levels in Fig. 5.15 we had already taken into

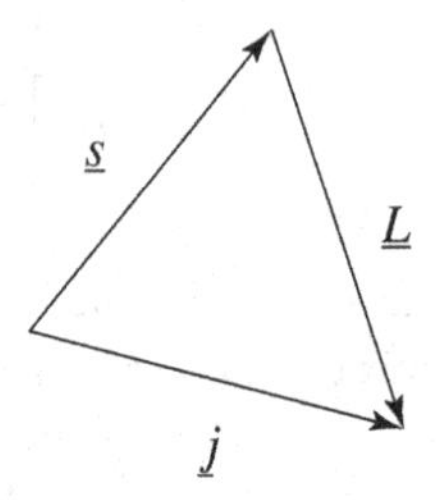

Fig. 5.14 Addition of spin $\underline{s}$ and orbital $\underline{L}$ angular momenta to give the total angular momentum $\underline{j}$

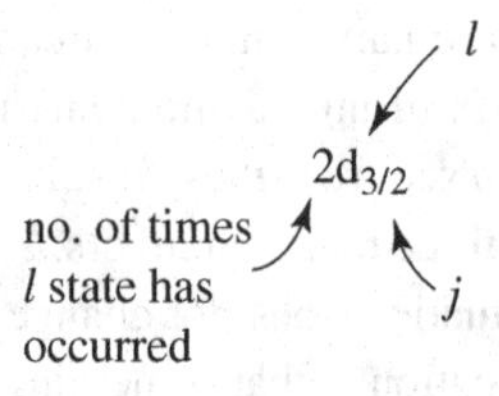

Fig. 5.15 spectroscopic notation for labeling each energy level with a specific orbital angular momentum ℓ and total angular momentum j.

account the fact that the nucleon has an orbital angular momentum (it is in a state with a specified ℓ), and that it has an intrinsic spin of $\frac{1}{2}$ (in unit of $\hbar$). For this reason the number of nucleons that we can put into each level has been counted correctly. For example, in the 1s ground state one can put two nucleons, for zero orbital angular momentum and two spin orientations (up and down). The reader can verify that for a state of given ℓ, the number of nucleons that can go into that state is $2(2\ell + 1)$. This comes about because the eigenfunctions we are using to describe the system is a representation that *diagonalizes* the square of the orbital angular momentum operator L^2, its z-component, L_z, the square of the intrinsic spin angular momentum operator s^2, and its *z-component* s_z. Let us use the following notation to label these eigenfunctions (or representation),

$$|\ell, m_\ell, s, m_s\rangle \equiv Y_\ell^{m_\ell} \chi_s^{m_s} \tag{5.46}$$

where $Y_\ell^{m_\ell}$ is the spherical harmonic we first encountered in Sec. 5.2, and we know it is the eigenfunction of the square of the orbital angular momentum operator L^2 (it is also the eigenfunction of L_z). The function $\chi_s^{m_s}$ is the spin eigenfunction with the expected properties,

$$s^2 \chi_s^{m_s} = s(s+1)\hbar^2 \chi_s^{m_s}, \quad s = 1/2 \tag{5.47}$$

$$s_z \chi_s^{m_s} = m_s \hbar \chi_s^{m_s}, \quad -s \le m_s \le s \tag{5.48}$$

The properties of $\chi_s^{m_s}$ with respect to operations by s^2 and s_z completely mirror the properties of $Y_\ell^{m_\ell}$ with respect to L^2 and L_z. Going back to our representation Eq. (5.46) we see that the eigenfunction is a "ket" with indices which are the good quantum numbers for the problem, namely, the orbital angular momentum and its projection (sometimes called the magnetic quantum number m, but here we use a subscript to denote that it goes with the orbital angular momentum), the spin (which has the fixed value of $\frac{1}{2}$) and its projection (which can be $+\frac{1}{2}$ or $-\frac{1}{2}$).

The representation given in Eq. (5.46) is no longer a good representation when the spin-orbit coupling term is added to the Hamiltonian. It turns out that the good representation is a linear combination of the old representation. It is sufficient for our purpose to just know this, without going into the details of how to actually construct the linear combination. To understand the properties of the new representation we now discuss angular momentum addition [Rose, 1957].

The two angular momenta we want to add are obviously the orbital angular momentum operator $\underline{L}$ and the intrinsic spin angular momentum operator $\underline{s}$, since they are the only angular momentum operators in our problem. Why do we want to add them? The reason lies in Eq. (5.45). Notice that if we define the total angular momentum as

$$\underline{j} = \underline{s} + \underline{L} \tag{5.49}$$

we then have by squaring $\underline{j}$,

$$\underline{s} \cdot \underline{L} = (j^2 - s^2 - L^2)/2 \tag{5.50}$$

so the problem of diagonalizing Eq. (5.45) is the same as diagonalizing j^2, s^2, and L^2. This is then the basis for choosing our new representation. In analogy to Eq. (5.46) we denote the new eigenfunctions by $|jm_j\ell s\rangle$, which has the properties

$$j^2|jm_j\ell s\rangle = j(j+I)\hbar^2|jm_j\ell s\rangle, \quad |\ell - s| \leq j \leq \ell + s \tag{5.51}$$

$$j_z|jm_j\ell s\rangle = m_j\hbar|jm_j\ell s\rangle, \qquad j \leq m_j \leq j \tag{5.52}$$

$$L^2|jm_j\ell s\rangle = \ell(\ell+1)\hbar^2|jm_j\ell s\rangle, \quad \ell = 0, 1, 2, \ldots \tag{5.53}$$

$$s^2|jm_j\ell s\rangle = s(s+1)\hbar^2|jm_j\ell s\rangle, \quad s = 1/2 \tag{5.54}$$

In Eq. (5.51), we indicate the values that j can take for a given ℓ and s ($= 1/2$ in our discussion), the lower (upper) limit corresponds to when $\underline{s}$ and $\underline{L}$ are antiparallel (parallel) as shown in Fig. 5.14.

Returning now to the energy levels of the nucleons in the shell model with spin-orbit coupling we can understand the conventional spectroscopic notation where the value of j is shown as a subscript, as indicated in Fig. 5.15.

This is then the notation in which the shell-model energy levels are displayed in Fig. 5.16. For a given (n, ℓ, j) level, the nucleon occupation number is $2j+1$. It would appear that having $2j+1$ identical nucleons occupying the same level would violate the Pauli exclusion principle. But this is not the case since each nucleon would have a distinct value of m_j (this is why there are $2j+1$ values of m_j for a given j).

We see in Fig. 5.16 the shell model with spin-orbit coupling gives a set of energy levels having breaks at the seven magic numbers. This is considered a major triumph of the model, for which Mayer and Jensen were awarded the Noble prize in physics (the 1963 Prize was shared by E. P. Wigner, cited for contributions to the theory of the atomic nucleus, and Mayer and Jensen). For our purpose we will use the results of the shell model to predict the ground-state spin and parity of nuclei. Before going into this discussion we leave the reader with the following comments. The shell model is most useful when applied to closed-shell or near closed-shell nuclei. Away from closed-shell nuclei collective models taking into account the rotation and vibration of the nucleus are more appropriate. Simple versions of the shell model do not take into account pairing forces, the effects of which are to make

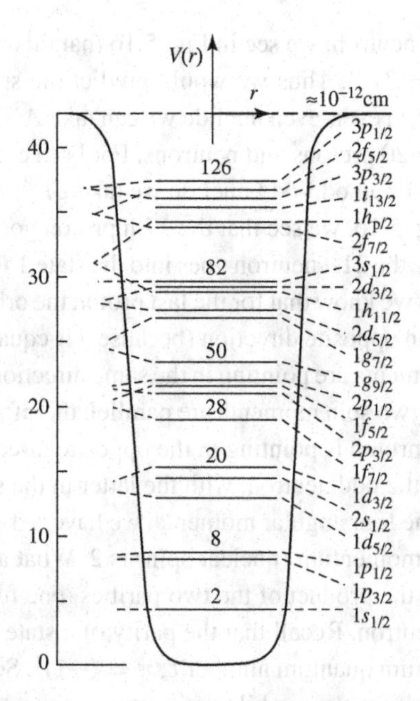

Fig. 5.16 Energy levels of nucleons in a smoothly varying potential well with a strong spin-orbit coupling term [adapted from Meyerhof, 1967]. Vertical axis gives the energy scale in MeV (see Fig. 5.13). Dashed lines on the left show the splitting of levels after the addition of spin-orbit coupling to the Hamiltonian. Dashed lines on the right show the resulting labeling of the levels explained in Fig. 5.15.

two like-nucleons combine to give zero orbital angula momentum. Shell model does not treat distortion effects (deformed nuclei) due to the attraction between one or more outer nucleons and the closed-shell core. When the nuclear core is not spherical, it can exhibit "rotational" spectrum.

Prediction of ground-state spin and parity

There are three general rules for using the shell model to predict the total angular momentum (spin) and parity of a nucleus in the ground state. These do not always work, especially in predictions away from the major shell breaks. First, the angular momentum of odd-A nuclei is determined by the angular momentum of the last nucleon in the species (neutron or proton) that is odd. Secondly, even-even nuclei have zero ground-state spin, because the net angular momentum associated with even N and even Z is zero, and even parity. Lastly, in odd-odd nuclei the last neutron couples to the last proton with their intrinsic spins in parallel orientation.

To illustrate how these rules work, we consider an example for each rule. Consider the odd-A nuclide Be^9 which has 4 protons and 5 neutrons. With the last nucleon being the fifth neutron, we see in Fig. 5.16 that this nucleon goes into the state $1p_{3/2}(\ell = 1, j = 3/2)$. Thus we would predict the spin and parity of this nuclide to be $3/2^-$. For an even-even nuclide we can take A^{36}, with 18 protons and neutrons, or Ca^{40}, with 20 protons and neutrons. For both cases we would predict spin and parity of 0^+. For an odd-odd nuclide we take Cl^{38}, which has 17 protons and 21 neutrons. In Fig. 5.16, we see that the 17th proton goes into the state $1d_{3/2}$ $(\ell = 2, j = 3/2)$, while the 21st neutron goes into the state $1f_{7/2}$ $(\ell = 3, j = 7/2)$. From the ℓ and j values we know that for the last proton the orbital and spin angular momenta are pointing in opposite direction (because j is equal to $\ell - 1/2$). For the last neutron the two momenta are pointing in the same direction $(j = \ell + 1/2)$. Now the rule tells us that the two spin momenta are parallel, therefore the orbital angular momentum of the odd proton is pointing in the opposite direction from the orbital angular momentum of the odd neutron, with the latter in the same direction as the two spins. Adding up the four angular momenta, we have $+3 + 1/2 + 1/2 - 2 = 2$. Thus the total angular momentum (nuclear spin) is 2. What about the parity? The parity of the nuclide is the product of the two parities, one for the last proton and the other for the last neutron. Recall that the parity of a state is determined by the orbital angular momentum quantum number ℓ, $\pi = (-1)^\ell$. So with the proton in a state with $\ell = 2$, its parity is even, while the neutron in a state with $\ell = 3$ has odd parity. The parity of the nucleus is therefore odd. Our prediction for Cl^{38} is then 2^-. The student can verify, using for example the Nuclide Chart, the foregoing predictions are in agreement with experiment.

Potential wells for neutrons and protons

We summarize the qualitative features of the potential wells for neutrons and protons. If we exclude the Coulomb interaction for the moment, then the well for a proton is known to be deeper than that for a neutron. The reason is that in a given nucleus usually there are more neutrons than protons, especially for the heavy nuclei, and the $n-p$ interactions can occur in more ways than either the $n-n$ or $p-p$ interactions on account of the Pauli exclusion principle. The difference in well depth ΔV_s is called the symmetry energy; it is approximately given by

$$\Delta V_s = \pm 27 \frac{(N - Z)}{A} \text{ MeV} \tag{5.55}$$

where the (+) and (−) signs are for protons and neutrons respectively. If we now consider the Coulomb repulsion between protons, its effect is to raise the potential for a proton. In other words, the Coulomb effect is a positive contribution to the nuclear potential which is larger at the center than at the surface.

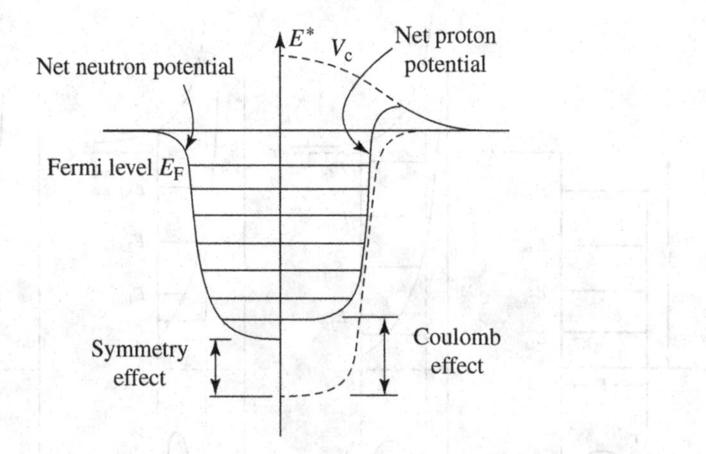

Fig. 5.17 Schematic showing the effects of symmetry and Coulomb interactions on the potential for a neutron and a proton. [Adapted from Marmier, 1969.]

Combining the symmetry and the Coulomb effects we have a sketch of the potential for a neutron and a proton as indicated in Fig. 5.17. One can also estimate the well depth in each case using the *Fermi Gas* model. One assumes the nucleons of a fixed kind behave like a fully degenerate gas of *fermions* (degeneracy here means that the states are filled continuously starting from the lowest energy state and there are no unoccupied states below the occupied ones), so that the number of states occupied is equal to the number of nucleons in the particular nucleus. This calculation is carried out separately for neutrons and protons. The highest energy state that is occupied is called the *Fermi level*, and the magnitude of the difference between this state and the ground state is called the *Fermi energy* E_F. It turns out that E_F is proportional to $n^{2/3}$, where n is the number of nucleons of a given kind, therefore E_F (neutron) > E_F (proton). The sum of E_F and the separation energy of the last nucleon provides an estimate of the well depth. (The separation energy for a neutron or proton is about 8 MeV for many nuclei, see Chap. 4 and Fig. 5.11.)

We have so far considered only a spherically symmetric nuclear potential well. We know there is in addition a centrifugal contribution of the form $\ell(\ell+1)\hbar^2/2mr^2$ and a spin-orbit contribution. As a result of the former the well becomes narrower and shallower for the higher orbital angular momentum states. The effects are illustrated in Fig. 5.18. Since the spin-orbit coupling is attractive, its effect depends on whether $\underline{S}$ is parallel or anti-parallel to $\underline{L}$.

We conclude this chapter with the remark that in addition to the bound states in the nuclear potential well, there exist also virtual states (levels) which are states

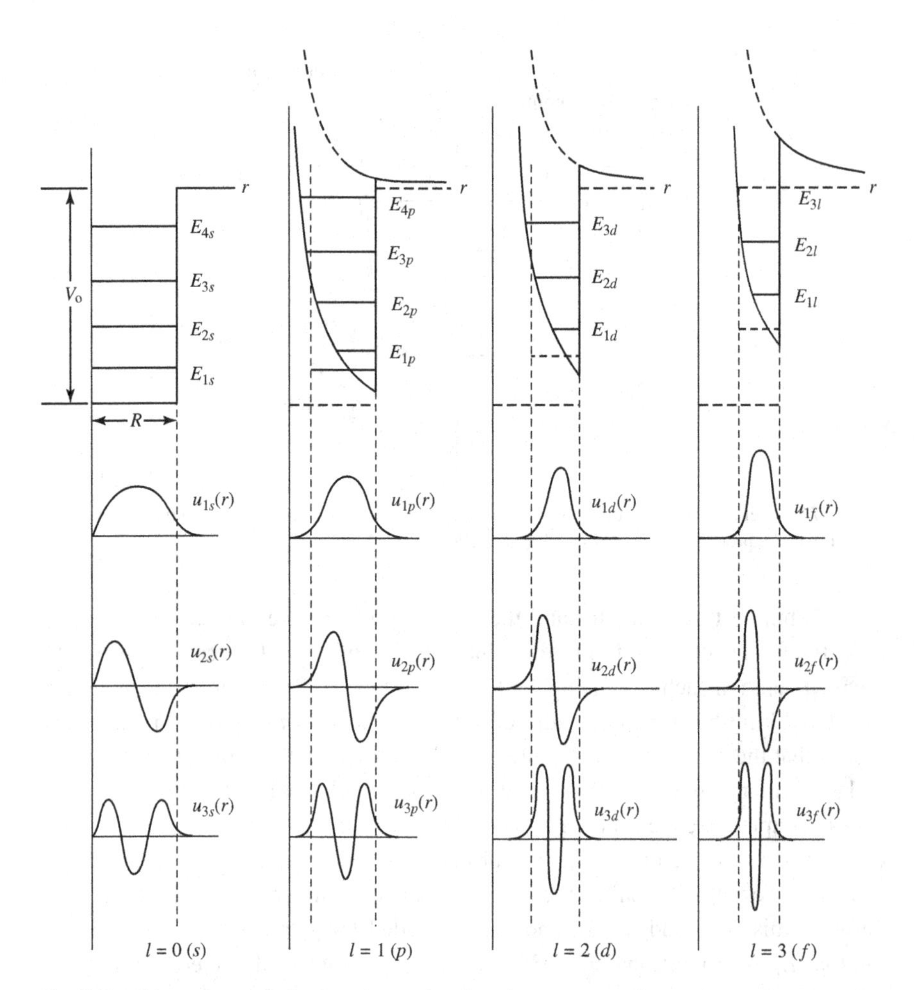

Fig. 5.18 Schematic energy levels and wave functions for a square well potential for the four partial waves $\ell = 0, 1, 2$, and 3, also referred to as *s*, *p*, *d*, and *f* states. [Adapted from Cohen, 1971.]

with positive energies (unbound) in which the wave function is large within the potential well. This can happen if the *deBroglie* wavelength is such that approximately standing waves are formed within the well. (Correspondingly, the reflection coefficient at the edge of the potential is large.) A virtual level is not a bound state; on the other hand, there is a non-negligible probability that inside the nucleus a nucleon can be found in such a state. The presence of virtual states is illustrated in Fig. 5.19. We will encounter an example of a virtual state for the deuteron in

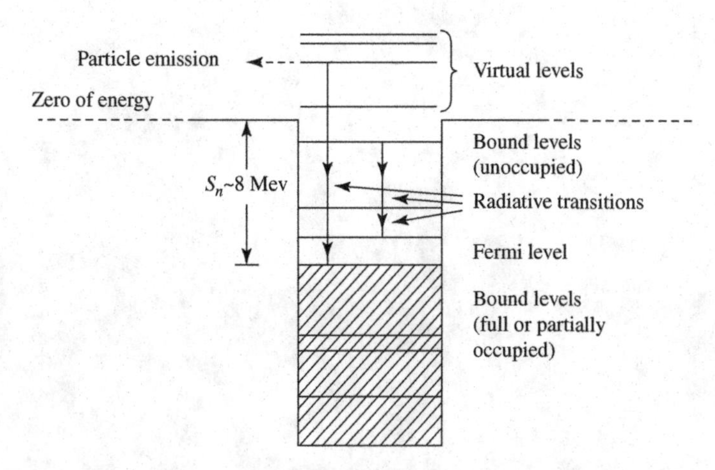

Fig. 5.19 Schematic illustration of bound and virtual energy levels of a nucleus relative to the *Fermi* level and the separation energy of a particle (neutron in this case).

Chap. 9 when we discuss neutron-proton scattering. In the discussion of neutron resonances in Chap. 12 we will introduce the concept of resonance levels which are formed during a neutron-induced reaction, levels which have sufficiently long lifetimes to be measureable and have significant technological implications.

REFERENCES

Bernard L. Cohen, *Concepts of Nuclear Physics* (McGraw-Hill, New York, 1971).

H. E. Duckworth, "Evidence for Nuclear Shells from Atomic Measurements", *Nature* **170**, 158 (1952).

O. Haxel, H. D. Jensen, H. E. Suess, "On the 'Magic Numbers' in Nuclear Structure", *Physical Review* **75**, 1766 (1949).

R. L. Liboff, *Introductory Quantum Mechanics* (Holden Day, New York, 1980).

P. Marmier and E. Sheldon, *Physics of Nuclei and Particles* (Academic Press, New York, 1969), vol. II, Chap. 15.2.

M. G. Mayer, "On Closed Shells in Nuclei", *Physical Review* **74**, 235 (1948).

W. E. Meyerhof, *Elements of Nuclear Physics* (McGraw-Hill, New York, 1967), Chap. 2.

P. M. Morse and H. Feshbach, *Methods of Theoretical Physics* (McGraw-Hill, New York, 1953).

L. I. Schiff, *Quantum Mechanics* (McGraw-Hill, New York, 1955).

6

Nuclear Disintegrations and Decays

Radioactivity is a topic that belongs in every nuclear physics text. As a result of the extensive studies of nuclear stability and energy levels we now have an understanding of the different ways in which a nucleus can exist and of various energy levels with specification of orbital angular momentum, intrinsic spin, etc. Once these levels are known from measurement or calculations using an appropriate model, one can speak of transitions from one level to another, either among levels of the same nucleus, or between levels that belong to neighboring nuclides. If the transitions occur spontaneously, they are called *radioactive disintegrations* or *nuclear decays*.

In this chapter we will first consider the process of radioactive decay without regard to the nuclear particle (radiation) that is emitted in the transitions. This will give us an understanding of *series decay*, the transition from a parent nuclide to its daughter which is itself unstable and therefore undergoes a second transition to the granddaughter, etc. The treatment will allow us to determine the concentration of all the nuclear species involved in the series decay. In the second part of the chapter, we consider the processes of *alpha*, *beta*, and *gamma* decays individually. In each case, our primary purpose is to estimate the *decay constant*, the reciprocal of the probability per unit time that the particular decay will occur.

6.1 RADIOACTIVE-SERIES DECAY

We begin with an experimental observation in radioactive decay that the probability of a decay during a small time interval Δt, which we will denote as $P(\Delta t)$, is proportional to Δt. Taking this as given one can write

$$P(\Delta t) = \lambda \Delta t \tag{6.1}$$

where λ is the proportionality constant which we will call the *decay constant*. Notice this expression is meaningful only when $\lambda \Delta t < 1$, a condition which defines what we mean by a small time interval. In other words, the time interval of measurement should be $\Delta t < 1/\lambda$, where $1/\lambda$ will turn out to be the mean life time of the radioisotope.

Suppose we are interested in the survival probability $S(t)$, the probability that the radioisotope does not decay during an arbitrary time interval t. To calculate $S(t)$ using Eq. (6.1) we divide the time interval t into many small, equal segments, each one of magnitude Δt. For a given t the number of such segments will be $t/\Delta t = n$. To survive the entire time interval t, we need to survive the first segment $(\Delta t)_1$, then the next segment $(\Delta t)_2$, ..., all the way up to the nth segment $(\Delta t)_n$. Thus we can write

$$\begin{aligned} S(t) &= \prod_{i=1}^{n}[1 - P((\Delta t)_i)] \\ &= [1 - \lambda(t/n)]^n \rightarrow e^{-\lambda t} \end{aligned} \tag{6.2}$$

where the arrow indicates the limit of $n \rightarrow \infty$, $\Delta t \rightarrow 0$. Unlike Eq. (6.1), Eq. (6.2) is valid for any t. When λt is sufficiently small compared to unity, it reduces to Eq. (6.1) as expected. Stated another way, Eq. (6.2) is the extension of $1 - P(t)$ for arbitrary t. One should also notice a close similarity between Eq. (6.2) and the probability that a particle will go a distance x without collision, $e^{-\Sigma x}$, where Σ is the macroscopic collision cross section (recall Sec. 3.4). The role of the decay constant λ in the probability of no decay in a time t is the same as the macroscopic cross section Σ in the probability of no collision in a distance x. The exponential attenuation in time or space is quite a general result, which one encounters frequently. There is another way to derive it. Suppose the radioisotope has not decayed up to a time interval of t_1. For it to survive the next small segment Δt the probability is just $1 - P(\Delta t) = 1 - \lambda\Delta t$. The probability of no decay up to the time $t_1 + \Delta t$ is then

$$S(t_1 + \Delta t) = S(t_1)[1 - \lambda\Delta t] \tag{6.3}$$

which can be rearranged to read

$$\frac{S(t + \Delta t) - S(t)}{\Delta t} = -\lambda S(t) \tag{6.4}$$

Taking the limit of small Δt, we get

$$\frac{dS(t)}{dt} = -\lambda \tag{6.5}$$

which is readily integrated to give Eq. (6.2), since the initial condition in this case is $S(t = 0) = 1$.

The decay of a single radioisotope is described by $S(t)$ involving a single physical constant λ. Instead of λ one can speak of two equivalent quantities, the

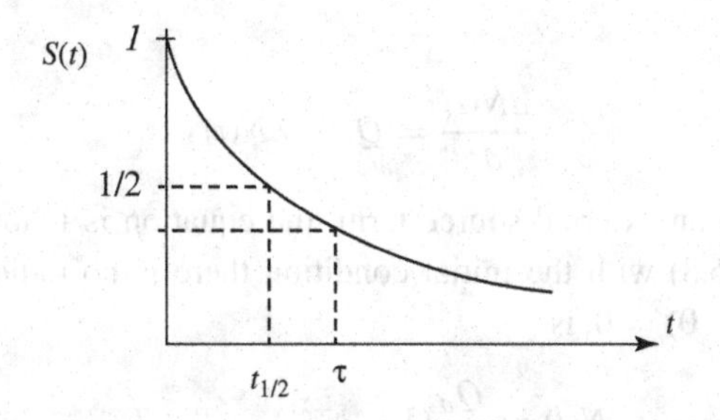

Fig. 6.1 The relation of half-life and mean life to a survival probability $S(t)$.

half life $t_{1/2}$ and the mean life τ. They are defined as

$$S(t_{1/2}) = 1/2 \rightarrow t_{1/2} = \ell n2/\lambda = 0.693/\lambda \tag{6.6}$$

and

$$\tau = \frac{\int_0^\infty dt' t' S(t')}{\int_0^\infty dt' S(t')} = \frac{1}{\lambda} \tag{6.7}$$

Figure 6.1 shows the relationship between these quantities and $S(t)$.

Radioactivity is measured in terms of the *rate of radioactive decay*. The quantity $\lambda N(t)$, where $N(t)$ is the number of radioisotope atoms at time t, is called *activity*. A standard unit of radioactivity has been the *curie*, 1 Ci $= 3.7 \times 10^{10}$ *disintegrations/sec*, which is roughly the activity of 1 gram of Ra^{226}. Now it is replaced by the *becquerel* (Bq), 1 Bq $= 2.7 \times 10^{-11}$ Ci. An old unit which is not often used is the *rutherford* (10^6 *disintegrations/sec*).

Radioisotope production by bombardment

Generally there are two ways of producing radioisotopes. One is activation by particle bombardment using an accelerator or a nuclear reactor. The other is through the decay of a radioactive series. Both methods can be treated using a differential equation that governs the number of radioisotopes at time t, $N(t)$. This is a first-order linear differential equation with constant coefficients, to which the solution can be readily written down. Although there are many different situations to which one can apply this equation, the analysis is fundamentally the same and quite straightforward. We will discuss the activation problem first. Let $Q(t)$, the rate of production of the radioisotope, be a constant Q_o during a time interval $(0, T)$, after which the production ceases. During production, $t < T$, the equation governing

$N(t)$ is

$$\frac{dN(t)}{dt} = Q_o - \lambda N(t) \tag{6.8}$$

Because we have an external source term, the equation is inhomogeneous. The solution to Eq. (6.8) with the initial condition there is no radioisotope prior to production, $N(t = 0) = 0$, is

$$N(t) = \frac{Q_o}{\lambda}(1 - e^{-\lambda t}), \quad t < T \tag{6.9}$$

For $t > T$, the governing equation is Eq. (6.8) without the source term. The solution is

$$N(t) = \frac{Q_o}{\lambda}(1 - e^{-\lambda T})e^{-\lambda(t-T)} \tag{6.10}$$

A sketch of the solutions Eqs. (6.9) and (6.10) is shown in Fig. 6.2. One sees a build up of $N(t)$ during production which approaches the asymptotic value of Q_o/λ, and after production is stopped $N(t)$ undergoes an exponential decay, so that if $\lambda T \gg 1$,

$$N(t) \approx \frac{Q_o}{\lambda}e^{-\lambda(t-T)} \tag{6.11}$$

Radioisotope production in series decay

Radioisotopes are also produced as the product(s) of a series of sequential decays. Consider the case of a three-member chain,

$$N_1 \xrightarrow{\lambda_1} N_2 \xrightarrow{\lambda_2} N_3 \text{ (stable)}$$

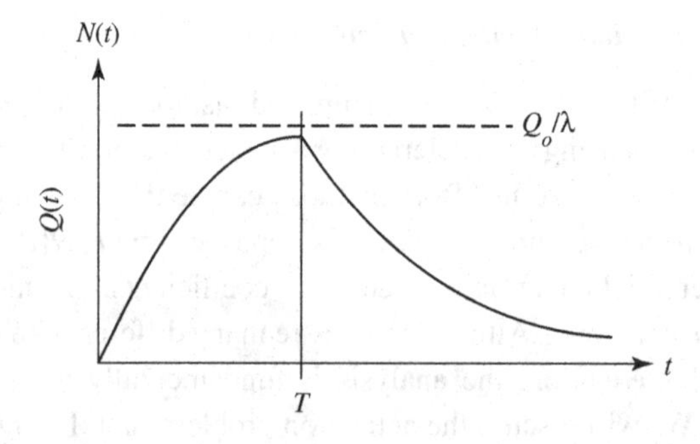

Fig. 6.2 Time variation of the number of radioisotope atoms produced at a constant rate Q_o for a time interval of T after which the system is left to decay.

where λ_1 and λ_2 are the decay constants of the parent (N_1) and the daughter (N_2) respectively. The governing equations are

$$\frac{dN_1(t)}{dt} = -\lambda_1 N_1(t) \tag{6.12}$$

$$\frac{dN_2(t)}{dt} = \lambda_1 N_1(t) - \lambda_2 N_2(t) \tag{6.13}$$

$$\frac{dN_3(t)}{dt} = \lambda_2 N_2(t) \tag{6.14}$$

For the initial conditions we assume there are N_{10} nuclides of species 1 and no nuclides of species 2 and 3. The solutions to Eqs. (6.12)–(6.14) then become

$$N_1(t) = N_{10} e^{-\lambda_1 t} \tag{6.15}$$

$$N_2(t) = N_{10} \frac{\lambda_1}{\lambda_2 - \lambda_1} (e^{-\lambda_1 t} - e^{-\lambda_2 t}) \tag{6.16}$$

$$N_3(t) = N_{10} \frac{\lambda_1 \lambda_2}{\lambda_2 - \lambda_1} \left(\frac{1 - e^{-\lambda_1 t}}{\lambda_1} - \frac{1 - e^{-\lambda_2 t}}{\lambda_2} \right) \tag{6.17}$$

Equations (6.15) through (6.17) are known as the *Bateman equations*. One can use them to analyze situations when the decay constants λ_1 and λ_2 take on different relative values. We consider two such scenarios, the case where the parent is short-lived, $\lambda_1 \gg \lambda_2$, and the opposite case where the parent is long-lived, $\lambda_2 \gg \lambda_1$.

One should notice from Eqs. (6.12)–(6.14) the sum of these three differential equations is zero. This means that $N_1(t) + N_2(t) + N_3(t) = constant$ for any t. We also know from our initial conditions that this constant must be N_{10}. One can use this information to find $N_3(t)$ given $N_1(t)$ and $N_2(t)$, or use it as a check that the solutions given by Eqs. (6.15)–(6.17) are indeed correct.

Series decay with short-lived parent

In this case one expects the parent to decay quickly and the daughter to build up quickly. The daughter then decays more slowly which means the grand daughter will build up slowly, eventually approaching the initial number of the parent. Figure 6.3 shows schematically the behavior of the three isotopes. The initial values of $N_2(t)$ and $N_3(t)$ can be readily deduced from an examination of Eqs. (6.16) and (6.17).

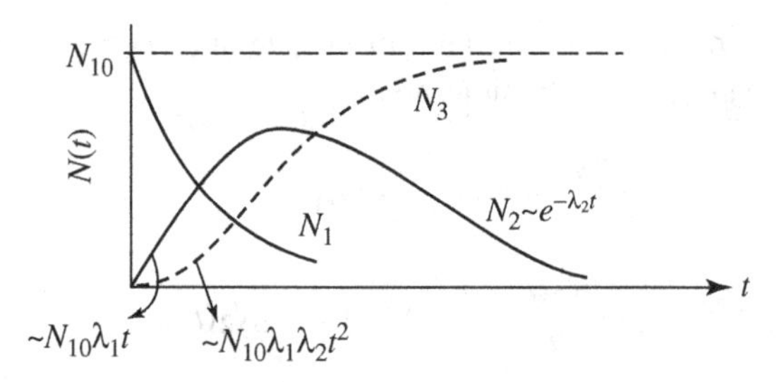

Fig. 6.3 Schematic of time variation of a three-member decay chain for the case $\lambda_1 \gg \lambda_2$, short-lived parent.

Series decay with long-lived parent

When $\lambda_1 \ll \lambda_2$, we expect the parent to decay slowly so the daughter and grand daughter will build up slowly. Since the daughter decays quickly the long-time behavior of the daughter follows that of the parent. Figure 6.4 shows the general behavior (admittedly the N_2 behavior is not sketched accurately). In this case we find

$$N_2(t) \approx N_{10}\frac{\lambda_1}{\lambda_2}e^{-\lambda t} \tag{6.18}$$

or

$$\lambda_2 N_2(t) \approx \lambda_1 N_1(t) \tag{6.19}$$

Between the two extremes of short- and long-lived parent is the condition of approximately equal activities, called *secular equilibrium*. Generalizing this to an arbitrary chain, we can say for the series

$$N_1 \rightarrow N_2 \rightarrow N_3 \rightarrow \cdots$$

if

$$\lambda_2 \gg \lambda_1, \ \lambda_3 \gg \lambda_1, \ldots$$

then

$$\lambda_1 N_1 \approx \lambda_2 N_2 \approx \lambda_3 N_3 \approx \cdots \tag{6.20}$$

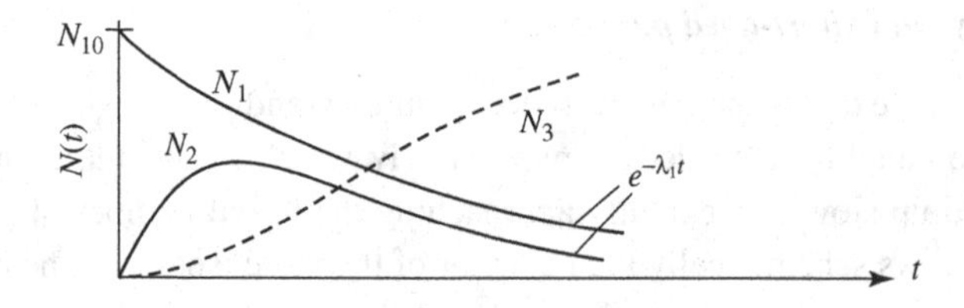

Fig. 6.4 Schematic of time variation of a three-member chain with a long-lived parent.

This condition can be used to estimate the half-life of a very long-lived radioisotope. An example is U^{238} whose half-life is so long that it is difficult to determine the value by directly measuring its decay. It is known that $U^{238} \rightarrow Th^{234} \rightarrow \cdots \rightarrow Ra^{226} \rightarrow \cdots$, and in uranium mineral the ratio of $N(U^{238})/N(Ra^{226}) = 2.8 \times 10^6$ has been measured, with $t_{1/2}(Ra^{226}) = 1620$ yr. Using these data we can write

$$\frac{N(U^{238})}{t_{1/2}(U^{238})} = \frac{N(Ra^{226})}{t_{1/2}(Ra^{226})} \text{ or } t_{1/2}(U^{238}) = 2.8 \times 10^6 \times 1620 = 4.5 \times 10^9 \text{ yr.}$$

In so doing we assume all the intermediate decay constants are larger than that of U^{238}. It turns out that this is indeed true, and that the above estimate is a good result. For an extensive treatment of radioactive series decay, the student should consult Evans [Evans, 1955].

6.2 NUCLEAR DECAYS

A nucleus in an excited state is unstable because it can always undergo a transition (decay) to a lower-energy state of the *same nucleus*. Such a transition will be accompanied by the emission of gamma radiation. A nucleus in either an excited or ground state also can undergo a transition to a lower-energy state of *another nucleus*. This decay is accomplished by the emission of a particle such as an *alpha*, electron or positron, with or without subsequent gamma emission. A nucleus which undergoes a transition *spontaneously*, that is, without being supplied with additional energy as in bombardment, is said to be *radioactive*. It is found experimentally that naturally occurring radioactive nuclides emit one or more of the three types of radiations, α-particles, β-particles, and γ-rays. Measurements of the energy of the nuclear radiation provide the most direct information on the energy-level structure of nuclides. One of the most extensive compilations of radioisotope data and detailed nuclear level diagrams is the *Table of Isotopes* [Lederer, 1978] introduced in Sec. 3.3.

In this chapter, we supplement our previous discussions of beta decay and radioactive decay by briefly examining the study of decay constants, selection rules, and some aspects of α-, β-, and γ-decay energetics.

Alpha decay

Most radioactive substances are α-emitters. Most nuclides with $A > 150$ are unstable against α-decay. On the other hand, α-decay is very unlikely for light nuclides. The decay constant is known to decrease exponentially with decreasing Q-value, here called the decay energy, $\lambda_\alpha \sim \exp(-c/v)$, where c is a constant and

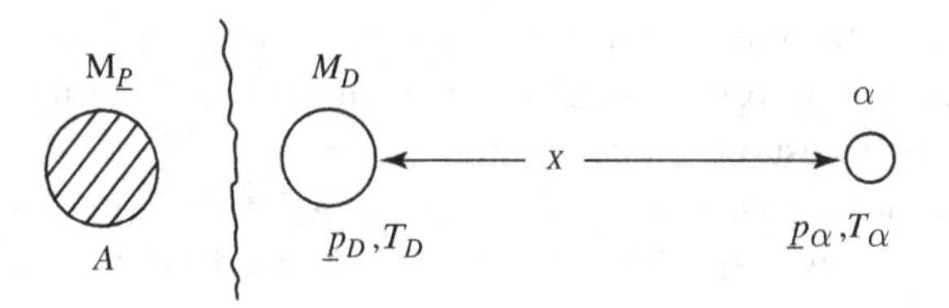

Fig. 6.5 Schematic of *α-decay* showing the excited parent nucleus P before the decay, and the emission of an α-particle with momentum $\underline{p}_\alpha$ and energy T_α along with recoil of the daughter nucleus D.

v the speed of the α-particle, $v \propto \sqrt{Q_\alpha}$. The momentum and energy conservation equations are quite straightforward in this case, as can be seen in Fig. 6.5.

$$\underline{p}_D + \underline{p}_\alpha = 0 \tag{6.21}$$

$$M_P c^2 = (M_D c^2 + T_D) + (M_\alpha c^2 + T_\alpha) \tag{6.22}$$

Both kinetic energies are small enough that non-relativistic energy-momentum relations may be used,

$$T_D = p_D^2/2M_D = p_\alpha^2/2M_D = (M_\alpha/M_D)T_\alpha \tag{6.23}$$

Treating the decay as a reaction the corresponding Q-value becomes

$$\begin{aligned} Q_\alpha &= [M_P - (M_D + M_\alpha)]c^2 \\ &= T_D + T_\alpha \\ &= \frac{M_D + M_\alpha}{M_D} T_\alpha \approx \frac{A}{A-4} T_\alpha \end{aligned} \tag{6.24}$$

This shows that the kinetic energy of the α-particle is always less than Q_α. Since $Q_\alpha > 0$ (T_α is necessarily positive), it follows that α-decay is an *exothermic* process. The various energies involved in the decay process can be displayed in an energy-level diagram shown in Fig. 6.6. One can see at a glance how the rest masses and the kinetic energies combine to ensure energy conservation. We will see in Chap. 12 energy-level diagrams are also useful in depicting collision-induced nuclear reactions. The separation energy S_α is the work necessary to separate an α-particle from the nucleus,

$$\begin{aligned} S_\alpha &= [M(A-4, Z-2) + M_\alpha - M(A, Z)]c^2 \\ &= B(A, Z) - B(A-4, Z-2) - B(4, 2) = -Q_\alpha \end{aligned} \tag{6.25}$$

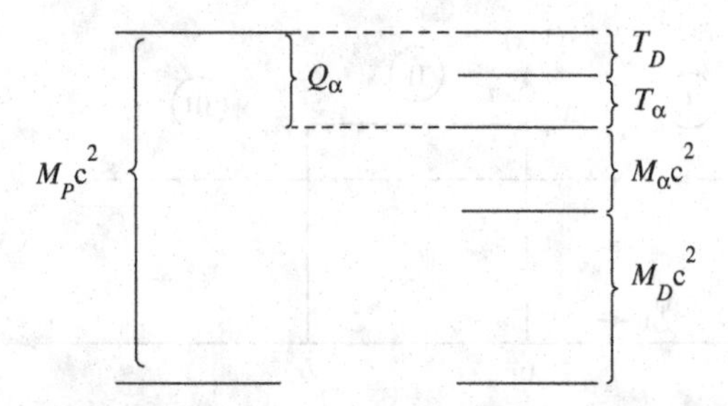

Fig. 6.6 Schematic energy-level diagram for α-decay showing the decomposition of the rest mass of the parent nucleus into the rest masses of the daughter and the α-particle plus the kinetic energies of the α-particle and the recoiling daughter. Compare this diagram with Fig. 6.5.

One can use the semi-empirical mass formula to determine whether a nucleus is stable against α-decay. In this way one finds $Q_\alpha > 0$ for $A > 150$. Equation (6.25) also shows that when the daughter nucleus is magic, $B(A\text{-}4,Z\text{-}2)$ is large, and Q_α is large. Conversely, Q_α is small when the parent nucleus is magic.

Barrier peneteration

Our next topic is to discuss a systematic way to estimate the decay constant for α-decay. However, because this approach is based on the quantum mechanical description of a particle going through a potential barrier, it would be appropriate to digress and connect with the discussion in Chap. 5 on the different solutions of the Schrödinger wave equation. In contrast to bound-state problems where we look for standing-wave solutions, we can consider traveling-wave solutions which reflect the passing of a particle over a potential well or barrier. For α-decay, the relevant potential would be a barrier because our picture of α-emission is an α-particle inside the parent nucleus trying to tunnel out.

We now look for positive-energy solutions as in a scattering problem (to be discussed in detail in the next chapter). We consider a one-dimensional system where a particle with mass m and energy $E > 0$ is incident upon a potential barrier with width L and height $V_o > E$, see Fig. 6.7. With the particle approaching from the left, the problem separates into three regions, left of the barrier (region I), inside the barrier (region II), and right of the barrier (region III).

In regions I and III the potential is zero, so the wave equation Eq. (3.1) is of the form

$$\frac{d^2\psi(x)}{dx^2} + k^2\psi(x) = 0, \quad k^2 = 2mE/\hbar^2 \tag{6.26}$$

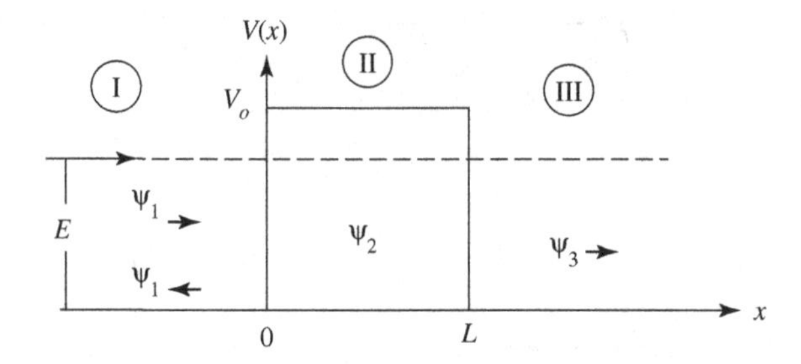

Fig. 6.7 Particle with energy E penetrating a square barrier of height V_o ($V_o > E$) and width L.

where k^2 is positive. The wave functions in these two regions are therefore

$$\psi_1 = a_1 e^{ikx} + b_1 e^{-ikx} \equiv \psi_{1\rightarrow} + \psi_{1\leftarrow} \tag{6.27}$$

$$\psi_3 = a_3 e^{ikx} + b_3 e^{-ikx} \equiv \psi_{3\rightarrow} \tag{6.28}$$

(A.4.3)

where we have set $b_3 = 0$ by imposing the boundary condition that there is no particle in region III traveling to the left (since there is nothing in this region that can reflect the particle). In contrast, in region *I* we allow for reflection of the incident particle by the barrier; this means b_1 will be nonzero. The subscripts $\rightarrow$ and $\leftarrow$ denote the wave functions traveling to the right and to the left respectively.

In region II, the wave equation is

$$\frac{d^2\psi(x)}{dx^2} - \kappa^2\psi(x) = 0, \quad \kappa^2 = 2m(|V_o| - E)/\hbar^2 \tag{6.29}$$

So we write the solution in the form

$$\psi_2 = a_2 e^{\kappa x} + b_2 e^{-\kappa x} \tag{6.30}$$

Notice in region II the kinetic energy, $E - V_o$, is *negative*, so the wavenumber is imaginary in a propagating wave (another way of saying the wave function is monotonically decaying rather than oscillatory). What this means is there is no wave-like solution in this region. By introducing κ we can think of it as the wavenumber of a hypothetical particle whose kinetic energy is positive, $V_o - E$.

Having obtained the wave function in all three regions we proceed to discuss how to organize this information into a useful form, namely, the transmission and reflection coefficients. We recall that given the wave function ψ, we know immediately the particle density (number of particles per unit volume, or the probability of the finding the particle in an element of volume d^3r about $\underline{r}$), $|\psi(\underline{r})|^2$, and the net current, given by (see Eq. (7.9)),

$$\underline{j} = \frac{\hbar}{2mi}(\psi^* \underline{\nabla} \psi - \psi \underline{\nabla} \psi^*) \tag{6.31}$$

Using the wave functions in regions I and III we obtain

$$j_1(x) = v[|a_1|^2 - |b_1|^2] \tag{6.32}$$

$$j_3(x) = v|a_3|^2 \tag{6.33}$$

where $v = \hbar k/m$ is the particle speed. We see from (6.32) that j_1 is the net current in region I, the difference between the current going to the right and that going to the left. Also, in region III there is only the current going to the right. Notice that current is like a flux in that it has the dimension of number of particles per unit area per second. This is consistent with (6.32) and (6.33) since $|a|^2$ and $|b|^2$ are particle densities with the dimension of number of particles per unit volume. From here on we can regard a_1, b_1, and a_3 as the amplitudes of the incident, reflected, and transmitted waves, respectively. With this interpretation we define transmission T and reflection R coefficients

$$T = \left|\frac{a_3}{a_1}\right|^2, \quad R = \left|\frac{b_1}{a_1}\right|^2 \tag{6.34}$$

Since particles cannot be absorbed or created in region II and there is no reflection in region III, the net current in region I must be equal to the net current in region III, or $j_1 = j_3$. It then follows that the condition

$$T + R = 1 \tag{6.35}$$

is always satisfied (as one should expect). The transmission coefficient is sometimes also called the *Penetration Factor* and denoted as P.

To calculate a_1 and a_3, we apply the boundary conditions at the interfaces, $x = 0$ and $x = L$,

$$\psi_1 = \psi_2, \quad \frac{d\psi_1}{dx} = \frac{d\psi_2}{dx} \quad x = 0 \tag{6.36}$$

$$\psi_2 = \psi_3, \quad \frac{d\psi_2}{dx} = \frac{d\psi_3}{dx} \quad x = L \tag{6.37}$$

These 4 conditions allow us to eliminate 3 of the 5 integration constants. For the purpose of calculating the transmission coefficient we need to keep a_1 and a_3. Thus we will eliminate b_1, a_2, and b_2 and in the process arrive at the ratio of a_1 to a_3 (after about a page of algebra),

$$\frac{a_1}{a_3} = e^{(ik-\kappa)L}\left[\frac{1}{2} - \frac{i}{4}\left(\frac{\kappa}{k} - \frac{k}{\kappa}\right)\right] + e^{(ik+\kappa)L}\left[\frac{1}{2} + \frac{i}{4}\left(\frac{\kappa}{k} - \frac{k}{\kappa}\right)\right] \tag{6.38}$$

This result then leads to (after another half-page of algebra)

$$\frac{|a_3|^2}{|a_1|^2} = \left|\frac{a_3}{a_1}\right|^2 = \frac{1}{1 + \frac{V_o^2}{4E(V_o-E)}\sinh^2 \kappa L} \equiv P \tag{6.39}$$

with $\sinh x = (e^x - e^{-x})/2$. A sketch of the variation of P with κL is shown in Fig. 6.8.

Using the leading expression of $\sinh(x)$ for small and large arguments, one can readily obtain simpler expressions for P in the limit of thin and thick barriers,

$$P \sim 1 - \frac{V_o^2}{4E(V_o - E)}(\kappa L)^2 = 1 - \frac{(V_o L)^2}{4E}\frac{2m}{\hbar^2} \qquad \kappa L \ll 1 \tag{6.40}$$

$$P \sim \frac{16E}{V_o}\left(1 - \frac{E}{V_o}\right)e^{-2\kappa L} \qquad \kappa L \gg 1 \tag{6.41}$$

Thus the transmission coefficient decreases monotonically with increasing V_o or L, relatively slowly for thin barriers and more rapidly for thick barriers.

Which limit is more appropriate for our interest? Consider a 5 MeV proton incident upon a barrier of height 10 MeV and width 10 F. This gives $\kappa \sim 5 \times 10^{12}\ \mathrm{cm}^{-1}$,

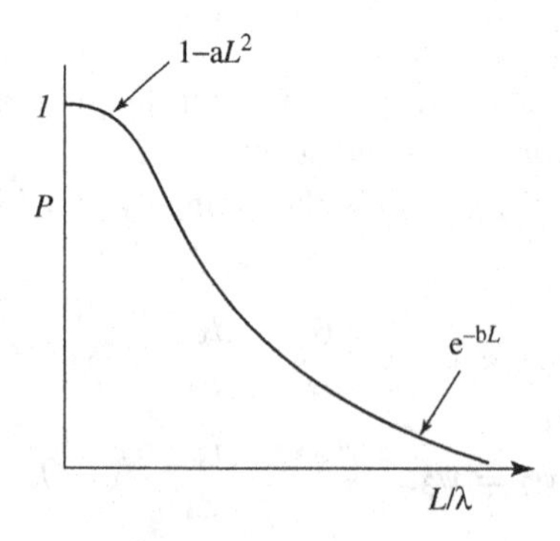

Fig. 6.8 Variation of transmission coefficient (*Penetration Factor*) with the ratio of barrier width L to λ, the effective wavelength of the incident particle where a and b are constants.

or $\kappa L \sim 5$. Using (6.41) we find

$$P \sim 16 \times \frac{1}{2} \times \frac{1}{2} \times e^{-10} \sim 2 \times 10^{-4}$$

As a further simplification, one sometimes even ignores the prefactor in Eq. (6.41) and takes

$$P \sim e^{-\gamma} \tag{6.42}$$

with

$$\gamma = 2\kappa L = \frac{2L}{\hbar}\sqrt{2m(V_o - E)} \tag{6.43}$$

We show in Fig. 6.9 a schematic of the wave function in each region. In regions I and III, ψ is complex, so we plot its real or imaginary part. In region II ψ is not oscillatory. Although the wave function in region II is nonzero, it does not appear in either the transmission or the reflection coefficient.

When the potential varies continuously in space, one can show that the attenuation coefficient γ is given approximately by the expression

$$\gamma \cong \frac{2}{\hbar}\int_{x_1}^{x_2} dx[2m\{V(x) - E\}]^{1/2} \tag{6.44}$$

where the limits of integration are indicated in Fig. 6.10; they are known as the "classical turning points". This result is for 1D. For a spherical barrier ($\ell = 0$ or s-wave solution), one has

$$\gamma \approx \frac{2}{\hbar}\int_{r_1}^{r^2} dr[2m\{V(r) - E\}]^{1/2} \tag{6.45}$$

We will now return to the discussion of α-decay keeping in mind the results of Eqs. (6.42) and (6.45).

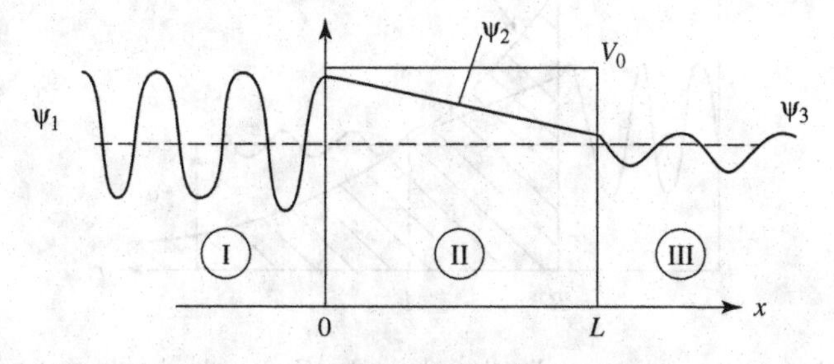

Fig. 6.9 Particle penetration through a square barrier of height V_o and width L at energy E ($E < V_o$) showing schematic behavior of wave functions in the three regions (see Fig. 6.7).

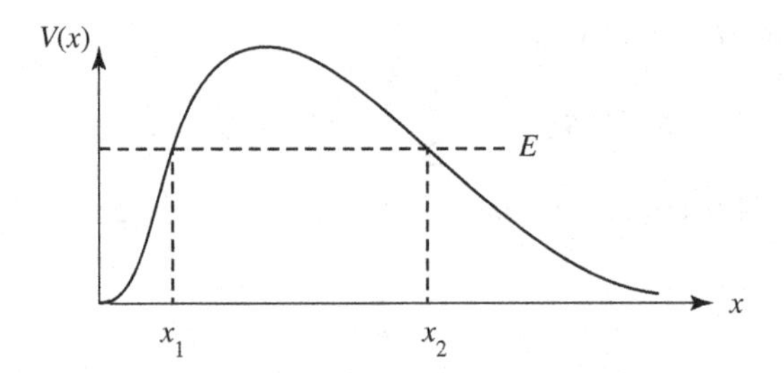

Fig. 6.10 Region of integration in Eq. (6.44) for a variable potential barrier.

Estimating α-decay Constant

An estimate of the decay constant can be made by treating the decay as a barrier penetration problem, an approach proposed by Gamow [Gamow, 1928] and also by Gurney and Condon [Gurney, 1928]. The idea is to assume the α-particle already exists as a particle inside the daughter nucleus where it is confined by the Coulomb potential, as illustrated in Fig. 6.11. The decay constant is then the probability per unit time that it can tunnel through the potential,

$$\lambda_\alpha \sim \left(\frac{v}{R}\right) P \tag{6.46}$$

where v is the relative speed of the α and the daughter nucleus, R is the radius of the daughter nucleus, and P the transmission coefficient. Equation (6.46) is a standard expression for describing tunneling probability in the form of a rate. The prefactor v/R is the attempt frequency, the rate at which the particle tries to tunnel through the barrier, and P is the probability of tunneling for each try. Recall from the

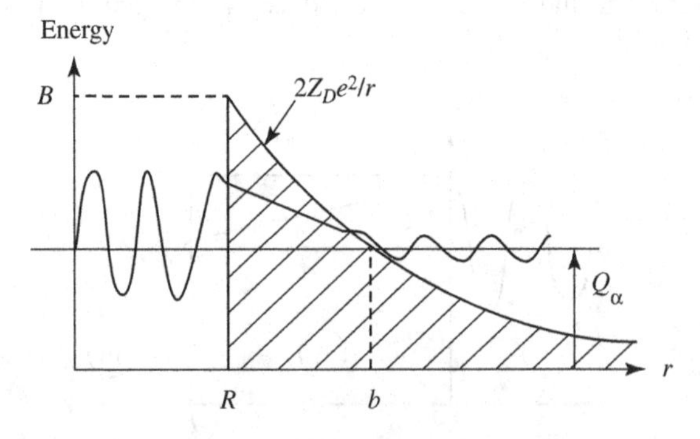

Fig. 6.11 Schematic of α-particle tunneling through a nuclear Coulomb barrier with energy Q_α. Charge of daughter nucleus is Z_D.

above discussion of barrier penetration the transmission coefficient can be written in the form

$$P \sim e^{-\gamma} \tag{6.47}$$

$$\gamma = \frac{2}{\hbar}\int_{r_1}^{r_2} dr(2m[V(r) - E])^{1/2}$$

$$= \frac{2}{\hbar}\int_{R}^{b} dr\left[2\mu\left(\frac{2Z_D e^2}{r} - Q_\alpha\right)\right]^{1/2} \tag{6.48}$$

with $\mu = M_\alpha M_D/(M_\alpha + M_D)$. The integral can be evaluated,

$$\gamma = \frac{8Z_D e^2}{\hbar v}[\cos^{-1}\sqrt{y} - \sqrt{y}(1-y)^{1/2}] \tag{6.49}$$

where $y = \mathrm{R/b} = Q_\alpha/\mathrm{B}$, $\mathrm{B} = 2Z_D e^2/\mathrm{R}$, $Q_\alpha = \mu v^2/2 = 2Z_D e^2/b$. Typically B is a few tens or more MeV, while Q_α~a few MeV. One can therefore invoke the thick barrier approximation, in which case $b \gg R$ (or $Q_\alpha \ll \mathrm{B}$), and $y \ll 1$. Then

$$\cos^{-1}\sqrt{y} \sim \frac{\pi}{2} - \sqrt{y} - \frac{1}{6}y^{3/2} - \ldots \tag{6.50}$$

the square bracket in Eq. (6.49) becomes

$$[\,] \sim \frac{\pi}{2} - 2\sqrt{y} + O(y^{3/2}) \tag{6.51}$$

and

$$\gamma \approx \frac{4\pi Z_D e^2}{\hbar v} - \frac{16 Z_D e^2}{\hbar v}\left(\frac{R}{b}\right)^{1/2} \tag{6.52}$$

So the expression for the decay constant becomes

$$\lambda_\alpha \approx \frac{v}{R}\exp\left[-\frac{4\pi Z_D e^2}{\hbar v} + \frac{8}{\hbar}\left(Z_D e^2 \mu R\right)^{1/2}\right] \tag{6.53}$$

where μ is the reduced mass. The exponent is sometimes known as the *Gamow factor G*.

To illustrate the application of Eq. (6.53) we consider estimating the decay constant of the 4.2 MeV α-particle emitted by U^{238}. Ignoring the small recoil effects, we can write

$$T_\alpha \sim \frac{1}{2}\mu v^2 \rightarrow v \sim 1.4 \times 10^9 \text{ cm/s}, \ \mu \sim M_\alpha$$

$$R \sim 1.4(234)^{1/3} \times 10^{-13} \sim 8.6 \times 10^{-13} \text{ cm}$$

$$-\frac{4\pi Z_D e^2}{\hbar v} = -173, \quad \frac{8}{\hbar}(Z_D e^2 \mu R)^{1/2} = 83$$

Thus

$$P = e^{-90} \sim 10^{-39} \tag{6.54}$$

As a result our estimate is

$$\lambda_\alpha \sim 1.7 \times 10^{-18} \, s^{-1}, \quad \text{or} \quad t_{1/2} \sim 1.3 \times 10^{10} \text{ yrs}$$

The experimental half-life is $\sim 0.45 \times 10^{10}$ *yrs*. Considering our estimate is very rough, the agreement is rather remarkable. In general one should not expect to predict λ_α to be better than the correct order of magnitude (say a factor of 5 to 10). Notice in our example, B $\sim$30 MeV and $Q_\alpha = 4.2$ MeV. Also $b = RB/Q_\alpha = 61 \times 10^{-13}$ cm. So the thick barrier approximation, B $\gg Q_\alpha$ or $b \gg R$, is indeed well justified.

The theoretical expression for the decay constant provides a basis for an empirical relation between the half-life and the decay energy. Since $t_{1/2} = 0.693/\alpha$, we have from Eq. (6.53)

$$\ell n(t_{1/2}) = \ell n(0.693R/v) + 4\pi Z_D e^2/\hbar v - \frac{8}{\hbar}(Z_D e^2 \mu R)^{1/2}. \tag{6.55}$$

We note $R \sim A^{1/3} \sim Z_D^{1/3}$, so the last term varies with Z_D like $Z_D^{2/3}$. Also, in the second trerm $v \propto \sqrt{Q_\alpha}$. Therefore Eq. (6.55) suggests the following relation,

$$\log(t_{1/2}) = a + \frac{b}{\sqrt{Q_\alpha}} \tag{6.56}$$

with a and b being parameters depending only on Z_D. A relation of this form is known as the *Geiger–Nuttall rule*.

We conclude our consideration of α-decay at this point. For further discussions the reader may wish to consult [Meyerhof, 1967, Chap. 4] and [Evans, 1955, Chap. 16].

Beta decay

Beta decay is considered to be a *weak* interaction since the interaction potential is $\sim 10^{-6}$ that of nuclear interactions, which are generally regarded as strong.

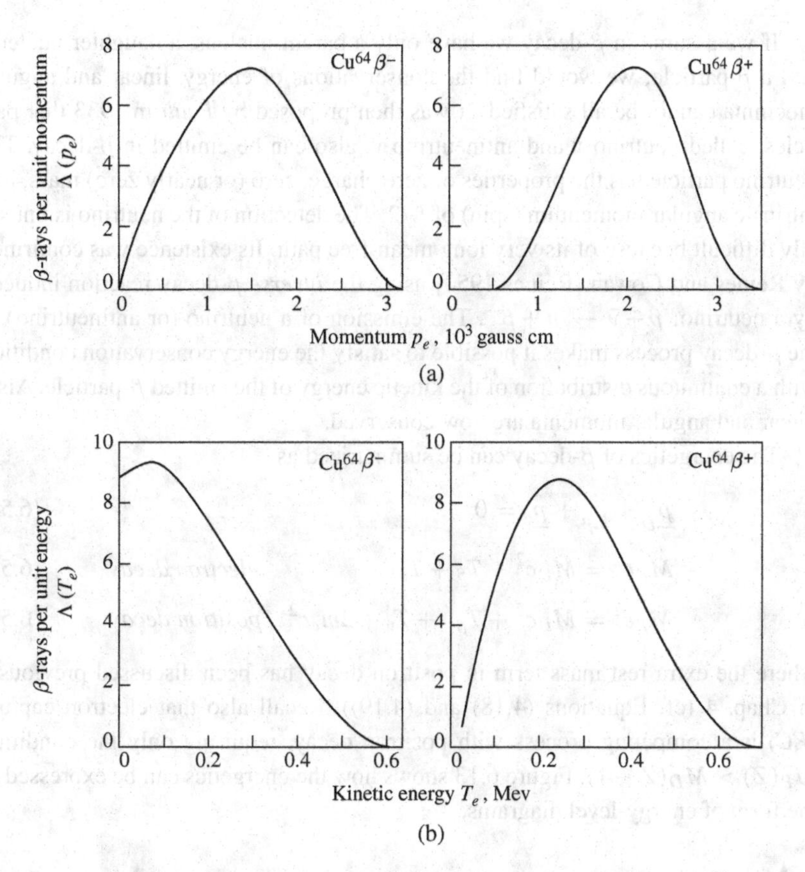

Fig. 6.12 Momentum (a) and energy (b) distributions of electron and positron decay in Cu^{64}. [Adapted from Meyerhof, 1967.]

Electromagnetic and gravitational interactions are intermediate in this sense. β-decay is the most common type of radioactive decay, all nuclides not lying in the "valley of stability" are unstable against this transition. The positrons or electrons emitted in β-decay have a continuous energy distribution, as illustrated in Fig. 6.12 for the decay of Cu^{64},

$$ {}_{29}\mathrm{Cu}^{64} \rightarrow {}_{30}\mathrm{Zn}^{64} + \beta + \bar{\nu}, \qquad T_{-}(max) = 0.57\,\mathrm{MeV} $$

$$ \rightarrow {}_{28}\mathrm{Ni}^{64} + \beta^{+} + \nu, \qquad T_{+}(max) = 0.66\,\mathrm{MeV} $$

The values of $T_{\pm}(\max)$ are characteristic of the particular radionuclide; they may be regarded as signatures.

If we assume in β-decay we have only a parent nucleus, a daughter nucleus, and a β-particle, we would find the conservations of energy, linear and angular moemnta cannot be all satisfied. It was then proposed by *Pauli* in 1933 that particles, called neutrino ν and antineutrino $\bar{\nu}$, also can be emitted in β-decay. The neutrino particle has the properties of zero charge, zero (or nearly zero) mass, and intrinsic angular momentum (spin) of $\hbar/2$. The detection of the neutrino is unusually difficult because of its very long mean-free path. Its existence was confirmed by Reines and Cowan [Reines 1953] using the *inverse* β-decay reaction induced by a neutrino, $p+\bar{\nu} \rightarrow n+\beta^-$. The emission of a neutrino (or antineutrino) in the β-decay process makes it possible to satisfy the energy conservation condition with a continuous distribution of the kinetic energy of the emitted β-particle. Also, linear and angular momenta are now conserved.

The energetics of β-decay can be summarized as

$$\underline{p}_D + \underline{p}_\beta + \underline{p}_{\bar{\nu}} = 0 \tag{6.57}$$

$$M_P c^2 = M_D c^2 + T_\beta + T_{\bar{\nu}} \qquad \textit{electron decay} \tag{6.58}$$

$$M_P c^2 = M_D c^2 + T_{\beta^+} + T_\nu + 2m_e c^2 \quad \textit{positron decay} \tag{6.59}$$

where the extra rest mass term in positron decay has been discussed previously in Chap. 4 (cf. Equations (4.18) and (4.19)). Recall also that electron capture (*EC*) is a competing process with positron decay, requiring only the condition $M_P(Z) > M_D(Z-1)$. Figure 6.13 shows how the energetics can be expressed in the form of energy-level diagrams.

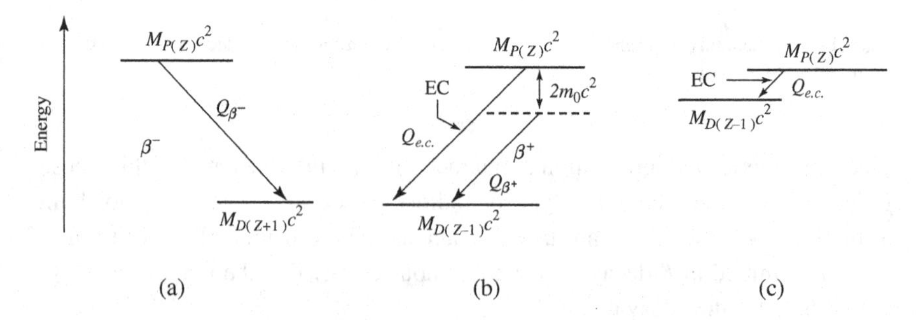

Fig. 6.13 Energetics of β-decay processes. [Adapted from Meyerhof, 1967.]

Typical decay schemes for β-emitters are shown in Fig. 6.14. For each nuclear level there is an assignment of spin and parity. This information is essential for determining whether a transition is allowed according to certain selection rules, as we will discuss below.

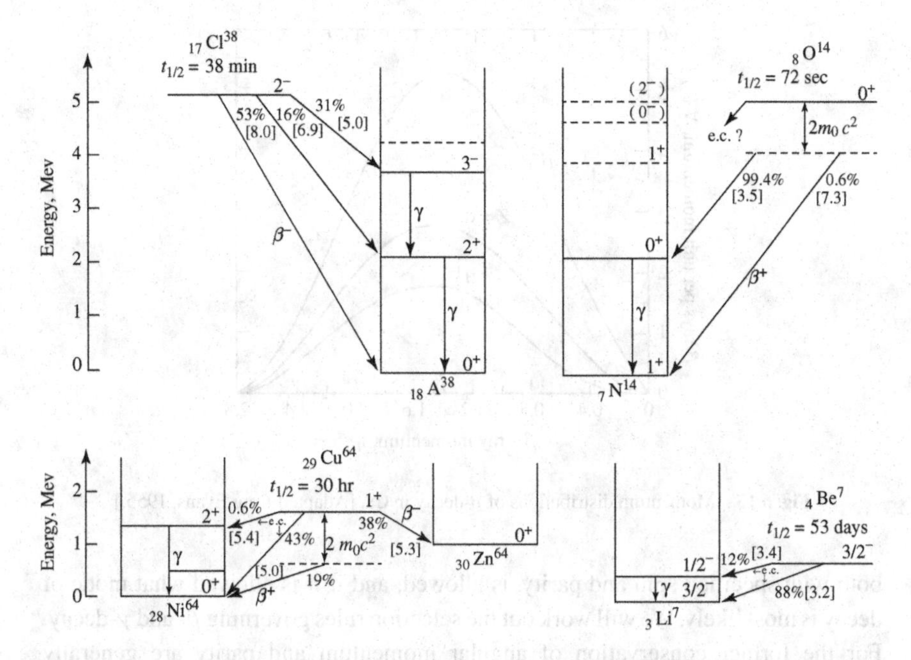

Fig. 6.14 Energy-level diagrams depicting nuclear transitions involving beta decay. [Adapted from Meyerhof, 1967.]

Experimental half-lives of β-decay have values spread over a very wide range, from 10^{-3} sec to 10^{16} yrs. Generally, $\lambda_\beta \sim Q_\beta^5$. The decay process cannot be explained classically. The theory of β-decay was developed by Fermi in 1934 in analogy with the quantum theory of electromagnetic decay. For a discussion of the elements of this theory one can begin with Meyerhof [Meyerhof, 1967] and follow the references given therein. We will be content to mention just one aspect of the theory, that concerning the statistical factor describing the momentum and energy distributions of the emitted β particle. Figure 6.15 shows the nuclear coulomb effects on the momentum distribution in β-decay in *Ca* ($Z = 20$). One can see an enhancement in the case of β^--decay and a suppression in the case of β^+-decay at low momenta. Coulomb effects on the energy distribution are even more pronounced.

Selection rules for beta decay

Besides energy and linear momentum conservation, a nuclear transition must also satisfy angular momentum and parity conservation. This gives rise to *selection rules* which specify whether a particular transition between initial and final states,

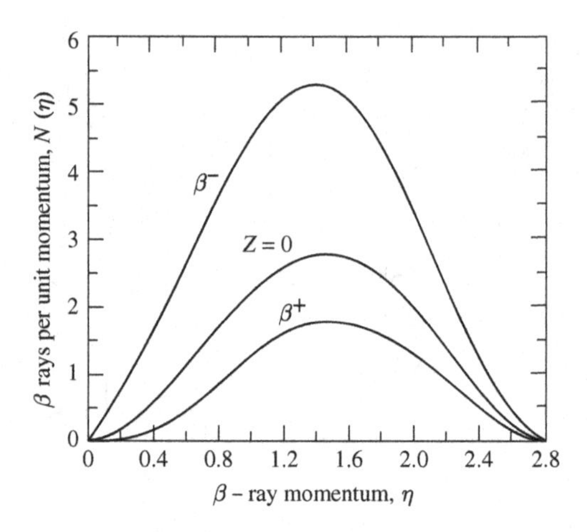

Fig. 6.15 Momentum distributions of β-decay in Ca. [Adapted from Evans, 1955.]

both with specified spin and parity, is allowed, and if it is allowed what mode of decay is most likely. We will work out the selection rules governing β- and γ-decay. For the former conservation of angular momentum and parity are generally expressed as

$$\underline{I}_P = \underline{I}_D + \underline{L}_\beta + \underline{S}_\beta \tag{6.60}$$

$$\pi_P = \pi_D(-1)^{L_\beta} \tag{6.61}$$

where L_β is the orbital angular momentum and S_β the intrinsic spin of the electron-antineutrino system. The magnitude of angular momentum vector can take integral values, 0, 1, 2, ... whereas the spin can take on values of 0 and 1 which would correspond to the antiparallel and parallel coupling of the electron and neutrino spins. These two orientations will be called *Fermi* and *Gamow–Teller* respectively in what follows.

In applying the conservation conditions, our goal is to find the lowest value of L_β that will satisfy Eq. (6.61) for which there is a corresponding value of S_β that is compatible with Eq. (6.60). This then identifies the most likely transition among all the allowed transitions. In other words, all the other allowed transitions will have higher values of L_β, which make them less likely to occur. The reason is the decay constant is governed by the square of a transition matrix element, which in turn can be written as a sum of contributions, one for each L_β (see the discussion of partial wave expansion in cross section calculation, Chap. 7, where we also argue that the higher order partial waves are less likely than the low order ones, ending up with only the s-wave),

$$\lambda_\beta \propto |M|^2 = |M(L_\beta = 0)|^2 + |M(L_\beta = 1)|^2 + |M(L_\beta = 2)|^2 + \cdots \quad (6.62)$$

Transitions with $L_\beta = 0, 1, 2, \ldots$ are called *allowed, first-forbidden, second-forbidden,*... etc. The magnitude of the matrix element squared decreases from one order to the next higher one by at least a factor of 10^2. For this reason we are interested only in the lowest order transition that is allowed.

To illustrate how the selection rules are determined, we consider the transition

$${}_2\mathrm{He}^6(0^+) \rightarrow {}_3\mathrm{Li}^6(1^+)$$

To determine the combination of L_β and S_β for the first transition that is allowed, we begin by noting parity conservation requires L_β to be even. Then we see that $L_\beta = 0$ plus $S_\beta = 1$ would satisfy both Eqs. (6.60) and (6.61). Thus the most likely transition is the transition designated as *allowed, G-T*. Following the same line of argument, one can arrive at the following assignments.

${}_8\mathrm{O}^{14}(0^+) \rightarrow {}_7\mathrm{N}^{14}(0^+)$	*allowed, F*
${}_0\mathrm{n}^1(1/2^+) \rightarrow {}_1\mathrm{H}^1(1/2^+)$	*allowed, G-T and F*
${}_{17}\mathrm{C}\ell^{38}(2^-) \rightarrow {}_{18}\mathrm{A}^{38}(2^+)$	*first-forbidden, GT and F*
${}_4\mathrm{Be}^{10}(3^+) \rightarrow {}_5\mathrm{B}^{10}(0^+)$	*second-forbidden, GT*

Parity non-conservation

The presence of neutrino in β-decay leads to a certain type of non-conservation of parity. It is known that neutrinos have intrinsic spin antiparallel to their velocity, whereas the spin orientation of the antineutrino is parallel to their velocity (keeping in mind that ν and $\bar{\nu}$ are *different* particles). Consider the mirror experiment where a neutrino is moving toward the mirror from the left, Fig. 6.16. Applying the inversion

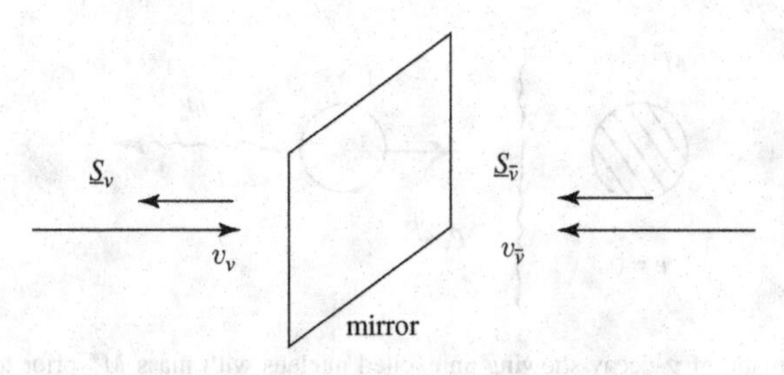

Fig. 6.16 Mirror reflection demonstrating parity non-sonserving property of neutrino. [Adapted from Meyerhof, 1967.]

symmetry operation ($x \to -x$), the velocity reverses direction, while the angular momentum (spin) does not. Thus, on the other side of the mirror we have an image of a particle moving from the right, but its spin is now parallel to the velocity so it has to be an antineutrino instead of a neutrino. This means that the property of ν and $\bar{\nu}$, namely definite spin direction relative to the velocity, is not compatible with parity conservation (symmetry under inversion).

For further discussions of beta decay we refer the reader to Segré and Krane.

Gamma decay

An excited nucleus can always decay to a lower energy state by either γ-emission or a competing process called internal conversion. In the latter the excess nuclear energy is given directly to an atomic electron which is ejected with a certain kinetic energy. In general, complicated rearrangements of nucleons occur during γ-decay.

The energetics of γ-decay is rather straightforward. As shown in Fig. 6.17 a γ is emitted while the nucleus recoils.

$$\hbar \underline{k} + \underline{p}_a = 0 \tag{6.63}$$

$$M^* c^2 = Mc^2 + E_\gamma + T_a \tag{6.64}$$

The recoil energy is usually quite small,

$$T_a = p_a^2/2M = \hbar^2 k^2/2M = E_\gamma^2/2Mc^2 \tag{6.65}$$

Typically, $E_\gamma \sim 2\,\text{MeV}$, so if $A \sim 50$, then $\text{T}_a \sim 40\,\text{eV}$. This is generally negligible.

Decay constants and selection rules

Nuclear excited states have half-lives for γ-emission ranging from 10^{-16} sec to >100 yrs. A rough estimate of λ_γ can be made using semi-classical ideas. From

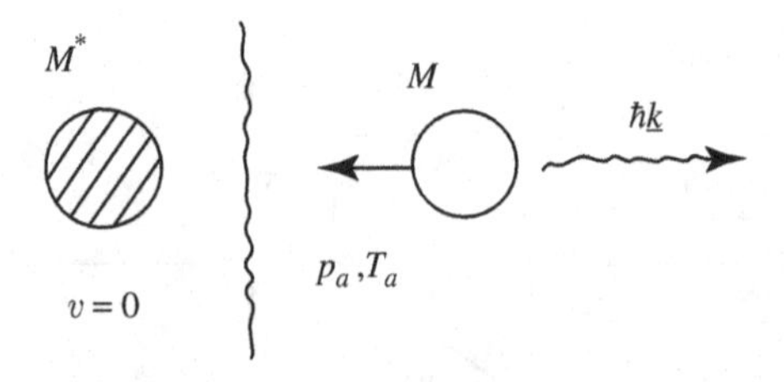

Fig. 6.17 Schematic of γ-decay showing an excited nucleus with mass M^* prior to emission of a high-energy photon with momentum $\hbar \underline{k}$ while the daughter nucleus recoils with momentum $\underline{p}_a$ and energy T_a.

Maxewell's equations one finds that an accelerated point charge e radiates electromagnetic radiation at a rate given by the *Lamor formula* [Jackson, 1962, Chap. 17],

$$\frac{dE}{dt} = \frac{2}{3}\frac{e^2 a^2}{c^3} \tag{6.66}$$

where a is the acceleration of the charge. Suppose the radiating charge has a motion like the simple oscillator,

$$x(t) = x_o \cos \omega t \tag{6.67}$$

where we take $x_o^2 + y_o^2 + z_o^2 = R^2$, R being the radius of the nucleus. From Eq. (6.67) we have

$$a(t) = R\omega^2 \cos \omega t \tag{6.68}$$

To get an average rate of energy radiation, we average Eq. (6.66) over a large number of oscillation cycles,

$$\left(\frac{dE}{dt}\right)_{\text{avg}} = \frac{2}{3}\frac{R^2\omega^4 e^2}{c^3}\left(\cos^2 \omega t\right)_{\text{avg}} \approx \frac{R^2\omega^4 e^2}{3c^3} \tag{6.69}$$

Now we assume each photon is emitted during a time interval τ (having the physical significance of a mean lifetime). Then,

$$\left(\frac{dE}{dt}\right)_{\text{avg}} = \frac{\hbar\omega}{\tau} \tag{6.70}$$

Equating this with Eq. (6.69) gives

$$\lambda_\gamma \approx \frac{e^2 R^2 E_\gamma^3}{3\hbar^4 c^3} \tag{6.71}$$

If we were to apply this result to a process in atomic physics, namely the de-excitation of an atom by electromagnetic emission, we would take $R \sim 10^{-8}$ cm and $E_\gamma \sim 1$ eV, in which case Eq. (6.71) gives

$$\lambda_\gamma \sim 10^6 \text{ sec}^{-1}, \quad \text{or} \quad t_{1/2} \sim 7 \times 10^{-7} \text{ sec}$$

On the other hand, if we apply Eq. (6.71) to nuclear decay, where typically $R \sim 5 \times 10^{-13}$ cm, and $E_\gamma \sim 1$ MeV, we would obtain

$$\lambda_\gamma \sim 10^{15} \text{ sec}^{-1}, \quad \text{or} \quad t_{1/2} \sim 3 \times 10^{-16} \text{ sec}$$

These results only indicate typical orders of magnitude. What Eq. (6.71) does not explain is the wide range of values of the half-lives that have been observed.

Turning to the question of selection rules for γ-decay, we can write down the conservation of angular momenta and parity in a form similar to Eqs. (6.61) and (6.62),

$$\underline{I}_i = \underline{I}_f + \underline{L}_\gamma \tag{6.72}$$

$$\pi_i = \pi_f \pi_\gamma \tag{6.73}$$

Notice in contrast to Eq. (6.71) the orbital and spin angular momenta are incorporated in $\underline{L}_\gamma$, which plays the role of the *total* angular momentum. Since the photon has spin $\hbar$ [Davydov 1965, pp. 306 and 578], the possible values of L_γ are 1 (corresponding to the case of zero orbital angular momentum), 2, 3, ... For the conservation of parity we know the parity of the photon depends on the value of L_γ. We now encounter two possibilities because in photon emission, which is the process of electromagnetic multipole radiation, one can have either electric or magnetic multipole radiation,

$$\pi_\gamma = (-1)^{L_\gamma} \quad \textit{electric multipole}$$
$$-(-1)^{L_\gamma} \quad \textit{magnetic multipole}$$

Thus we can set up the following table,

Radiation	**Designation**	**Value of L_γ**	π_γ
Electric dipole	E1	1	−1
Magnetic dipole	M1	1	+1
Electric quadrupole	E2	2	+1
Magnetic quadrupole	M2	2	−1
Electric octupole	E3	3	−1
etc.			

Similar to the case of β-decay, the decay constant can be expressed as a sum of contributions from each multipole [Blatt, 1952, p. 627],

$$\lambda_\gamma = \lambda_\gamma(E1) + \lambda_\gamma(M1) + \lambda_\gamma(E2) + \cdots \tag{6.74}$$

provided each contribution is allowed by the selection rules. We are again interested only in the lowest order allowed transition, and if both E and M transitions are allowed, E will dominate. Take, for example, a transition between an initial state with spin and parity of 2^+ and a final state of 0^+. This transition requires the photon parity to be positive, which means that for an electric multipole radiation L_γ would have to be even, and for a magnetic radiation it has to be odd. In view of the initial and final spins, we see that angular momentum conservation Eq. (6.52) requires L_γ to be 2. Thus, the most likely mode of γ-decay for this transition is $E2$. A few other examples are:

$$1^+ \to 0^+ \qquad M1$$

$$\frac{1}{2}^- \to \frac{1}{2}^+ \qquad E1$$

$$\frac{9}{2}^+ \to \frac{1}{2}^- \qquad M4$$

$$0^+ \to 0^+ \qquad \text{no } \gamma\text{-decay allowed}$$

We conclude this discussion of nuclear decays by the remark that internal conversion (IC) is a competing process with γ-decay. The atomic electron ejected has a kinetic energy given by (ignoring nuclear recoil)

$$T_e = E_i - E_f - E_B \tag{6.75}$$

where $E_i - E_f$ is the energy of de-excitation, and E_B is the binding energy of the atomic electron. If we denote by λ_e the decay constant for internal conversion, then the total decay constant for de-excitation is

$$\lambda = \lambda_\gamma + \lambda_e \tag{6.76}$$

REFERENCES

J. M. Blatt and V. F. Weisskopf, *Theoretical Nuclear Physics* (Wiley, New York, 1952), p. 627.

A. S. Davydov, *Quantum Mechanics* (Pergamon Press, London, 1965), pp. 306 and 578.

R. D. Evan, *The Atomic Nucleus* (McGraw-Hill, New York, 1955).

E. Fermi, *Nuclear Physics*, Lecture Notes by J. Orear, A. H. Rosenfeld and R. A. Schluter (University of Chicago Press, 1949).

G. Gamow, *Z. Phys.* **51**, 204 (1928).

R. W. Gurney and E. U. Condon, "Wave Mechanics and Radioactive Disintegration", *Nature* **122**, 439 (1928).

J. D. Jackson, *Classical Electrodynamics* (Wiley, New York, 1962), Chap. 17.

K. S. Krane, *Introductory Nuclear Physics* (Wiley, New York, 1987).

C. M. Lederer *et al*, eds., *Table of Isotopes* (Wiley, New York, 1978).
W. E. Meyerhof, *Elements of Nuclear Physics* (McGraw-Hill, New York, 1967).
F. Reines and C. L. Cowan, "Detection of the Free Neutrino", *Physical Review* **92**, 830 (1953).
E. Segrè, *Nuclei and Particles* (W. A. Benjamin, New York, 1965).

7

Collision Cross Sections

Among all the radiation interactions, the two-body scattering collisions are the simplest and most fundamental. Based on the understanding of this process one can extend the analysis to phenomena where the interaction involves a general reaction between two colliding particles. The quantities of interest in all cases are the cross sections which specify the likelihood of an interaction. There are several types of cross sections depending on the variety of interactions under consideration. For each type of interaction, there is a cross section σ which typically is a function of an appropriate energy of the collision. For collisions where a particle is emitted (including the scattering collision) there is an angular differential cross section $d\sigma/d\Omega$ which describes the angular distribution of emitted particle, and an energy differential cross section $d\sigma/dE$ for the energy distribution of the particle. In the case of scattering resulting in changes in energy and direction of motion, one can even define a double differential cross section $d^2\sigma/d\Omega dE$ to describe the measurements of sophisticated inelastic scattering experiments (see Sec. 13.4).

The purpose of this chapter is to focus only on the interaction of elastic scattering. We begin by introducing the physical concept of a cross section and then describe a basic method of calculating the angular differential scattering cross section. The method, called phase-shift analysis, is quite standard in many textbooks on quantum mechanis [Liboff, 1980; Schiff, 1955]. It is designed to solve the Schrödinger wave equation for the scattering (positive energy) solution. We recall from Chap. 5 the discussion on solving the wave equation to obtain bound-state (negative energy) solutions by matching appropriate boundary conditions at the interface of the interaction potential. Similarly, for the scattering solutions, one also applies boundary conditions at the interface, in the process thereby introducing a phase shift between the interior and exterior waves. We will proceed in two steps. First we formulate the cross section in terms of the phase shift without specifying the interaction potential. Then we consider the case of low-energy scattering where only the s-wave solution, which is very simple, is needed. The results of the cross section derivation are also useful as the basis to extend the cross section calculation to nuclear reactions in the following chapter. For first-time readers we recommend studying first Sec. 7.3, where the reduction of the two-body collision problem to an effective one-body calculation is discussed, as a general orientation. This chapter marks the beginning of the part of the book that we would regard as the fundamentals of *Nuclear Radiation Interactions*, the Part II that was mentioned in Chap. 2.

7.1 CONCEPT OF SCATTERING CROSS SECTIONS

The physical meaning of a cross section σ, is a measure of the probability of a certain interaction, whether the process is a collision (no change in reactants or products) or a reaction. We will speak about cross section for a particle interacting with a nucleus, where the particle can be a neutron, gamma, or charged particle. Imagine a beam of particles, all with the same energy, incident on a thin sample of thickness Δx covering an area A as sketched in Fig. 7.1. Let the intensity of the beam incident on the sample be I, having the unit of number of particles per second. The incident flux is therefore I/A, with unit of *particles per cm*^{2}-*second.*

If the nuclear density of the sample is N nuclei/cm^3, then the number of nuclei exposed is $NA\Delta x$. We are assuming no shadowing effects, i.e., the nuclei do not cover each other with respect to interacting with the incoming particles. For this situation the probability for a collision-induced reaction is

$$\{\text{reaction probability}\} = \Theta/I = \left(\frac{NA\Delta x}{A}\right) \bullet \sigma \tag{7.1}$$

where Θ is the number of reactions occurring per sec. Notice the quantity σ, with dimension of area or length squared, simply appears in our definition of reaction probability as *a proportionality constant*, with no further justification. (Sometimes this simple fact is not emphasized.) There are other ways to introduce or motivate

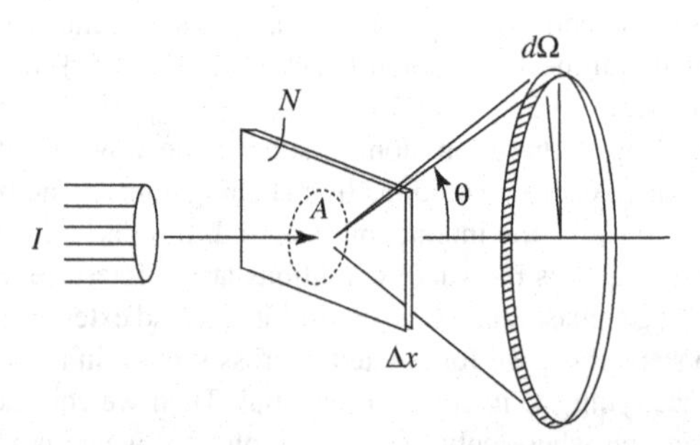

Fig. 7.1 Schematic of a particle beam of intensity I covering an area A incident upon a thin target of thickness Δx. After interaction the particles emerge into a cone subtending an angle θ (the scattering angle) relative to the direction of incidence. The element of solid angle $d\Omega$ is a small area on the surface of the cone (see Fig. 7.2 below).

the meaning of the cross section. They are essentially all equivalent when one imagines the physical set-up of a beam of particles colliding with a target of atoms. Equation (7.1) can be rearranged to give

$$\sigma = \{\text{reaction probability}\}/\{\text{no. exposed per unit area}\}$$

$$= \frac{\Theta}{IN\Delta x} = \frac{1}{I}\left[\frac{\Theta}{N\Delta x}\right]_{\Delta x \to 0} \tag{7.2}$$

Moreover, we define $\Sigma = N\sigma$, called the *macroscopic cross section*, a quantity having the dimension of inverse length. Then (7.2) becomes

$$\Sigma \Delta x = \frac{\Theta}{I}, \tag{7.3}$$

or

$$\Sigma \equiv \{\text{probability per unit path for small path that a reaction will occur}\} \tag{7.4}$$

Both the *microscopic cross section* σ (unit of σ is the **barn** which is 10^{-24} cm^2 as already noted in Chap. 3), and its counterpart, the macroscopic cross section Σ, which has the dimension of reciprocal length, are fundamental to our study of radiation interactions. Our discussion thus far can be applied to any radiation or particle.

We extend the concept of cross section to that of **angular differential** cross section $d\sigma/d\Omega$. Now we imagine counting the reactions per second in an angular cone subtended at angle θ with respect to the direction of incidence (incoming particles), as shown previously in Fig. 7.1. Again $d\Omega$ is the element of solid angle, which is the small area through which the unit vector $\underline{\Omega}$ passes through (see Fig. 7.2). Thus, $d\Omega = \sin\theta d\theta d\varphi$.

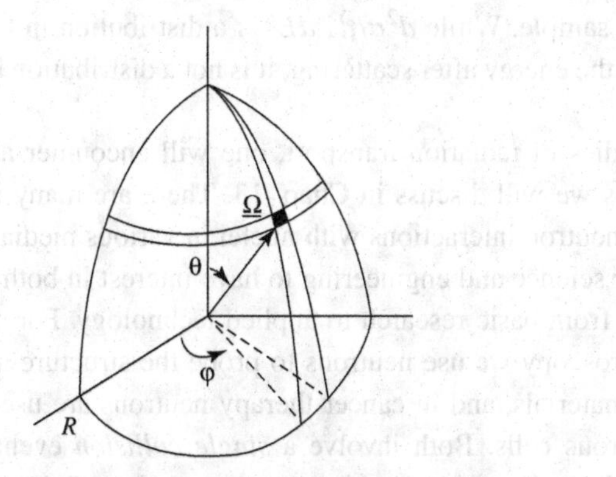

Fig. 7.2 The unit vector $\underline{\Omega}$ in spherical coordinates, with θ and φ being the polar and azimuthal angles respectively (R would be unity if $\underline{\Omega}$ ends on the unit sphere). The element of solid angle $d\Omega$ is the small dark area on the unit sphere.

We can write

$$\frac{1}{I}\left(\frac{d\Theta}{d\Omega}\right) = N\Delta x\left(\frac{d\sigma}{d\Omega}\right) \tag{7.5}$$

In analogy to Eq. (7.1), $d\sigma/d\Omega$ appears also as a proportionality constant between the reaction rate per unit solid angle and a product of two simple factors specifying the interacting system — the incident flux and the number of nuclei exposed (or the number of nuclei available for reaction). The normalization condition of the angular differential cross section is $\int d\Omega\,(d\sigma/d\Omega) = \sigma$, which shows the relation between $d\sigma/d\Omega$ and σ and explains why the former is called the *angular differential cross section*, and the latter is often called the *total* cross section in the sense of an integrated quantity.

There is another differential cross section which we should introduce. Suppose we consider the incoming particles to have energy E and the particles after reaction to have energy in dE' about E'. One can define in a similar way as above an *energy differential cross section*, $d\sigma/dE'$, which is a measure of the probability of an incoming particle with energy E will have outgoing energy E' as a result of the reaction. Both $d\sigma/d\Omega$ and $d\sigma/dE'$ are *distribution functions*, the former is a distribution in the variable $\underline{\Omega}$, the solid angle, whereas the latter is a distribution in E', the energy after scattering. Their dimensions are *barns per steradian* and *barns per unit energy*, respectively.

Combining the two extensions from cross section to differential cross sections, we can further extend to a *double differential cross section* $d^2\sigma/d\Omega dE'$, which is a quantity that has been studied extensively in thermal neutron scattering (see the section on neutron thermalization in Chap. 13). This particular cross section contains the most fundamental information about the structure and dynamics of the scattering sample. While $d^2\sigma/d\Omega dE'$ is a distribution in two variables, the solid angle and the energy after scattering, it is not a distribution in E, the energy before scattering.

In studies of radiation transport, one will encounter all three types of cross sections as we will discuss in Chap. 13. There are many important applications based on neutron interactions with nuclei in various media. It is within the scope of nuclear science and engineering to have interest in both the cross sections and their uses from basic research to applied technology. For example, in diffraction and spectroscopy we use neutrons to probe the structure and dynamics of many types of materials, and in cancer therapy neutrons are used to preferentially kill the cancerous cells. Both involve a *single collision* event between the neutron and a nucleus. For these problems, a knowledge of the cross sections is sufficient. In contrast, for nuclear reactors and other nuclear systems applications, one is interested in the effects of *a sequence of collisions or multiple collisions*, in which case knowing only the cross sections is *not* sufficient. One needs to follow the neutrons as they undergo many collision sequences in the media of interest. This then requires the study of **neutron transport** — *the distribution of neutrons in configuration space, direction of travel, and energy*, as we will discuss in Chap. 13.

7.2 CROSS SECTION CALCULATION — METHOD OF PHASE SHIFT

Since the unit process in radiation interactions can be completely characterized by the cross sections, clearly the next step is to discuss how this central quantity can be formulated and calculated. It is to be expected that σ is a dynamical quantity which depends on the nature of the interaction forces between the radiation particle and the target nucleus. Since nuclear forces are very complicated, it follows any calculation of σ will be complicated as well. It is therefore fortunate that methods for calculating σ have been developed which we expect to be within the grasp of the reader. The method we follow is widely known in the literature. Here we discuss the key steps involved without belaboring all the details, from the introduction of the *scattering amplitude* $f(\theta)$ to the derivation of the expression for the angular differential cross section $\sigma(\theta) = d\sigma/d\Omega$.

Expressing $\sigma(\theta)$ in terms of the Scattering Amplitude $f(\theta)$

We consider a scattering scenario sketched in Fig. 7.3, where the incident plane wave as (see Sec. 7.3 for the reduction to an effective one-body potential scattering problem) is taken to be,

$$\Psi_{in} = be^{i(\underline{k}\cdot\underline{r}-\omega t)} \tag{7.6}$$

where the wavenumber k is set by the energy of the incoming effective particle E, and the scattered spherical outgoing wave is written as

$$\Psi_{sc} = f(\theta)b\frac{e^{i(kr-\omega t)}}{r} \tag{7.7}$$

where $f(\theta)$ is the scattering amplitude. The angular differential cross section for scattering through $d\Omega$ about $\underline{\Omega}$ is

$$\sigma(\theta) = \frac{\underline{J_{sc}} \cdot \underline{\Omega}}{J_{in}} = |f(\theta)|^2 \tag{7.8}$$

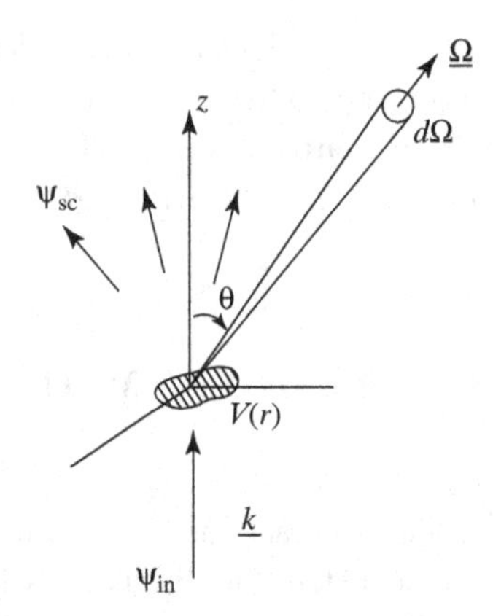

Fig. 7.3 Schematic of potential scattering where the incident particle, represented by an incoming plane wave ψ_{in} with wave vector $\underline{k}$, is scattered by a potential field $V(r)$ giving rise to spherical outgoing wave ψ_{sc}. The scattered current crossing an element of surface area $d\Omega$ about the direction $\underline{\Omega}$ is used to define the angular differential cross section $d\sigma/d\Omega \equiv \sigma(\theta)$, where the scattering angle θ is the angle between the direction of incidence and direction of scattering.

where we have used the expression for the current (see (6.31) and any text on quantum mechanics [Liboff, 1980; Scchiff, 1955]),

$$\underline{J} = \frac{\hbar}{2\mu i}[\Psi^*(\underline{\nabla}\Psi) - \Psi(\underline{\nabla}\Psi^*)] \tag{7.9}$$

Calculating f (θ) from the Schrödinger wave equation

The *Schrödinger* equation to be solved is of the form (see Eq. (5.1) and Sec. 7.3)

$$\left(-\frac{\hbar^2}{2\mu}\nabla^2 + V(r)\right)\psi(\underline{r}) = E\psi(\underline{r}) \tag{7.10}$$

where $\mu = m_1 m_2/(m_1 + m_2)$ is the reduced mass, and $E = \mu v^2/2$, with v being the relative speed, is positive. Notice Eq. (7.10) is the equation of motion for an effective particle with mass equal to the reduced mass at position r which is the relative coordinate between the two colliding particles. The reduction from a two-body collision with a central force potential $V(r)$ is is quite standard in classical or quantum mechanics. Because it is conceptually an important step in the understanding of cross section calculations, the details of this reduction are explained in Sec. 7.3 at the end of the chapter.

To obtain a solution to our particular scattering set-up, we will impose the boundary condition

$$\psi_k(\underline{r}) \rightarrow_{r \gg r_o} e^{ikz} + f(\theta)\frac{e^{ikr}}{r} \tag{7.11}$$

where r_o is the range of force, $V(r) = 0$ for $r > r_o$. In the region beyond the force range the wave equation describes a free particle. This free-particle solution is what we want to match up with the RHS of Eq. (7.11). The most convenient form of the free-particle wave function is an expansion in terms of partial waves (see Eq. (5.29)),

$$\psi(r, \theta) = \sum_{\ell=0}^{\infty} R_\ell(r) P_\ell(\cos\theta) \tag{7.12}$$

where $P_\ell(\cos\theta)$ is the *Legendre polynomial of order* ℓ. Inserting Eq. (7.12) into Eq. (7.10), and setting $u_\ell(r) = rR_\ell(r)$, we obtain

$$\left(\frac{d^2}{dr^2} + k^2 - \frac{2\mu}{\hbar^2}V(r) - \frac{\ell(\ell+1)}{r^2}\right) u_\ell(r) = 0, \tag{7.13}$$

One can compare this with Eq. (5.30). Equation (7.13) describes the wave function everywhere. Its solution clearly depends on the form of $V(r)$. Outside of the interaction region, $r > r_o$, Eq. (7.13) reduces to the radial wave equation for a free particle,

$$\left(\frac{d^2}{dr^2} + k^2 - \frac{\ell(\ell+1)}{r^2}\right) u_\ell(r) = 0 \tag{7.14}$$

with general solution

$$u_\ell(r) = B_\ell r j_\ell(kr) + C_\ell r n_\ell(kr) \tag{7.15}$$

where B_ℓ and C_ℓ are integration constants, and j_ℓ and n_ℓ are *spherical Bessel and Neumann* functions respectively [Liboff, 1980; Schiff, 1955]. The latter are tabulated functions; for our purposes it is sufficient to note they have the following properties.

$$j_o(x) = \sin x/x, \qquad n_o(x) = -\cos x/x$$

$$j_1(x) = \frac{\sin x}{x} - \frac{\cos x}{x}, \qquad n_1(x) = -\frac{\cos x}{x^2} - \frac{\sin x}{x}$$

$$j_\ell(x) \underset{x\to 0}{\rightarrow} \frac{x^\ell}{1\cdot 3\cdot 5\ldots(2\ell+1)} \qquad n_\ell(x) \underset{x\to 0}{\rightarrow} \frac{1\cdot 3\cdot 5\ldots(2\ell-1)}{x^{\ell+1}}$$

$$j_\ell(x) \underset{x\gg 1}{\rightarrow} \frac{1}{x}\sin(x-\ell\pi/2) \qquad n_\ell(x) \underset{x\gg 1}{\rightarrow} -\frac{1}{x}\cos(x-\ell\pi/2)$$

Introduction of the phase shift δ_ℓ

We rewrite the general solution Eq. (7.15) as

$$u_\ell(r) \underset{kr \gg 1}{\rightarrow} (B_\ell/k)\sin(kr - \ell\pi/2) - (C_\ell/k)\cos(kr - \ell\pi/2) \tag{7.16}$$

$$= (a_\ell/k)\sin[kr - (\ell\pi/2) + \delta_\ell] \tag{7.17}$$

where we have replaced B and C by two other constants, a and δ, the latter is seen to be a *phase shift*. The partial-wave expansion of the free-particle wave function in the asymptotic region becomes

$$\psi(r,\theta) \underset{kr \gg 1}{\rightarrow} \sum_\ell a_\ell \frac{\sin[kr - (\ell\pi/2) + \delta_\ell]}{kr} P_\ell(\cos\theta) \tag{7.18}$$

This is the LHS of Eq. (7.12). Now we prepare the RHS of Eq. (7.12) to have the same form of partial wave expansion by writing

$$f(\theta) = \sum_\ell f_\ell P_\ell(\cos\theta) \tag{7.19}$$

and

$$\begin{aligned} e^{ikr\cos\theta} &= \sum_\ell i^\ell (2\ell+1) j_\ell(kr) P_\ell(\cos\theta) \\ &\underset{kr \gg 1}{\rightarrow} \sum_\ell i^\ell (2\ell+1) \frac{\sin(kr - \ell\pi/2)}{kr} P_\ell(\cos\theta) \end{aligned} \tag{7.20}$$

Inserting both Eqs. (7.19) and (7.20) into the RHS of Eq. (7.12), we match the coefficients of $\exp(ikr)$ and $\exp(-ikr)$ to obtain

$$f_\ell = \frac{1}{2ik}(-i)^\ell [a_\ell e^{i\delta_\ell} - i^\ell(2\ell+1)] \tag{7.21}$$

$$a_\ell = i^\ell (2\ell+1) e^{i\delta_\ell} \tag{7.22}$$

Combing Eqs. (7.21) and (7.19) we obtain

$$f(\theta) = (1/k)\sum_{\ell=0}^{\infty} (2\ell+1) e^{i\delta_\ell} \sin\delta_\ell P_\ell(\cos\theta) \tag{7.23}$$

Final Expressions for $\sigma(\theta)$ *and* σ

In view of Eq. (7.23), the angular differential cross section Eq. (7.8) becomes

$$\sigma(\theta) = \lambda\!\!\!^{-2} \left| \sum_{\ell=0}^{\infty} (2\ell+1) e^{i\delta_\ell} \sin\delta_\ell P_\ell(\cos\theta) \right|^2 \tag{7.24}$$

where$\lambdabar = 1/k$ is the *reduced* wavelength. Correspondingly, the total cross section is

$$\sigma = \int d\Omega\sigma(\theta) = 4\pi\lambdabar^2 \sum_{\ell=0}^{\infty} (2\ell+1)\sin^2\delta_\ell \tag{7.25}$$

S-wave scattering

We have seen that if kr_o is appreciably less than unity, then only the $\ell = 0$ term contributes in Eqs. (7.24) and (7.25) (see the discussion in Chap. 5 in connection with Fig. 5.6(b)). The differential and total cross sections for *s-wave* scattering are therefore

$$\sigma(\theta) = \lambdabar^2 \sin^2\delta_o(k) \tag{7.26}$$

$$\sigma = 4\pi\lambdabar^2 \sin^2\delta_o(k) \tag{7.27}$$

Notice *s*-wave scattering is spherically symmetric, or $\sigma(\theta)$ is independent of the scattering angle. This is true in CMCS, but not in LCS (see Chap. 9 for further discussions). From Eq. (7.21) we see $f_o = (e^{i\delta_o}\sin\delta_o)/k$. Since the cross section must be finite at low energies, as $k \to 0$ f_o has to remain finite, or $\delta_o(k) \to 0$. We can set

$$\lim_{k\to 0}[e^{i\delta_o(k)}\sin\delta_o(k)] = \delta_o(k) = -ak \tag{7.28}$$

where the constant a is called the *scattering length.* Thus for low-energy scattering, the differential and total cross sections depend only on knowing the scattering length of the target nucleus,

$$\sigma(\theta) = a^2 \tag{7.29}$$

$$\sigma = 4\pi a^2 \tag{7.30}$$

Physical significance of the sign of the scattering length

Figure 7.4 shows two sine waves, one is the reference wave sin kr which has not had any interaction (unscattered) and the other one is the wave $\sin(kr + \delta_o)$ which has suffered a phase shift by virtue of the scattering. The entire effect of the scattering is represented by the phase shift δ_o, or equivalently the scattering length through Eq. (7.28). In the vicinity of the potential, we take kr_o to be small (again invoking the condition of low-energy scattering), so that $u_o \sim k(r - a)$, in which case a *becomes the distance at which the wave function extrapolates to zero* from its value and slope at $r = r_o$. There are two situations in which this extrapolation can take place, depending on the value of kr_o. As shown in Fig. 7.5 below, when $kr_o > \pi/2$, the wave function has reached more than a quarter of its wavelength at $r = r_o$.

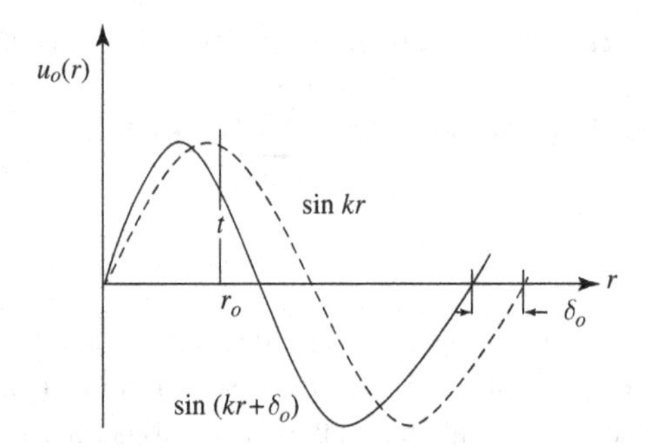

Fig. 7.4 Schematic of comparison of *unscattered* (dashed curve) and *scattered* (solid curve) waves showing a phase shift δ_o in the asymptotic region as a result of the scattering. The range of the scattering potential is r_o.

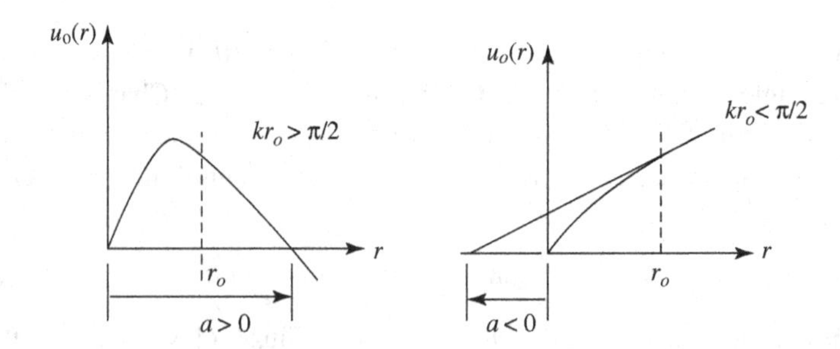

Fig. 7.5 Geometric interpretation of *positive* (left panel) and *negative* (right panel) scattering lengths as the extrapolation distances of the wave function at the interface r_o between interior and exterior solutions. The two situations correspond to potentials for which a *bound* state can exist (left panel) and for which only *virtual* states are possible (right panel), respectively. The range of the potential is r_o.

So its slope is *downward* and the extrapolation gives a distance a which is *positive*. If on the other hand, $kr_o < \pi/2$, then the extrapolation gives a distance a which is *negative*. The significance of $a > 0$ is the potential is such that it can have a *bound* state, whereas $a < 0$ means that the potential can only give rise to a *virtual* state (recall the discussions surrounding Fig. 5.19).

The interface boundary condition

Before closing this section we wish to derive a general result that will be useful later for the actual determination of the phase shift δ_ℓ. Recall δ_ℓ was introduced in Eq. (7.17) in writing the partial-wave expansion of the free-particle wave function in the asymptotic region. This result also shows how the effects of scattering can be expressed in one part pertaining to interaction at the surface of the nucleus and another part describing the interaction in the interior region. The separation of the two parts forms the basis for extending the present description to treat nuclear reactions, a topic we will take up in the next chapter, Sec. 8.3.

For an interaction potential of a definite range, the phase shift δ_ℓ is determined by applying the boundary condition at this potential range, where one imposes continuity on the wave function and its derivative (the current). We had previously encountered such interface boundary conditions, namely, Eqs. (5.6) and (5.7), which were introduced to determine the bound states of a given potential. We have also seen in Chap. 5 an efficient way to impose continuity at the interface is to require the equality of the logarithmic derivative to be evaluated using the interior and exterior wave functions,

$$q_\ell^{\rm int} = q_\ell^{ext} \tag{7.31}$$

where

$$q_\ell \equiv \left[\frac{r du_\ell}{u_\ell dr}\right]_{r=r_o} \tag{7.32}$$

The superscripts, *int* and *ext*, denote the radial wave function in the interior ($r < r_o$) and exterior ($r > r_o$) regions, respectively. We will leave $q_\ell^{\rm int}$ to be specified later when we consider a particular calculation. For now we continue to manipulate the exterior region solution to obtain a particular expression for $e^{i\delta_\ell} \sin\delta_\ell$, the quantity that appears in the angular differential scattering cross section, Eq. (7.24). Going back to Eq. (7.16) we see for the chosen normalization of u_ℓ, the two integration constants must satisfy $B_\ell = \cos\delta_\ell$, and $C_\ell = -\sin\delta_\ell$. Inserting these relations into Eq. (7.16), we find the radial wave function in the exterior region becomes

$$u_\ell^{ext} = Im\lfloor u_\ell^{(+)} e^{-i\delta_\ell}\rfloor \tag{7.33}$$

where we have defined a conjugate pair of quantities

$$u_\ell^{(\pm)} = kr[n_\ell(kr) \pm i j_\ell(kr)] \tag{7.34}$$

with j_ℓ and n_ℓ being the *spherical Bessel* and *Neumann* functions respectively (see Eq. (7.15)). The right hand side of Eq. (7.31) then becomes

$$q_\ell^{ext} = \xi \frac{\mathrm{Im}\{e^{-i\delta_\ell}[du_\ell^{(+)}/d\xi]\}}{\mathrm{Im}\{e^{-i\delta_\ell}u_\ell^{(+)}\}} \tag{7.35}$$

with $\xi = kr_o$. The usefulness of this formal expression for the logarithmic derivative involving the exterior wave, is admittedly not yet apparent. To avoid writing the imaginary part of a complex quantity in Eq. (7.35) we introduce yet another conjugate pair of functions,

$$q_\ell^{(\pm)} \equiv \frac{\xi}{u_\ell^{(\mp)}} \frac{du_\ell^{(\mp)}}{d\xi} \tag{7.36}$$

Notice the sign reversal in the superscripts between $q_\ell^{(\pm)}$ and $u_\ell^{(\pm)}$ which we follow as a matter of convention. After some algebra we find Eq. (7.35) becomes

$$q_\ell^{ext} = \frac{u_\ell^{(+)} q_\ell^{(-)} - e^{2i\delta_\ell} u_\ell^{(-)} q_\ell^{(+)}}{u_\ell^{(+)} - e^{2i\delta_\ell} u_\ell^{(-)}} \tag{7.37}$$

Equation (7.37) shows a certain symmetry in the appearance of the quantities $u_\ell^{(\pm)}$. Since they are a conjugate pair, their ratio can be expressed in terms of a phase angle, τ_ℓ,

$$u_\ell^{(\pm)} = \frac{1}{\sqrt{v_\ell}} \exp(\mp i\tau_\ell) \tag{7.38}$$

or equivalently,

$$\tau_\ell = -\tan^{-1}(j_\ell/n_\ell) = \tan^{-1}(j_\ell/n_\ell) \tag{7.39}$$

which shows the sign invariance of the phase angle. In Eq. (7.39), the amplitude $1/\sqrt{v_\ell}$ is of no interest in this discussion. In the literature on nuclear reactions, it is sometimes called the penetration factor. Using Eq. (7.38) we can eliminate $u_\ell^{(\pm)}$ in favor of τ_ℓ in the expression Eq. (7.37). Then equating the result to q_ℓ^{int} which is still unspecified, we obtain

$$\eta_\ell \equiv e^{2i\delta_\ell} = e^{-2i\tau_\ell} \frac{q_\ell^{\text{int}} - q_\ell^{(-)}}{q_\ell^{\text{int}} - q_\ell^{(+)}} \tag{7.40}$$

This is a classic result in the quantum theory of scattering [Roman 1965, Foderaro 1971]. In Eq. (7.40) we define the quantity η_ℓ which is related to the well-known scattering matrix. It is sometimes called the *reflection factor* [Foderaro, 1971, p. 284]. Considering the definitions and manipulations it takes to arrive at this compact expression, we note this way of defining the phase shift suggests a useful interpretation, namely, the effects of scattering can be separated into two contributions, one pertains to scattering at the potential surface, represented by the factor $\exp(2i\tau_\ell)$, and the other pertains to interactions within the potential, thus involving q_ℓ^{int}. We will see in Chap. 8, Sec. 8.3 it is the second part that allows us to extend the present analysis of scattering to treat nuclear reactions.

Returning to the angular differential scattering cross section given in Eq. (7.24), we point out that the particular factor, $\exp(i\delta_\ell)\sin\delta_\ell$, can be rewritten as $[\exp(2i\delta_\ell) - 1]/2i$, and by using Eq. (7.40) one obtains

$$e^{i\delta_\ell}\sin\delta_\ell = e^{-2i\tau_\ell}\left[\frac{s_\ell}{q_\ell^{\text{int}} - \Delta_\ell - is_\ell} - e^{i\tau_\ell}\sin\delta_\ell\right] \tag{7.41}$$

where Δ_ℓ and s_ℓ are the real and imaginary parts of $q_\ell^{(\pm)}$,

$$q_\ell^{(\pm)} \equiv \Delta_\ell \pm is_\ell \tag{7.42}$$

Equation (7.41) is the expression to be use in actual calculation of the angular differential scattering cross section, as given by Eq. (7.24). The significance of the two separate contributions just mentioned is even more evident in Eq. (7.41). The first term clearly has a resonant form. If the wave function does not penetrate inside the potential, which would be the case for scattering by a hard sphere, we would expect this term to be absent. Then the scattering is entirely described by the second term. For later discussions in Chap. 12 we remark here that the two terms in Eq. (7.41) should be interpreted as the contributions from resonance and potential (or shape elastic) scattering, respectively.

Hard sphere scattering

In a quantum mechanical treatment this is a degenerate case in that physically one requires the wave function to vanish at the range cutoff $r = r_o$ and there is no penetration inside the potential. If one is considering an actual collision between two identical hard spheres, then r_o is just the hard sphere diameter. Since the logarithmic derivative Eq. (7.32) is now infinite, Eq. (7.40) reduces to $\delta_\ell = -\tau_\ell$, which confirms our earlier interpretation that to within a sign, τ_ℓ is the phase shift due to the interaction at the potential surface. Conversely, one can say *shape elastic scattering* is the same as scattering from a hard-sphere potential. In the limit of low energy, $kr_o \ll 1$, only the $\ell = 0$ term, the s-wave contribution, is important in Eq. (7.24),

$$\sigma(\theta) = \sigma(\theta)|_{\ell=0} = r_o^2 \tag{7.43}$$

The total cross section at the low-energy limit is therefore

$$\sigma = 4\pi r_o^2, \quad kr_o \ll 1 \tag{7.44}$$

At the opposite limit of high energy, one finds [Mott and Massey, 1949, pp. 38]

$$\sigma \cong 2\pi\lambda\!\!\!^{-2}\sum_{\ell=0}^{kr_o}(2\ell+1) \cong \frac{2\pi}{k^2}\int_0^{kr_o} 2\ell d\ell = 2\pi r_o^2, \quad kr_o \gg 1 \tag{7.45}$$

We see the quantum mechanical cross section for hard sphere scattering is dependent on the incident energy E, and the cross sections at the limits of low and high energy are greater than the classical cross section which has the value of πr_o^2. This difference between the quantum and classical results lies in the effects of diffraction, sometimes known as "shadow scattering", which are present in the quantum mechanical treatment. These effects are most pronounced in the forward scattering direction in $\sigma(\theta)$.

We can test the energy-dependent limiting behavior of the total cross section predicted by the hard-sphere model against experimental measurements that have been averaged over different ranges of energy [Foderaro, 1971, p. 289]. Figure 7.6 shows three sets of data, averaged over low-energy range (1,100) eV, intermediate-energy range around 10 MeV, and high-energy range around 100 MeV. These show the variation of the averaged cross sections with mass number. The two curves are the predictions of Eqs. (7.44) and (7.45), with $r_o = 1.5 \times 10^{-13} A^{1/3}$ cm. We see the high-energy data indeed behave like Eq. (7.45) for all mass numbers, indicating that scattering at high energies can be described in the sense of energy averaging by a hard-sphere interaction with a range that scales like the mass number to the one-third power. In contrast, the low-energy data follow the hard-sphere result reasonably well only for the heavy nuclides. For mass number less than ~80 the data show pronounced fluctuations at values consistently above the hard-sphere limit.

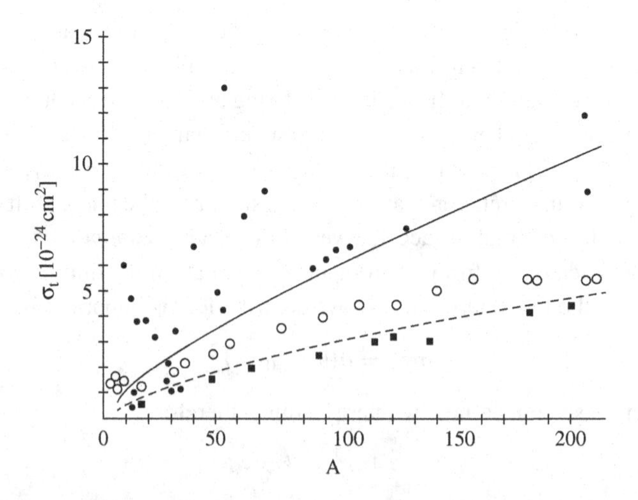

Fig. 7.6 Variation with mass number of measured total neutron cross sections [10^{-24} cm^2] averaged over three energy ranges, 1–100 eV (closed circles) 1–100 eV, around 10 MeV (open circles), and around 100 MeV (closed squares). The low-energy limit of $4\pi r_0^2$ is represented by the solid line, while the high-energy limit of $2\pi r_0^2$ is given by the dashed line. [Adapted from Foderaro, 1971, p. 289.]

This is not surprising because we expect at low energies there will be reactions and resonances which do not behave like hard-sphere scattering. Furthermore, for the light nuclei the resonances will be spaced more widely in energy (compared to resonances in the heavy nuclei) and therefore their effects are less likely to be averaged out. These considerations suggest the discrepancy between the data and Eq. (7.44) is to be expected based on the considerations we have just discussed. At intermediate energy, around 10 MeV, one can further analyze the data in terms of an effective-range model that treats the total cross section as *k-dependent* [Foderaro 1971, p. 289].

s- and p-wave scattering

In Chap. 5, we have made a simple semiclassical argument to show that for a given incident wavenumber k, with $k = \sqrt{2\mu E}/\hbar$, only those partial waves with $\ell < kr_o$ will make significant contributions to the scattering (see the discussions pertaining to Fig. 5.6(b)). We repeat that argument here expecting the reader at this point will have had a chance to look over Sec. 7.3 below. Consider a two-body collision at an impact parameter b in relative coordinates. According to Sec. 7.3, the problem can be expressed as that of an effective particle scattered by a spherically symmetric interaction potential with an angular momentum $\hbar\ell = pb$, where $p = \hbar k$ is the linear momentum, all in relative coordinates. Depending on the value of b compared to the range of the interaction potential r_o, one can say whether or not there will be a significant interaction. One expects scattering will take place only if $b < r_o$, or only the ℓ values satisfying $b = \ell/k < r_o$, or

$$\ell < kr_o \tag{7.46}$$

will contribute to scattering. This is a generally useful criterion in treating the results derived from partial wave expansions. In particular, it means that under the condition of low incident energy, namely, $kr_o \ll 1$, only the $\ell = 0$, *s*-wave, contribution needs to be considered. We have already discussed the *s*-wave scattering results above. Recall that in going from Eq. (7.27) to Eq. (7.30) one is assuming that $\sin\delta_o$ can be replaced by $-ak$, where a turns out to be the scattering length. This assumption can be violated if the scattering involves a resonance reaction in the vicinity of the energy under consideration. The implication of Eq. (7.30) is that both the angular differential and the total cross sections are energy-independent.

It is useful to have an indication of when $\ell = 1$, *p-wave*, contribution in the partial-wave expansion starts to become important. Including the first two terms in the expression for the scattering amplitude, Eq. (7.23), we have

$$f(\theta)|_{s,p} = \lambda\!\!\!^{-} \lfloor e^{i\delta_o} \sin\delta_o + 3e^{i\delta_1} \sin\delta_1 \cos\theta \rfloor \tag{7.47}$$

which then gives

$$\sigma(\theta)|_{s,p} = \lambda\!\!\!^{-2}(A + B\cos\theta + C\cos^2\theta) \tag{7.48}$$

with $A = \sin^2\delta_o$, $B = 6\cos(\delta_o - \delta_1)\sin\delta_o \sin\delta_1$, $C = 9\sin^2\delta_1$. Usually $\delta_1 \ll \delta_o$, then C is negligible and $B \approx 3\delta_1 \sin 2\delta_o$. Moreover, $|\delta_o| < \pi/2$, so B is positive. This means $\sigma(\theta)$ will show a forward-peaking component when the p-wave contribution needs to be considered.

One can integrate Eq. (7.48) to obtain the total cross section,

$$\sigma|_{s,p} = 4\pi\lambda\!\!\!^{-2}(\sin^2\delta_o + 3\sin^2\delta_1) \approx \sigma|_s + O(\delta_1^2) \tag{7.49}$$

so the p-wave contribution is of order $\delta_1$in $\sigma(\theta)$, but of order δ_1^2 in σ. This can be understood physically by the way the different partial waves interfere with each other in giving the differential and total cross sections. The waves act coherently (square of the sum over ℓ) in $\sigma(\theta)$, whereas they act incoherently (sum of squares) in σ.

Optical theorem

Before ending this section we note a simple relation between the scattering amplitude at $\theta = 0$ and the total cross section. From Eq. (7.23) we have the identity

$$Im[f(\theta = 0)] = \lambda\!\!\!^{-} \sum_{\ell} (2\ell + 1)\sin^2\delta_\ell \tag{7.50}$$

Comparing this result with Eq. (7.25) we obtain

$$\sigma = 4\pi\lambda\!\!\!^{-2} Im[f(\theta = 0)] \tag{7.51}$$

This relation is called *the optical theorem* in analogy with a similar expression in optics between the absorption coefficient and the imaginary part of the complex index of refraction. It is a direct consequence of the conservation of probability [cf. Roman, 1965, p. 337]. Equation (7.50) holds even when inelastic scattering is present. At any energy the scattering amplitude for elastic scattering in the forward direction determines the total scattering cross section.

7.3 THE EFFECTIVE ONE-BODY PROBLEM

We have discussed how to solve the Schrödinger equation for an effective particle scattered by a central potential $V(r)$ in Eq. (7.10), as well as in Chap. 5.

Throughout the analyses it was always understood the two-body collision problem has been reduced to an effective one-body problem for a particle with a mass that is the reduced mass and moving in a relative coordinate system that depends only the separation distance between the two colliding particles. This reduction is quite standard in either classical [Goldstein, 2001] or quantum mechanics [Meyerhof, 1967, pp. 21; Roman, 1965]. Because it is conceptually important for the reader to keep in mind this simplification — the assumption of a central potential of interaction — we provide here the details of the reduction. By a central force potential $V(r)$ we mean the interaction potential is only a function of the relative coordinate. We will first go through the argument in classical mechanics. The equation describing the motion of particle 1 moving under the influence of particle 2 is the Newton's equation of motion,

$$m_1\ddot{\underline{r}}_1 = \underline{F}_{12} \tag{7.52}$$

where $\underline{r}_1$ is the position of particle 1 and $\underline{F}_{12}$ is the force on particle 1 exerted by particle 2. Similarly, the equation of motion for particle 2 is

$$m_2\ddot{\underline{r}}_2 = \underline{F}_{21} = -\underline{F}_{12} \tag{7.53}$$

where we have noted that the force exerted on particle 2 by particle 1 is exactly the opposite of $\underline{F}_{12}$. Now we transform from laboratory coordinate system to the center-of-mass coordinate system by defining the center-of-mass and relative positions,

$$\underline{r}_c = \frac{m_1\underline{r}_1 + m_2\underline{r}_2}{m_1 + m_2}, \quad \underline{r} = \underline{r}_1 - \underline{r}_2 \tag{7.54}$$

Solving for $\underline{r}_1$ and $\underline{r}_2$we have

$$\underline{r}_1 = \underline{r}_c + \frac{m_2}{m_1 + m_2}\underline{r}, \quad \underline{r}_2 = \underline{r}_c - \frac{m_1}{m_1 + m_2}\underline{r} \tag{7.55}$$

We can add and subtract Eqs. (7.52) and (7.53) to obtain equations of motion for $\underline{r}_c$ and $\underline{r}$. One finds

$$(m_1 + m_2)\ddot{\underline{r}}_c = 0 \tag{7.56}$$

$$\mu\ddot{\underline{r}} = \underline{F}_{12} = -dV(r)/d\underline{r} \tag{7.57}$$

with $\mu = m_1m_2/(m_1 + m_2)$ being the reduced mass. Thus the center-of-mass moves in a straight-line trajectory like a free particle, while the relative position obeys the equation of an effective particle with mass μ moving under the force generated by the potential $V(r)$. Equation (7.57) is the desired result of our reduction. It is manifestly the one-body problem of an effective particle scattered by a potential field. Far from the interaction field the particle has the kinetic energy$E = \mu(\dot{\underline{r}})^2/2$.

The quantum mechanical analogue of this reduction proceeds from the *Schrödinger* equation for the system of two particles,

$$\left(-\frac{\hbar^2}{2m_1}\nabla_1^2 - \frac{\hbar^2}{2m_2}\nabla_2^2 + V(|\underline{r}_1 - \underline{r}_2|)\right)\Psi(\underline{r}_1, \underline{r}_2) = (E_1 + E_2)\Psi(\underline{r}_1, \underline{r}_2) \tag{7.58}$$

Transforming the Laplacian operator ∇^2 from operating on $(\underline{r}_1, \underline{r}_2)$ to operating on $(\underline{r}_c, \underline{r})$, we find

$$\left(-\frac{\hbar^2}{2(m_1+m_2}\nabla_c^2 - \frac{\hbar^2}{2\mu}\nabla^2 + V(r)\right)\Psi(\underline{r}_c, \underline{r}) = (E_c + E)\Psi(\underline{r}_c, \underline{r}) \tag{7.59}$$

Since the Hamiltonian is now a sum of two parts, each involving either the center-of-mass position or the relative position, the problem is separable. Anticipating this, we have also divided the total energy, previously the sum of the kinetic energies of the two particles, into a sum of center-of-mass and relative energies. Therefore we can write the wave function as a product, $\Psi(\underline{r}_c, \underline{r}) = \psi_c(\underline{r}_c)\psi(\underline{r})$ so that Eq. (7.59) reduces to two separate problems,

$$-\frac{\hbar^2}{2(m_1+m_2)}\nabla_c^2\psi_c(\underline{r}_c) = E_c\psi_c(\underline{r}_c) \tag{7.60}$$

$$\left(-\frac{\hbar^2}{2\mu}\nabla^2 + V(r)\right)\psi(\underline{r}) = E\psi(\underline{r}) \tag{7.61}$$

It is clear that Eqs. (7.60) and (7.61) are the quantum mechanical analogues of Eqs. (7.56) and (7.57). The problem of interest is to solve either Eq. (7.57) or Eq. (7.61). Since we are describing the collisions using quantum mechanics, our concern has been the solution of Eq. (7.61), or Eq. (7.10).

REFERENCES

A. Foderaro, *The Elements of Neutron Interaction Theory* (MIT Press, 1971), Chap. 4.

H. Goldstein, C. P. Poole and J. L. Safko, *Classical Mechanics* (Addison-Wesley, Reading, 2001), Third Edition.

R. L. Liboff, *Introductory Quantum Mechanics* (Holden Day, New York, 1980).

W. E. Meyerhof, *Elements of Nuclear Physics* (McGraw-Hill, New York, 1967).

N. F. Mott and H. S. W. Massey, *The Theory of Atomic Collisions* (Oxford University Press, London, 1949).

P. Roman, *Advanced Quantum Theory* (Addison-Wesley, Reading, 1965), Chap. 3.

L. I. Schiff, *Quantum Mechanics* (McGraw-Hill, New York, 1955).

8

Nuclear Reactions Fundamentals

The study of nuclear reactions is a subject much of which is beyond the scope of this book. Yet there is a part of *Nuclear Radiation Interactions* that requires basic knowledge of the kinematics of nuclear reactions, of simple cross section behavior, and of some notions of reaction cross section calculations. To follow up on the treatment of collision cross sections in the previous chapter, we take up a discussion of the fundamentals of nuclear reactions, two-body interactions for which the Q-value is nonzero (recall Eq. (4.2)). We first consider the kinematics of two-body reactions which give us relations based purely on the conservation laws of energy and momentum. Then we survey the energy behavior of reactions using simple results of perturbation theory in quantum mechanics [Fermi, 1949; Foderaro, 1971] and relying only on kinematical factors in the cross section expressions. Lastly we give a brief discussion of elementary theory of nuclear reactions touching on the optical model and the R-matrix formalism in order to show how the method of partial waves discussed in Chap. 7 can be extended to treat nuclear reactions [Foderaro, 1971; Satchler, 1980].

8.1 KINEMATICS (*Q-EQUATION*)

Consider the two-body reaction depicted below. Incoming particle 1, from the set of *nuclear radiation particles*, (n, γ, CP), with a prescribed energy and momentum reacts with target particle 2, a particular nuclide with specified energy and momentum which are also specified. As a result of the reaction an outgoing particle 3, which can be any member of the set, is emitted with a certain energy and momentum, while the residual target, particle 4, recoils with energy and momentum. To keep matters as simple as possible, in Fig. 8.1 we assume the target nucleus is initially at rest. While this simplifies considerably the analysis, we should keep in mind the assumption is physically reasonable when the energy of the incoming particle is large compared to $k_B T$, where T is the target temperature. It will be seen in Chap. 9 where we apply the analysis to neutron elastic scattering, the assumption of stationary target nucleus is valid for neutrons during their slowing down from energies above the thermal energy of $k_B T$, but not for neutrons which have already reached energies in the thermal region.

The kinematics of a reaction is fully described by the velocity vectors of the particles. With three components to each velocity, the 12 velocity components constitute the 12 degrees of freedom of the problem. Generally we are interested in situations where very few degrees of freedom are left unspecified after the application of conservation laws. In the case of neutron elastic scattering, the number of independent degrees of freedom can be reduced to just one (cf. Chap. 9).

Figure 8.1 depicts the same general situation which we have considered previously in Chap. 4 (recall Fig. 4.2 and Eq. (4.2)), with a change of particle labeling. The equivalence is simply

$$1 + 2 \rightarrow 3 + 4$$
$$i + I \rightarrow f + F$$

The present system of labeling is more convenient for the analysis we are about to undertake. We are interested in the *kinematics* of two-body collision. By this we mean the energy-momentum, or energy-angle, relations that can be established by applying the principles of momentum and energy conservations. These relations are valid for all circumstances since no assumptions have been made about the reaction itself. Notice also these relations say nothing about the *dynamics* of the collision which is governed by the interaction between the particles. Dynamical relations are therefore subject to whatever assumptions go into specifying the

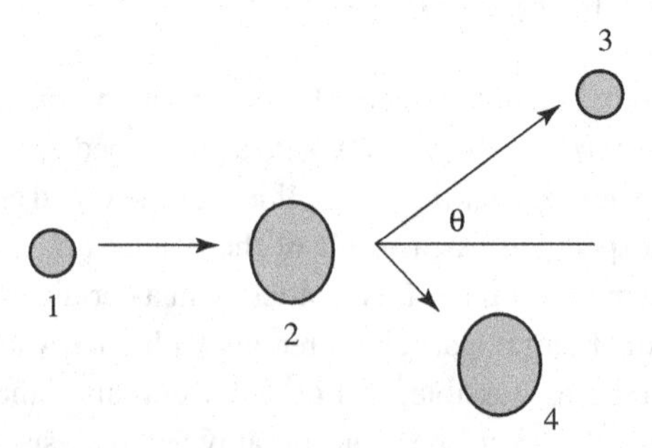

Fig. 8.1 Schematic of a two-body reaction. An incoming particle 1 is incident on the target particle 2, taken to be stationary for simplicity, before the reaction. After the reaction a particle 3 is emitted at an angle θ, while the residual particle 4 undergoes recoil.

interactions. Invariably assumptions are involved in specifying the particular interactions of interest, given that our knowledge of nuclear forces is still limited. These assumptions appear in the form of various models describing the energy of the system of interacting particles. As we have seen in Chap. 5, nuclear models can be quite crude (idealized) concerning the interaction potentials. Not all the knowledge we have about nuclear interactions (forces) are always incorporated in the simple models that we will discuss. We will use these idealized descriptions to calculate the consequences of simple collisions, such as neutron elastic scattering. For the more complicated nuclear reactions we will apply models which do not require explicit information about the interaction potentials, see for example, Chap. 12.

Returning to Fig. 8.1, we invoke the conservations of total energy and momentum by writing

$$(E_1 + M_1c^2) + M_2c^2 = E_3 + M_3c^2 + E_4 + M_4c^2 \tag{8.1}$$

where E_i is the kinetic energy and M_i the rest mass of particle i, with $i = 1, \ldots, 4$. Eq. (8.1) has already incorporated the condition that particle 2 is stationary. The corresponding momentum conservation is

$$\begin{aligned} p_4^2 &= \left(\underline{p}_1 - \underline{p}_3\right)^2 \\ &= p_1^2 + p_3^2 - 2p_1p_3\cos\theta \end{aligned} \tag{8.2}$$

Recalling the definition of Q-value in Eq. (4.2),

$$\begin{aligned} Q &= (M_1 + M_2 - M_3 - M_4)\,c^2 \\ &= E_3 + E_4 - E_1 \end{aligned} \tag{8.3}$$

we combine Eqs. (8.1)–(8.3) to obtain

$$Q = E_3(1 + M_3/M_4) - E_1(1 - M_1/M_4) - (2/M_4)\sqrt{M_1M_3E_1E_3}\cos\theta \tag{8.4}$$

This relation is known as the *Q-equation*. It is really not an equation to be used to calculate Q, rather it is an equation which relates E_3 and θ, with Q being a constant whose value is known once the reaction, $1 + 2 \rightarrow 3 + 4$, is specified. In Eq. (8.4), the energies E_i and θ are in the laboratory coordinate system (LCS) whereas Q is independent of coordinate system by virtue of its definition (see the first line of Eq. (8.3)). The *Q-equation* is typically used when E_1, the masses, and the Q-value are known. Then the purpose is to solve this equation for E_3 in terms of θ, or vice versa. Returning to the issue of reducing the number of degrees of freedom in the *Q-equation* by applying the conservation of energy and momentum, clearly not every single degree of freedom is an unknown in the situations of interest to us. Suppose we enumerate all the degrees of freedom to see which is given (known) and which is a variable. First, if the direction of travel and energy of the incoming particle are given, which is usually the case, this specifies 3 degrees of freedom. Secondly, unless one is specifically interested in thermal motion effects it is often customary to take the target nucleus to be stationary because the energy of the incoming particle is much larger than the thermal energies at the target temperature. This we have already done in arriving at Eq. (8.4). So another 3 degrees of freedom are specified. Since conservations of energy and momentum must hold in any collision (three conditions since momentum and energy are related), this eliminates another three degrees of freedom, now leaving only three unspecified degrees of freedom in the problem. If we further assume the emission of the outgoing particle (particle 3) is azimuthally symmetric (that is, emission into a cone subtended by the angle θ is equally probable), then only two degree of freedom are left. This way of counting shows that the outcome of the collision is completely determined if we just specify another degree of freedom. What variable should we take? Because we are often interested in knowing the energy or direction of travel of the outgoing particle, we can choose this last variable to be either E_3 or the scattering angle θ. In other words, if we know either E_3 or θ, then everything else (energy and direction) about the collision is determined. Keeping this in mind, it should come as no surprise that what we will do with Eq. (8.4) is to turn it into a relation between E_3 and θ.

Thus far we have used non-relativistic expressions for the kinematics. To turn Eq. (8.4) into the *relativistic Q-equation* we can simply replace the rest mass M_i by an effective mass, $M_i^{eff} = M_i + T_i/2c^2$, and use the expression $p^2 = 2MT + T^2/c^2$ instead of $p^2 = 2ME$. For photons, we take $M^{eff} = h\nu/2c^2$.

Seeing that Eq. (8.4) contains a term with $(E_3)^{1/2}$ we can regard it as a *quadratic* equation in $(E_3)^{1/2}$, which has two general solutions. Since energy E_3 is always positive or zero, only the solutions for $(E_3)^{1/2}$ that are real and positive are physically acceptable. A situation could arise when both solutions are acceptable which would mean there is no one-to-one correspondence between E_3 and θ (see, Evans, 1955, pp. 413–415 and Meyerhof, 1967, p. 178). This situation is more than just a matter of curiosity. It will be instructive to see what does a double-valued solution mean physically. We will return in Chap. 9 to give an illustration of such a solution.

Solutions to the Q-equation for reactions of definite Q

To see further the properties of the Q-equation, we examine the two general solutions of the quadratic equation in $(E_3)^{1/2}$

$$[\sqrt{E_3}]_\pm = s \pm \sqrt{s^2 + t} \tag{8.5}$$

with

$$s = \frac{\sqrt{M_1 M_3 E_1}}{M_3 + M_4} \cos\theta \tag{8.6}$$

$$t = \frac{M_4 Q + (M_4 - M_1) E_1}{M_3 + M_4} \tag{8.7}$$

Now we examine the consequences of the condition that $(E_3)^{1/2}$ must be *real and positive* with the masses fixed and Q is therefore known. Equation (8.5) can be regarded as a relation between the energy of the outgoing particle and the angle of emission, if we also take the energy of the incoming particle to be given, which is usually the case of interest. We distinguish between exothermic and endothermic reactions. In the latter situations, Eq. (8.5) becomes a condition on the critical value of E_1 for which the reaction is energetically barely possible.

Exothermic reaction ($Q > 0$), *or* $E_3 + E_4 > E_1$

An example of such a reaction would be $B^{10}(\alpha, p)C^{13}$, $Q = 4$ MeV. At very low incident energy E_1, we can set $s \sim 0$. Then, $E_3 = [M_4/(M_3 + M_4)]Q$ and correspondingly, $E_4 = [M_3/(M_3 + M_4)]Q$. One can show that in this case the outgoing particle and product nucleus go off in opposite directions, confirming our expectations based on momentum conservation. As E_1 increases, since $M_1 < M_4$ (if we choose C^{13} to be the product nucleus), then $t > 0$ and only the upper sign in Eq. (8.5) is acceptable. But the assignment of outgoing particle is arbitrary. Suppose we choose instead C^{13} as the outgoing particle 3, then Eq. (8.7) gives $t = (Q - 3E_1)/14$. For $E_1 > Q/3$, t will be negative, which means we now have two physical solutions for $(E_3)^{1/2}$. Another noteworthy consequence is that for $\cos\theta < 0$, $s < 0$, and with $E_1 > Q/3$, $(E_3)^{1/2}$ is always negative. This means that no C^{13} nucleus will be emitted at $\theta \geq \pi/2$.

Endothermic reaction ($Q < 0$, *threshold*)

The inverse of all exothermic reactions must be endothermic, for which the Q-value changes sign. From the foregoing example, we have $C^{13}(p, \alpha)B^{10}$, $Q = -4$ MeV. At low incident energy, $s \sim 0$, $t < 0$ so $(E_3)^{1/2}$ is imaginary. This indicates there is not enough kinetic energy supplied by particle 1 to make up the required mass increase. We define the threshold energy as

$$(E_1)_{thres} \equiv \text{lowest incoming energy at which reaction can occur}$$

As E_1 is increased, t will become less and less negative, while the most positive that s can be for a given E_1 is when the scattering is in the *forward* direction ($\theta = 0$), effectively no scattering. For increasing E_1 the reaction first becomes allowable when

$$s^2 + t = 0 \tag{8.8}$$

This condition then reflects back to give a critical value for E_1, with θ arbitrary. Let us denote this special value as $(E_1)_\theta$. From Eq. (8.8) we obtain

$$(E_1)_\theta = -Q\left[\frac{M_3 + M_4}{M_3 + M_4 - M_1 - (M_1 M_3/M_4)\sin^2\theta}\right] \tag{8.9}$$

Equation (8.9) shows the smallest value of E_1 occurs at $\theta = 0$, for which the denominator is largest. The threshold value is therefore

$$(E_1)_{thres} = -Q\frac{M_3 + M_4}{M_3 + M_4 - M_1} \tag{8.10}$$

Since $M_1 + M_2 = M_3 + M_4 + Q/c^2$, to a good approximation Eq. (8.10) simplifies to

$$(E_1)_{thres} \approx -Q\frac{M_1 + M_2}{M_2} \tag{8.11}$$

Equation (8.11) is an intuitively reasonable result inthat the threshold energy is seen to be always greater than the excitation energy or Q-value in a reaction. This is because a fraction of the incoming kinetic energy goes into moving the center-of-mass, which is therefore not available for driving the reaction. One needs an amount of kinetic energy that is larger than the Q value by approximately the ratio $(M_1 + M_2)/M_2$. In other words, the amount in excess of Q that is needed is that given to the center-of-mass, or $(M_1/M_2)Q \sim (M_1/M_2)E_1$.

Double-valued solutions to the Q-equation

If E_1 and θ are such that the condition

$$0 \leq s^2 + t \leq s^2 \tag{8.12}$$

is satisfied, which is equivalent to

$$-s^2 \leq t \leq 0 \tag{8.13}$$

is satisfied, this is sufficient for the *Q-equation* to have double-valued solutions. Notice the lower limit in Eq. (8.12) ensures E_3 is real, while the upper limit ensures that it is positive. These two limits imply corresponding limits on the incoming kinetic energy E_1. In Eq. (8.12) the upper limit gives

$$E_1 \leq -Q\frac{M_4}{M_4 - M_1} \tag{8.14}$$

while one obtains from the lower limit, after a bit of manipulation,

$$E_1 \geq -Q\left(\frac{M_3 + M_4}{M_3 + M_4 - M_1 - (M_1M_3/M_4)\sin^2\theta}\right) = (E_1)_\theta \tag{8.15}$$

At $\theta = \pi/2$,

$$(E_1)_{\pi/2} = -Q\frac{M_4(M_3 + M_4)}{M_3M_4 + M_4^2 - M_1M_3 - M_1M_4}$$
$$= -Q\frac{M_4}{M_4 - M_1} \tag{8.16}$$

Thus the condition for double-valued solutions can be put into the form,

$$(E_1)_\theta \leq E_1 \leq (E_1)_{\pi/2} \tag{8.17}$$

One consequence of Eq. (8.17) is that in the domain of double-valued solutions no outgoing particle can be emitted at an angle θ greater than $\pi/2$. Having an angle of emission greater than $\pi/2$ makes $t > 0$, which in turn puts E_3 into the single-valued domain. Physically this means the kinetic energy of the reaction products is large enough that M_3 can be projected into the backward direction in LCS. For this to occur, it is also necessary that $M_3 < M_2$. The "heavier fragment" (product nucleus) in an endothermic reaction can never be projected in the backward direction. Figure 8.2 summarizes our discussion of the different domains of physical solutions to the *Q-equation* in the case of endothermic reaction.

One way to appreciate the meaning of the double-valued solutions is illustrated in Fig. 8.3. For an endothermic reaction, these solutions correspond to a particle being emitted at a certain angle with two different energies in LCS. The construction

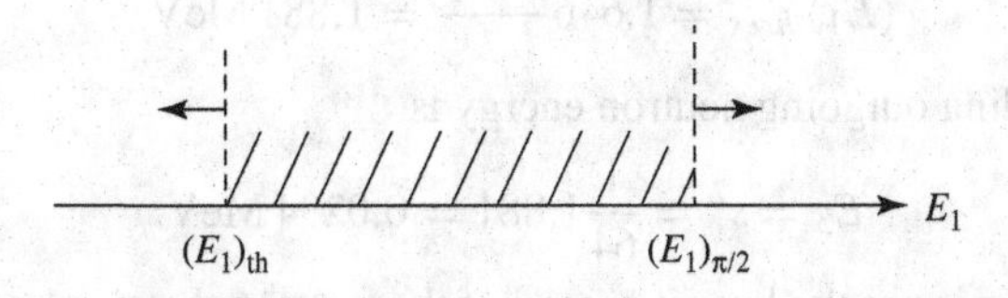

Fig. 8.2 Range of incident energy E_1 where double-valued solutions to the *Q-equation* are allowed (shaded region) in endothermic reactions. In the range below $(E_1)_{th}$, given by Eq. (8.11), (indicated by the left arrow) there are no solutions, and in the range above $(E_1)_{\pi/2}$, given by Eq. (8.16), (indicated by the right arrow) there are only singled-valued solutions.

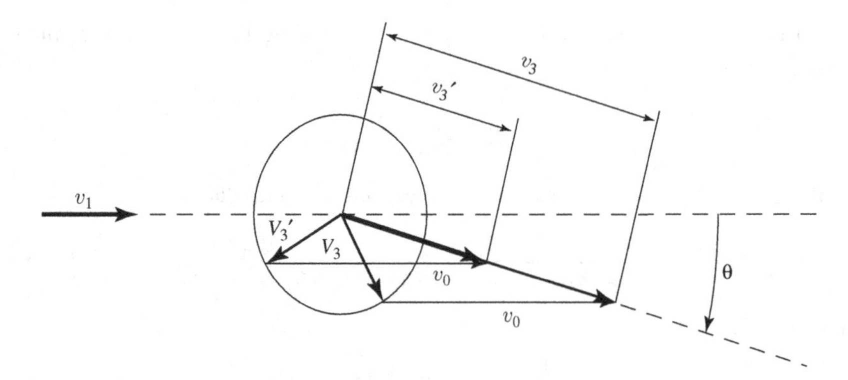

Fig. 8.3 Velocity diagram construction for an endothermic reaction showing how a doubled-valued solution can be physically realized. CMCS velocities of two outgoing particles (particle 3 in Fig. 8.1), depicted as V_3 and V_3' respectively, are combined with the *CM* speed v_o to give two LCS velocities, v_3 and v_3', at the same scattering LCS angle θ but with *different* magnitudes. [Adapted from Meyerhof, 1967, p. 178.]

depicted in Fig. 8.3 shows how this can be achieved. We see the construction requires the outgoing particle (particle 3, see Fig. 8.1) to be emitted at different angles but the same speeds in CMCS. When these are converted to LCS, one obtains a particle emitted with two different speeds but at the same angle in LCS. Thus the one-to-one correspondence between outgoing speed and angle does not hold in either CMCS or LCS. The reader should note this is not the case with elastic scattering ($Q = 0$). It will be seen in Chap. 9 the one-to-one correspondence between outgoing energy and angle, which holds explicitly, is a useful consequence of the kinematics.

We close this section with an example that illustrates further the relation among the three principal quantities in the present discussion, the incoming and outgoing particle energies, and the angle of emission. The example is the reaction $Li^7(p,n)Be^7$, $Q = -1.646\,\text{MeV}$ [Evans, 1955, p. 416]. The threshold energy, in this case the incoming proton energy, is

$$(E_1)_{thres} = 1.646\frac{1+7}{7} = 1.881\,\text{MeV}.$$

The corresponding outgoing neutron energy is

$$E_3 = s^2 = \frac{1}{64}1.881 = 0.0294\,\text{MeV}$$

Interestingly this is not the lowest energy of the neutron that can be emitted, since the proton energy at which the neutron is barely emitted, $E_3 = 0(s = 0, t = 0)$, is

$$(E_1)_{\pi/2} = 1.646\left(\frac{8}{7 - 1/8}\right) = 1.915\,\text{MeV}$$

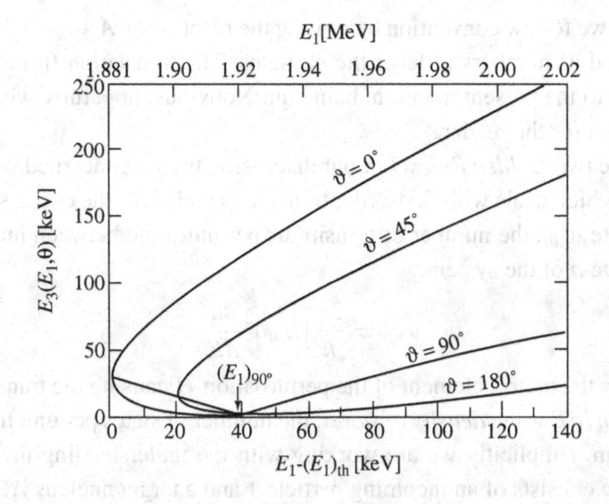

Fig. 8.4 Energetics of an endothermic reaction, $Li^7(p,n)Be^7$, $Q = -1.64$ MeV. All energies and angles are in LCS. Reaction first takes place at the threshold proton energy of 1.881 MeV, with neutron emitted at 0.0294 MeV. As proton energy is increased, neutron is emitted within a small forward cone, the extent of which increases with increasing proton energy. At proton energy of 1.920 MeV, all directions of neutron emission are allowed. One sees that doubled-valued solutions are allowed for emission at an angle less than 90°, and in the range between 90° and 180° only single-valued solution is possible. [Adapted from Evans, 1955, p. 416.]

In the range of proton energies between $(E_1)_{thres}$ and $(E_1)_{\pi/2}$, only certain emission angles are allowed. These lie in a cone which increases with increasing proton energy, as shown in Fig. 8.4 [Evans, 1955, p. 416].

8.2 CROSS SECTION SYSTEMATICS

In this section, we survey the energy variations of nuclear reaction cross sections, making use of simple results from perturbation theory in quantum mechanics. We have already discussed in Chap. 7 the concept of collision cross sections and their calculations using the method of phase-shift analysis. Here we apply the so-called Fermi's Golden Rules in a qualitative manner based on time-dependent perturbation theory [Fermi, 1949]. Our intent is to provide some appreciation of the variety of cross sections that have found useful applications in nuclear science and technology. In quantum mechanics the calculation of cross section $\sigma(a, b)$, where a and b designate the incoming and outgoing particle, generally entails applying time-dependent perturbation theory to the reaction $A(a, b)B$, where A and B designate the target and product nuclei, obtaining expressions known as Fermi's Golden

Rules. Here we follow convention by writing the reaction as $A+a \rightarrow B+b$, using letters instead of numbers to label the particles. The translation from Fig. 4.2 to Fig. 8.1 and to the present notation, being quite obvious, hopefully will not cause any confusion for the reader.

There are two *Golden Rules*. For our discussion we are concerned with *Golden Rule no. 2*, which deals with first-order transitions and gives the expression for the transition rate w_{ab}, the number of transitions per unit time between initial state a and final state b of the system,

$$w_{ab} = \frac{2\pi}{\hbar} |H_{ab}|^2 \frac{dn}{dE} \tag{8.18}$$

where H_{ab} is the matrix element of the perturbation H causing the transition from a to b, and dn/dE is the *density of states*, the number of states per unit final energy of the system. (Implicitly we are working with the understanding the system in initial state a consists of an incoming particle 1 and a target nucleus A, or particle 2. Similarly, the system in the final state b consists of outgoing particle 3 and a product nucleus B, or particle 4 in the scheme of Fig. 8.1.) For a particle in vacuum (free particle) with momentum $\underline{p}$, the density of states can be simply calculated. Let the vacuum have a finite volume Ω (we take the system volume to be finite for purpose of simple normalization of the wave function), the number of states with momentum in dp about $\underline{p}$ is

$$dn = 4\pi p^2 dp\Omega/(2\pi\hbar)^3 \tag{8.19}$$

If the outgoing particle b and the product nucleus B have spins I_b and I_B, respectively, then we need to include in Eq. (8.19) a statistical factor, $(2I_b + 1)(2I_B + 1)$ for the number of spin orientations one can have in the final state. For an outgoing particle with momentum $\underline{p}_b$,

$$dE = v_b dp_b \tag{8.20}$$

Strictly speaking, v_b and p_b should be the velocity and momentum of the final state $(3 + B)$ in the center of mass coordinate system, rather than of an outgoing particle 3. The difference is small when the product nucleus is heavy compared to the mass of the particle. We can rewrite Eq. (8.17) by combining Eqs. (8.19), (8.20) and the statistical factor,

$$w_{ab} = \frac{1}{\pi\hbar^4} \frac{p_b^2}{v_b} \Omega |H_{ab}|^2 (2I_b + 1)(2I_B + 1) \tag{8.21}$$

To go from the rate of transition to the cross section, we simply write

$$w_{ab} = n_a v_{rel} \sigma(a, b) \tag{8.22}$$

which defines the cross section. One can compare this argument with the definition introduced in Chap. 7. In Eq. (8.22), n_a is the density of the incoming particle and v_{rel} is the relative velocity in the initial state $(1 + A)$. Again, if the target nucleus is massive compared to the incoming particle, then to a good approximation v_{rel} is given by v_a, the velocity of the incoming particle in CMCS. Combining these results we obtain

$$\sigma(a, b) = \frac{1}{\pi \hbar^4} |\Omega H_{ab}|^2 \frac{p_b^2}{v_a v_b} (2I_b + 1)(2I_B + 1) \tag{8.23}$$

This formula is useful for deducing the qualitative variation of the cross section with energy for a number of typical reactions, without knowing much about the really complicated part of the calculation, namely. the matrix element of the perturbation H between initial and final states.

We now assume that there are two parts to the transition matrix element, one involves nuclear interactions among the nucleons and the other pertains to electrostatic (Coulomb) interactions between the incoming or outgoing particle, if it were a charged particle, and the nucleus (either the target or the product as the case may be). We further assume that the nuclear interactions, however complicated, may be taken to be a constant for the purpose of estimating the energy variation of $\sigma(a, b)$. What is left in H_{ab} is the Coulomb interaction. The effect of this can be estimated by using the model of a charged particle tunneling through a Coulomb barrier. Recall from Chap. 6 such a model was introduced to calculate the decay constant for α-decay. For a positively charged particle, charge ze with velocity v, to penetrate a nucleus, charge Ze, the transmission factor is $\exp(-G/2)$, where [Fermi, 1949, p. 143]

$$\frac{G}{2} \approx \frac{\pi Z z e^2}{\hbar v} \tag{8.24}$$

and G is the *Gamow factor* encountered in Chap. 6. Accordingly, if the incoming and outgoing particles are both charged, we will write

$$|H_{ab}|^2 \approx \overline{U}^2 \exp(-G_a - G_b) \tag{8.25}$$

with $\overline{U}$ denoting the nuclear interaction part, which we will take to be energy independent.

Now we are ready to examine the variation with energy of incoming particle of various neutron reactions at low energies, say eV range. Equation (8.23) shows that there are two parts which can vary with energy, the kinematical factor, $p_b^2/v_a v_b$, and the transmission factors, if present. We can distinguish four types of neutron interactions of interest.

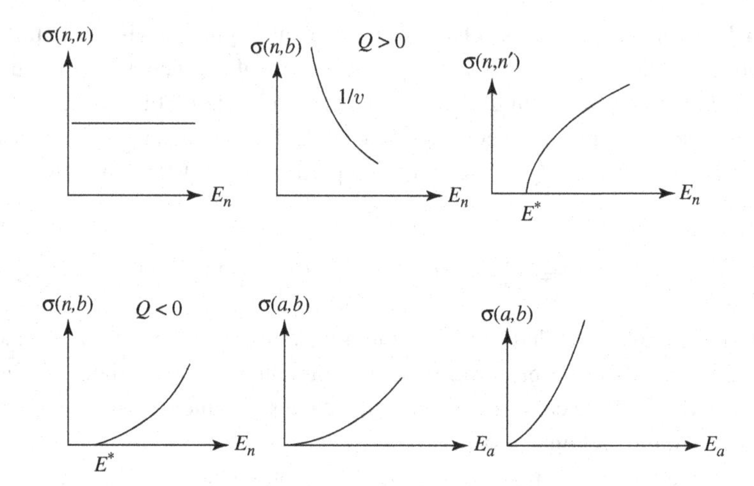

Fig. 8.5 Schematic energy variations of cross sections, the first four (left to right) correspond to neutron elastic scattering, neutron-induced reaction (exothermic), neutron inelastic scattering, and neutron-induced reaction (endothermic). The last two reactions are respectively charged particle in and uncharged particle out, and charged particles in and out, both reactions being exothermic. [Adapted from Fermi, 1949, p. 144.]

1. Elastic scattering (n, n)
   With incoming and outgoing particles being uncharged, the transmission factor is unity. Also, with $v_a = v_b$, the kinematical factor $p_b^2/v_a v_b$ is a constant. Equation (8.23) then predicts the cross section to be a constant, as shown in Fig. 8.5(a) below. Notice elastic scattering is, strictly speaking, not a proper reaction.
2. Charged-Particle Emission (exothermic)
   Since Q is typically a few MeV, while the incoming neutron energy is of order eV, $p_b^2/v_a v_b \approx 1/v_a$. $|H_{ab}|^2 \approx \overline{U}^2 \exp(-G_b) \approx$ constant. So $\sigma \approx 1/v_n$; this is the "1/v" behavior often seen in reactions like (n, α), (n, γ), etc.
3. Inelastic scattering (n, n')
   Since the product nucleus is left in an excited state, this is an endothermic reaction that requires a threshold value for the neutron energy, $E^* = -Q \sim 1$ MeV. Velocity of the outgoing neutron is $v_{n'}$, with $v_{n'}^2$ = (excess energy above threshold)/$(m_n/2) \sim (E_n - E^*)$. Thus,

$$\sigma(n, n') \propto v_{n'} \sim (E_n - E^*)^{1/2}$$

An example here is $O^{16}(n, \alpha)C^{13}$, with $Q = -2.215$ MeV.

4. Charged-Particle Emission (endothermic)
   This reaction is like the endothermic reaction of inelastic neutron scattering, except there is now a Coulomb factor,

$$\sigma(n,b) \sim e^{-G_b}(E_n - E^*)^{1/2}$$

   with $G_b \sim 1/v_b$.

In addition, we can consider exothermic reactions with incoming charged particles. If the outgoing particle is uncharged, then with $E_a \ll Q$, $p_b^2/v_a v_b \sim 1/v_a$, $p_b \sim$ constant, we have

$$\sigma(a,b) \sim \frac{1}{v_a} e^{-G_a}$$

An example would be the inverse reaction to that already mentioned, $C^{13}(\alpha, n)O^{16}$, $Q = 2.215$ MeV.

If the outgoing particle is also charged, as in (α, p), $p_b^2/v_a v_b \sim 1/v_a$, and

$$\sigma(a,b) \sim \frac{1}{v_a} e^{-(G_a+G_b)}$$

Both cases are also shown in Fig. 8.5.

8.3 ELEMENTARY NUCLEAR REACTION THEORY

Cross Sections for Nuclear Reactions

The concept of a reaction cross section can be introduced quite simply as an extension of the scattering cross section which we have studied in Chap. 7. The extension lies in allowing the interaction to be a complex energy with real and imaginary parts. For the square well potential this would mean taking $V(r)$ to be

$$V(r) = \begin{matrix} -V_o - iW_o \\ 0 \\ 1 \end{matrix} \qquad \begin{matrix} r < r_o \\ r > r_o \end{matrix} \tag{8.26}$$

The presence of an imaginary part in V(r) makes the phase shift δ_ℓ complex. Correspondingly, the scattering amplitude, Eq. (7.40), takes on an imaginary component,

$$\eta_\ell \rightarrow e^{2i(\rho_\ell + i\gamma_\ell)} \tag{8.27}$$

In Eq. (8.27), the real and imaginary parts of the phase shift are denoted as ρ_ℓ and γ_ℓ respectively. We will see that in the discussion of nuclear reaction theory, η_ℓ plays the role of the phase shift in the theory of elastic scattering in Chap. 7.

To prepare for this transition, we rewrite the angular differential elastic scattering cross section and the scattering cross section as

$$\sigma_{el}(\theta) = (\lambdabar^2/4)\left|\sum_{\ell=0}^{\infty}(2\ell+1)(1-\eta_\ell)P_\ell(\cos\theta)\right|^2 \tag{8.28}$$

$$\sigma_{el} = \pi\lambdabar^2\sum_{\ell=0}^{\infty}(2\ell+1)\left|1-\eta_\ell\right|^2 \tag{8.29}$$

We have added the subscript *el* to indicate the elastic scattering part of the cross section. (In the terminology of nuclear reactions the process of elastic scattering is not considered a *true* reaction, it is sometimes called an improper reaction.)

To find the reaction part of the cross section we adopt the definition,

$$\sigma_r \equiv (1/I_o)\left[J_{in}^R - J_{out}^R\right] \tag{8.30}$$

where $I_o = \hbar k/\mu$ is the incident unit flux, and J^R is the radial component of the current vector $\underline{J}$, with subscripts denoting incident and outgoing (previously we had used the term 'scattered') components. By the radial current we mean

$$J^R = r^2\int d\Omega\underline{\hat{r}}\cdot\underline{J} \tag{8.31}$$

the integral being taken over the surface of a large sphere of radius r. In Eq. (8.31), the current $\underline{J}$ can be calculated from Eq. (7.9) once the wave function is known. From Eqs. (7.18) and (7.22) we can write the wave function in the asymptotic region as

$$\begin{aligned}\psi(r,\theta) &= \sum_\ell \frac{i^\ell(2\ell+1)}{2ikr}P_\ell(\cos\theta)\left[\eta_\ell e^{i(kr-\ell\pi/2)} - e^{-i(kr-\ell\pi/2)}\right]\\ &= \psi_{out}+\psi_{in}\end{aligned} \tag{8.32}$$

Having identified the outgoing and incident wave functions we can find the corresponding currents from Eq. (7.9). The resulting expressions for the *reaction cross section* is

$$\sigma_r(k) = \pi\lambdabar^2\sum_\ell(2\ell+1)\left[1-\left|\eta_\ell\right|^2\right] \tag{8.33}$$

The sum of σ_{el} and σ_r now makes up the total cross section

$$\begin{aligned}\sigma_t(k) &= 2\pi\lambdabar^2\sum_\ell(2\ell+1)\left[1-Re\left\{\eta_\ell(k)\right\}\right]\\ &= \sigma_{el}(k)+\sigma_r(k)\end{aligned} \tag{8.34}$$

We have thus extended the previous formulation of elastic scattering to include a reaction component. In so doing we have obtained results for the angular differential elastic scattering cross section, given in Eq. (8.28), the *elastic* scattering cross section, Eq. (8.29), and the *reaction* cross section, Eq. (8.33), all involving the scattering amplitude η_ℓ for partial waves. In addition the total cross section depends only on the real part of η_ℓ for each partial wave.

Energy averaging

The purpose of our discussion of elementary theory of nuclear reactions is to introduce the idea that the gross features of the nuclear cross section variation with energy can be understood rather simply using the formulation of scattering theory. The energy variations of nuclear cross sections can be quite complicated in the presence of resonance reactions. But if one is willing to ignore the details of individual resonances and focus only on the broad energy behavior in an average sense, it is then possible to provide an interpretation of the cross section variations across the periodic table of elements. In other words, we will test our understanding of nuclear reactions essentially based on the formalism developed in Sec. 7.2 by considering the variation with mass number A of nuclear cross sections that have been averaged over an appropriate energy range, see for example, the discussion pertaining to Fig. 7.6. By appropriate we mean an energy range sufficiently wide that the detailed behavior of individual resonances is effectively smeared out. In so doing we are deferring the considerations of cross section behavior of well-isolated resonances to a later discussion, see Chap. 12.

Restricting our attention to only energy averaged cross sections brings a significant simplification to the theoretical analysis. Given that our challenge lies in determining the scattering amplitude η_ℓ, we are now motivated to consider an energy averaged quantity,

$$\langle \eta \rangle \equiv \frac{1}{\Delta E} \int_{E-\Delta E/2}^{E+\Delta E/2} dE' \eta(E') \tag{8.35}$$

The simplification is that while the energy variation of η_ℓ can be very complex, it may be that the energy averaged behavior is more amenable to being described by simple physical models. We will see that using physically reasonable models for the averaged scattering amplitude allows us to understand a number of gross features of the averaged nuclear cross section variation with mass number A. Applying the energy averaging procedure to the elastic scattering and reaction cross sections we obtain

$$\langle\sigma_{el}\rangle = \pi\lambda\!\!\!^{-2} \sum_{\ell} (2\ell + 1) \left[\left|1 - <\eta_\ell\right|^2 + \langle|\eta_\ell|^2\rangle - \left|\langle\eta_\ell\rangle\right|^2 \right]$$

$$= \sigma_{se} + \sigma_{ce} \tag{8.36}$$

$$\langle\sigma_r\rangle = \pi\lambda\!\!\!^{-2} \sum_{\ell} (2\ell + 1) \left[1 - \langle\left|\eta_\ell\right|^2\rangle \right] \tag{8.37}$$

In Eq. (8.36), we have identified two separate contributions to elastic scattering, the *shape elastic scattering* cross section σ_{se} and the *compound elastic scattering* cross section σ_{ce}. The latter is the combination of the second and third terms in the bracket in Eq. (8.36). It should be noted if one introduces a model of nuclear reaction that gives an approximation only for $\langle\eta\rangle$ but not $\langle|\eta|^2\rangle$, then the model can not give σ_{ce} and $\langle\sigma_r\rangle$ individually. This points out the special nature of the compound elastic scattering and reaction cross sections, namely, they involve fluctuations (correlations or interference) in the form of $\langle|\eta|^2\rangle$. On the other hand, the sum of these two quantities can be calculated. We therefore define

$$\sigma_c \equiv \sigma_{ce} + \langle\sigma_r\rangle = \pi\lambda\!\!\!^{-2} \sum_{\ell} (2\ell + 1) \left[1 - \left|\langle\eta_\ell\rangle\right|^2 \right] \tag{8.38}$$

which we can call the *compound formation* cross section. Summarizing our results we see once $\langle\eta_\ell\rangle$ is known, we can obtain the shape elastic scattering cross section as

$$\sigma_{se} = \pi\lambda\!\!\!^{-2} \sum_{\ell} (2\ell + 1) \left|1 - \langle\eta_\ell\rangle\right|^2 \tag{8.39}$$

Notice the sum of σ_{se} and $\sigma_c$is also the total cross section, $\sigma_t = \sigma_{se} + \sigma_c$, which can be separately calculated by knowing $\langle\eta_\ell\rangle$ from Eq. (8.34). This means an approximation for $\langle\eta_\ell\rangle$ is sufficient to give σ_{se} and σ_t, and $\sigma_c = \sigma_{ce}$ and $\langle\sigma_r\rangle$ is then simply determined by subtraction.

A similar set of relations can be established at the level of angular differential cross section for elastic scattering. We rewrite Eq. (8.28) as

$$\langle\sigma_{el}(\theta)\rangle = \langle\sigma_{se}(\theta)\rangle + \langle\sigma_{ce}(\theta)\rangle \tag{8.40}$$

with

$$\langle\sigma_{se}(\theta)\rangle = (\pi\lambda\!\!\!^{-2}/4) \left| \sum_{\ell} (2\ell + 1)(1 - \langle\eta_\ell\rangle) P_\ell(\cos\theta) \right|^2 \tag{8.41}$$

and

$$\langle\sigma_{ce}(\theta)\rangle = (\pi\lambda\!\!\!^{-2}/4) \sum_{\ell,\ell'} (2\ell + 1)(2\ell' + 1) P_\ell(\cos\theta) P_{\ell'}(\cos\theta)$$

$$\times \left[\langle\eta_\ell^* \eta_{\ell'}\rangle - \langle\eta_\ell^*\rangle\langle\eta_{\ell'}\rangle \right] \tag{8.42}$$

In Eq. (8.42), we see we need to know the energy average of a product of two scattering amplitudes. Thus knowing the energy average, Eq. (8.35), is not enough to calculate the compound elastic differential cross section. This will be the problem later when we consider the optical model, which is an approximate description that only gives $\langle \eta_\ell \rangle$.

If one has no further information about the phase shift one could simplify the calculation somewhat by assuming there is complete absence of interference between the states of different orbital angular momentum. This assumption can be expressed by inserting

$$\langle \eta_\ell^* \eta_{\ell'} \rangle - \langle \eta_\ell^* \rangle \langle \eta_{\ell'} \rangle \approx \left[\langle |\eta_\ell|^2 \rangle - |\langle \eta_\ell \rangle|^2 \right] \delta_{\ell\ell'} \tag{8.43}$$

into Eq. (8.42) which then becomes

$$\langle \sigma_{ce}(\theta) \rangle \approx \frac{1}{4} \sum_\ell (2\ell+1) \sigma_{ce}^{(\ell)}(k) P_\ell^2(\cos\theta) \tag{8.44}$$

with

$$\sigma_{ce}^{(\ell)}(k) = \text{ƛ}^{(2)}(2\ell+1) \left[\langle |\eta_\ell|^2 \rangle - |\langle \eta_\ell \rangle|^2 \right] \tag{8.45}$$

Equation (8.45) again reveals that compound elastic scattering is to be associated with fluctuations in the scattering amplitude, when average of the square is greater than the square of the average. While the assumption of no interference between states of different ℓ reduces the calculation of angular differential scattering cross section to the same level as the total cross section calculation, still one is unable to calculate $\langle \sigma_{ce}(\theta) \rangle$, Eq. (8.44), without further approximation. In other words, we still need to know the first term in Eq. (8.45) which is not given by Eq. (8.35). A simple way to proceed is to replace this term by unity, which amounts to assuming maximum fluctuations, or taking the upper bound on $\sigma_{ce}^{(\ell)}(k)$. Then the approximation is to write Eq. (8.38) as

$$\sigma_c = \pi \sum_\ell \sigma_c^{(\ell)} \tag{8.46}$$

where

$$\sigma_c^{(\ell)} = \text{ƛ}^2(2\ell+1) \left[1 - |\langle \eta_\ell \rangle|^2 \right] \tag{8.47}$$

Complex square-well potential

Before proceeding to the optical model description of nuclear reactions, we consider one more derivational aspect in the preparation for the numerical results that we will soon examine. This concerns the approximation of using a square-well potential to describe the interaction between an incoming particle and the nucleus. A resulting simplification is the availability of recurrence relations that will facilitate the actual computation of the quantities appearing in the scattering amplitude η_ℓ. Notice in Eq. (7.40) three quantities are involved in the determination of η_ℓ. They are the logarithmic derivative of the interior wave function q_ℓ^{int}, the phase angle τ_ℓ, and a part of the logarithmic derivative of the exterior wave function $q_\ell^{(\pm)}$. Both τ_ℓ and $q_\ell^{(\pm)}$ involve the spherical Bessel function $j_\ell(z)$ for which the following recurrence relations exist,

$$\begin{aligned}\frac{dj_\ell(z)}{dz} &= -\frac{\ell+1}{z}j_\ell(z) + j_{\ell-1}(z) \\ &= \frac{\ell}{z}j_\ell(z) - j_{\ell+1}(z)\end{aligned} \tag{8.48}$$

In the complex square-well potential model the potential is a constant in the interior region, $r < r_o$. So the interior radial wave equation has the same form as Eq. (7.14), with k^2 replaced by $K^2 = k^2 + (2\mu V_o/\hbar^2)(1 + i\varsigma)$, with $\varsigma = W_o/V_o$. This means the interior wave function for a complex square well potential is of the form given by Eq. (7.15). Moreover, we can discard the second term in Eq. (7.15) because the Neumann function n_ℓ is singular at the origin. Thus the interior wave becomes

$$u_\ell^{int}(Kr) = A_\ell r j_\ell(Kr) \quad r < r_o \tag{8.49}$$

With the interior wave solution determined, we can go back to Eq. (7.32) to find the corresponding interior logarithmic derivative,

$$q_\ell^{int} = 1 + Kr_o\frac{j_\ell'}{j_\ell} \tag{8.50}$$

We see that along with τ_ℓ and $q_\ell^{(\pm)}$, which pertain to the exterior wave solution, now the interior logarithmic derivative also depends on the spherical Bessel function. This means we can apply recurrence relations to all the quantities that appear in Eq. (7.40), which turns out to facilitate considerably the numerical computations. Applying Eqs. (8.48) to (8.50) we have (suppressing the superscript *int* for the moment)

$$q_\ell = -\ell + \frac{zj_{\ell-1}}{j_\ell} \tag{8.51}$$

$$q_{\ell-1} = 1 + (\ell+1) - \frac{zj_\ell}{j_{\ell-1}} \tag{8.52}$$

Solving for $j_{\ell-1}/j_\ell$ in Eq. (8.52) and inserting the result into Eq. (8.51) we obtain a useful relation between q_ℓ and $q_{\ell-1}$,

$$q_\ell = \frac{z^2}{\ell - q_{\ell-1}} - \ell \tag{8.53}$$

This relation is also applicable to $q_\ell^{(\pm)}$ through Eqs. (7.34) and (7.36), and including the separation into real and imaginary parts, Eq. (7.42). With some algebra one finds

$$\Delta_\ell = \frac{\xi^2(\ell - \Delta_{\ell-1})}{(\ell - \Delta_{\ell-1})^2 + s_{\ell-1}^2} - \ell \tag{8.54}$$

$$s_\ell = \frac{\xi^2 s_{\ell-1}}{(\ell - \Delta_{\ell-1})^2 + s_{\ell-1}^2} \tag{8.55}$$

Thus we can rewrite Eq. (7.40) as

$$\eta_\ell = e^{-2i\tau_\ell} \frac{q_\ell^{int} - \Delta_\ell + is_\ell}{q_\ell^{int} - \Delta_\ell - is_\ell} \tag{8.56}$$

while keeping in mind q_ℓ^{int} is complex through its argument $z = Kr_o$, with K being complex. If one wishes to display the real and imaginary parts of q_ℓ^{int} explicitly, then Eq. (8.56) becomes

$$\eta_\ell = e^{-2i\tau_\ell} \left[1 - \frac{2s_\ell}{M_\ell + iN_\ell} \right] \tag{8.57}$$

with

$$M_\ell = s_\ell - \text{Im} q_\ell^{int} \tag{8.58}$$

$$N_\ell = \text{Re} q_\ell^{int} - \Delta_\ell \tag{8.59}$$

Collecting all our results we return to computing the cross sections. For the compound formation cross section, the expression to use is

$$\sigma_c = \sum_\ell \sigma_c^{(\ell)} \equiv \pi ƛ^2 \sum_\ell (2\ell + 1) T_\ell(k) \tag{8.60}$$

with

$$T_\ell(k) = -4s_\ell \frac{\text{Im} q_\ell^{int}}{M_\ell^2 + N_\ell^2} \tag{8.61}$$

and for the total cross section,

$$\sigma_t = 4\pi ƛ^2 \sum_\ell (2\ell + 1) \left\{ \sin^2 \tau_\ell + s_\ell \frac{M_\ell \cos 2\tau_\ell - N_\ell \sin 2\tau_\ell}{M_\ell^2 + N_\ell^2} \right\} \tag{8.62}$$

The quantity $T_\ell(k)$, which is also equal to $1 - |\langle \eta_\ell \rangle|^2$, plays the role of a transmission coefficient, and $\sigma_c^{(\ell)}/\pi r_o^2$ is sometimes known as the penetrability. For the complex square-well potential model, Eqs. (8.61) and (8.62) are convenient expressions for computing σ_c and σ_t, and by difference one obtains σ_{se}. To find the starting values for the recurrence relations one can show using Eqs. (7.36) and (7.38)

$$q_\ell^{(\pm)} = \xi \left[-\frac{1}{2v_\ell} \frac{dv_\ell}{d\xi} \pm i \frac{d\tau}{d\xi} \right] \tag{8.63}$$

with $\xi = Kr_o$. Since $1/v_\ell = \xi^2[j_\ell^2 + n_\ell^2]$, one finds for $\ell = 0$, $v_o = 1$ and $\tau_o = \xi^2$. Putting

$$q_o^{(+)} = \Delta_o + is_o \tag{8.64}$$

in Eq. (8.57) we obtain the starting values for part of the the scattering amplitude, Eq. (8.60), that depends on the exterior solution,

$$\Delta_o = 0, \quad s_o = \xi, \quad \tau_o = \xi^2. \tag{8.65}$$

It might be noted that the apparent simplicity of what is actually a rather complex cross section calculation is largely the result of the several derivations we have gone through in order to arrive at expressions such as Eqs. (8.60), (8.61), and (8.62).

Comparison of optical model results with experiments

To gain an understanding of the extent to which the cross sections formulated here can describe significant features of actual cross sections, we resort to comparing numerical results obtained from the above analysis against selected experimental data. We consider two kinds of results, a few cross section values obtained from previous classroom exercises, and cross section calculations reported in the literature. In different respects both are instructive.

Table 8.1 shows the cross sections of three elements for neutrons with energies in the MeV range (fast neutrons). Each calculated cross section is the sum of partial waves for $\ell = 0$ to $\ell = \ell_{\max}$. The experimental data, given in parenthesis, are values for the reaction cross section rather than the compound formation cross section. Similarly, values are given for elastic scattering rather than for shape

Table 8.1 Fast neutron cross sections (in barns) calculated using the square-well optical model and the corresponding experimental data [BNL-325] shown in parenthesis.

Nuclide	Energy	$\ell_{\max}$	σ_t	$\sigma_c(\sigma_r)$	$\sigma_{se}(\sigma_{el})$
Al^{27}	4.1 MeV	0	0.59 (2.4)	0.016	0.576
Bi^{209}	7	6	6.43 (5.86)	1.35 (2.5)	5.07 (3.36)
U^{238}	2	4	6.35 (7.0)	0.65 (2.5)	5.7 (4.5)

elastic scattering in the next column. In the case of the light element Al^{27}, the underestimate of the total cross section is most likely due to the neglect of the higher partial wave contributions. Judging from the results for Bi^{209} and U^{238}, where the total cross sections are in reasonable agreement, one can infer that only a few higher order partial waves than $\ell = 0$ are needed to get meaningful results. Notice that for these two cases the reaction cross sections are underestimated and correspondingly the elastic scattering is overestimated. The explanation is that the square-well model, by virtue of the step change at the potential range r_o, causes the reflectivity at the surface of the nucleus to be too large. This is an inherent feature of any discontinuous potential.

Figure 8.6 shows the comparisons of the square-well optical model calculations with experiment data [Feshbach, 1954]. Figure 8.6(a) shows the model results for total cross section σ_t over the entire range of mass number and a range of energy, expressed in terms of $X^2 = \left[A^{5/3}/10(A+1)\right] E$, with E in MeV. The cross sections are given in $\sigma_t/\pi r_o^2$, where the potential range is $r_o = 1.45A^{1/3} \times 10^{-13}$ cm. The model potential parameter values are $V_o = 42$ MeV, $\varsigma = W_o/V_o = 0.03$ as defined in Eq. (8.26). The corresponding experimental data, shown in Fig. 8.6(b), were taken from many sources. They have been smoothed over in the sense of energy averaging discussed above. Both calculations and experimental data do not go below about 50 keV. The comparison shows the square-well optical model can account for a number of striking features of the experimental data. There is good correspondence in low cross section values at low energy around $A \approx 40$ and $100 < A < 140$. The calculation identifies the large cross sections at low energy around $A \sim 60$ and 150 to be s-wave resonances, and similarly at $A \sim 90$ p-wave resonances. Furthermore, "*D maxima*" can be seen for $x^2 \approx 3$ and $A \sim 40$ and 140.

Regarding sensitivity of the calculation to the parameter values, an earlier comparison using a smaller value of well depth, $V_o = 19$ MeV, gave less satisfactory agreement than that shown in Fig. 8.6. It appears the agreement is not sensitive to V_o so long as $r_o\sqrt{V_o}$ is kept constant. It is known that σ_t is quite sensitive to the value of ς, the depth of the imaginary part of the potential. An increase in ς tends to flatten out the maxima and minima, and using $\varsigma = 0.02$ and 0.05 gives poorer agreement.

Comparison with experiment also has been carried out for the angular differential scattering cross section $\sigma_{el}(\theta)$. Figure 8.7 shows two sets of calculations, one assuming that shape elastic scattering constitutes the only contribution, or $\sigma_{ce}(\theta) = 0$, and the other assuming a maximum contribution from the compound elastic component, $\sigma_{ce} = \sigma_c$. These two assumptions correspond to the opposite extremes of underestimate and overestimate respectively. The potential parameter values used are the same as those in Fig. 8.6(a). As in the comparison of σ_t, a number

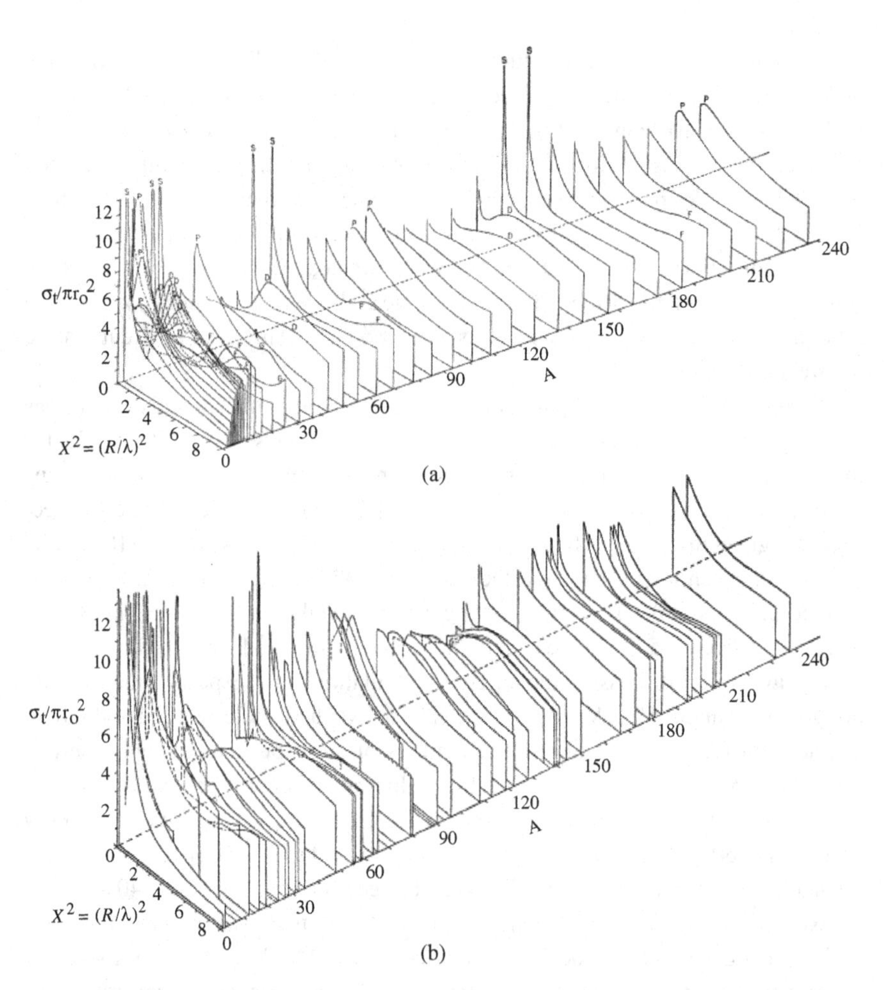

Fig. 8.6 (a) Variation of total cross section, expressed as $\sigma_t/\pi r_o^2$ with $r_o = 1.45\,A^{1/3} \times 10^{-13}$ cm, for nuclei of mass number A reacting with neutrons of energy E calculated using an optical model with parameters determined by using experimental data for neutron reaction in lead at 7 MeV. Optical model potential parameters are defined in Eq. (8.26) with $V_o = 42\,\text{MeV}$, $\varsigma = W_o/V_o = 0.03$. (b) Experimental total cross section averaged over energy range to smooth over the resonances, are plotted in the same way as the optical model calculations shown in (a). [Adapted from Feshbach, 1954.]

of features in the experimental data are also seen in the optical model results, in particular, strong forward and backward scattering intensities at $A \sim 140$, and a relatively flat distribution around $A \sim 60$. Judging from the fact that the corresponding results for $V_o = 19\,\text{MeV}$ are not very different, one may conclude that $\sigma_{el}(\theta)$ is not sensitive to the potential well depth.

Refinements of the optical model

The complex square-well model we have just discussed is the original version of the optical model concept of nuclear reaction theory. There exists a natural parallel between the present study of nuclear reactions and the preceding study of energy levels in a nucleus in Chap. 5. There we were concerned with solving the time-independent Schrödinger equation for negative energy (eigenvalue) solutions. The square-well model, Eq. (5.2), played an important role of simplifying the interaction in order to enable the solutions of the radial wave equation essentially by inspection. This simplification helps to explain the basic concept of quantization which involved matching the interior and exterior wave functions and derivatives at the interface without the mathematical details that would be involved if more realistic (and necessarily more complicated) interaction potential models were considered. On the other hand, it is also to be expected that there will be significant physical effects that would not be captured by the oversimplification of a square-well interaction for nuclear reactions such as Eq. (8.26). Moreover,

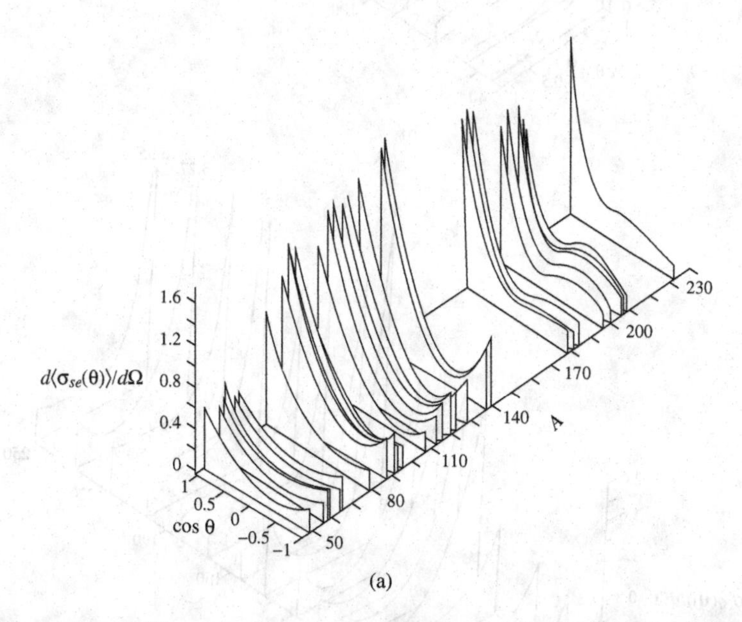

Fig. 8.7 (a) Angular differential scattering cross section of nuclei of mass number A calculated using the optical model of Fig. 8.6 by taking into account only the shape-elastic scattering contribution. Energy of incoming neutron is 1 MeV. The model potential parameters used are the same as those in Fig. 8.6(a). (b) Same as (a) except both shape-elastic and the maximum of compound elastic contributions are taken into account. (c) Experimental data on energy-averaged angular differential cross section for comparison with the calculations shown in (a) and (b). [Adapted from Feshbach, 1954.]

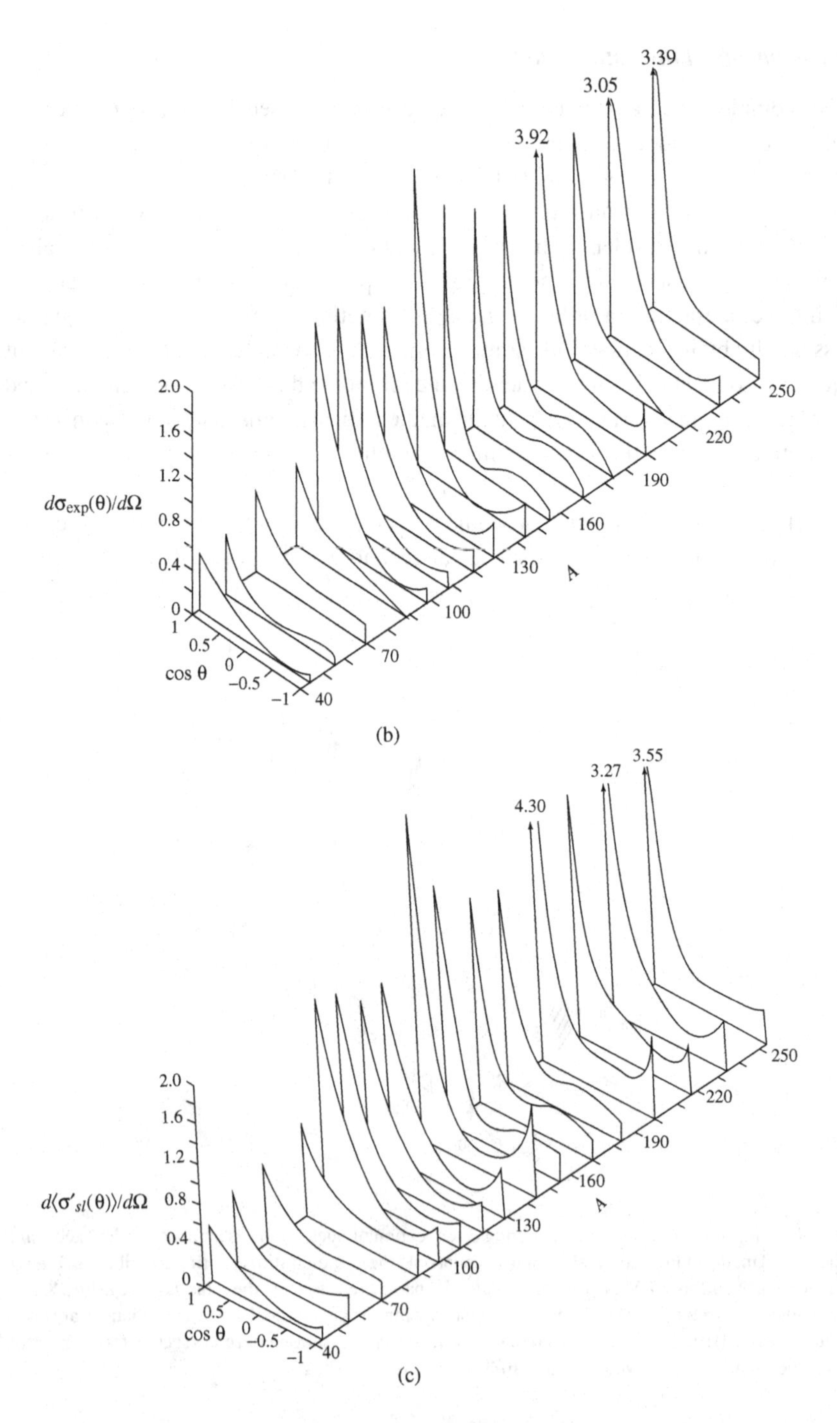

Fig. 8.7 (*Continued*).

beyond just the form of the interaction $V(r)$ there are additional interactions which need to be considered. In Chap. 5 such an example is the *spin-orbit* interaction which is necessary to explain adequately the shell structure of the energy levels of a nucleus. Regarding the complex square-well model we can therefore expect significant improvements also can be made by introducing refinements of various kinds. It is not our intention to go too deeply into this topic, while it is relevaant that the reader is aware of these developments [Feshbach, 1958; Hodgson, 1963; Satchler, 1980].

Diffuse boundary effects

The square-well potential is known to give a value of σ_c that is too low when compared to experiments, see Table 8.1. Intuitively one can expect this is due to an overestimate of the reflectivity at the surface of the nucleus by the step-change in the interaction. It is more reasonable to replace the square well by a potential with a continuous, diffuse boundary such as that suggested in Chap. 3 regarding the nuclear density distribution, Eq. (3.7) and Fig. 3.2 (the potential would have a form that is inverted). The potential that is often used is a four-parameter function,

$$V(r) = -(V_o + iW_o)\frac{1}{1 + e^{(r-R)/a}} \tag{8.66}$$

Here R is a measure of the nuclear radius, and a is a measure of the boundary thickness. Equation (8.66) is known as the Woods-Saxon potential, previously seen in Fig. (3.7). The main effect of the diffuse boundary is to lower the reflectivity at the surface, thus giving a larger value of σ_c.

Surface absorption

As the incident wave starts to penetrate into the nucleus, one can imagine a self-shielding effect which results in more absorption taking place at the surface compared to the interior region. This would suggest that the imaginary part of the potential should be more concentrated in the surface region. One way to express this surface absorption behavior without introducing another parameter is to modify the complex part of the interaction through a derivative,

$$V(r) = \begin{cases} -V_o & r < r_o \\ -(V_o + iW_o\frac{d}{dx})\frac{1}{e^x - x} & r > r_o \end{cases} \tag{8.67}$$

with $x = (r - r_o)/a$. We show an example of the effects of diffuse boundary and of surface absorption in Fig. 8.8.

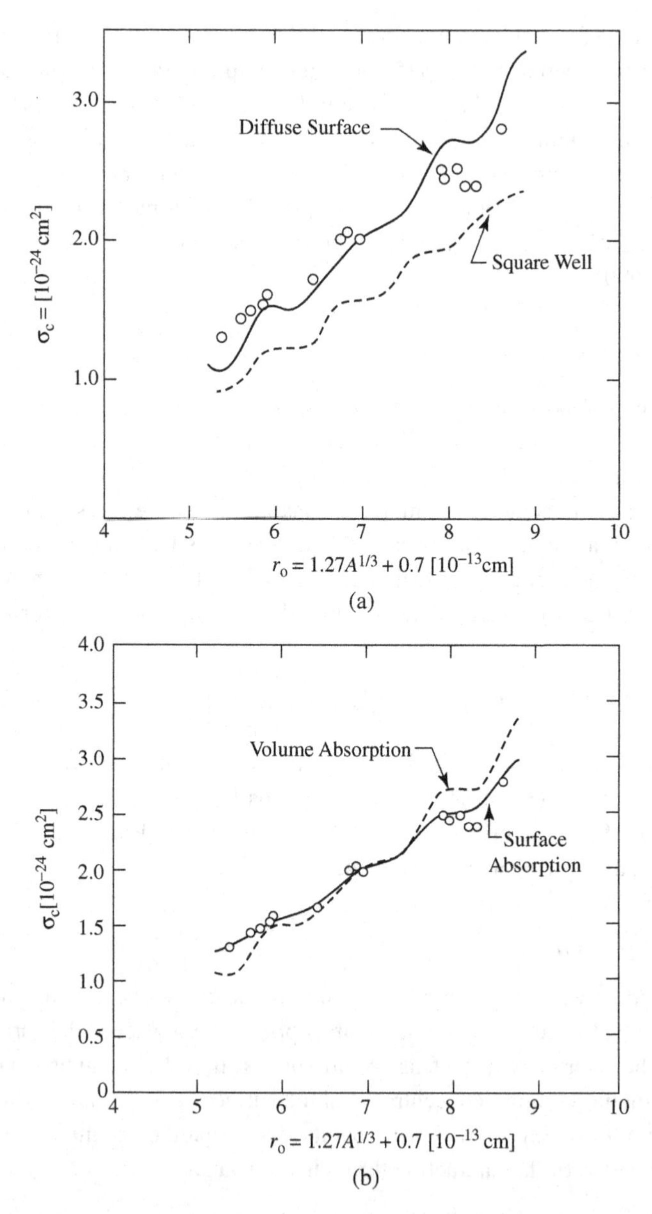

Fig. 8.8 (a) Variations of reaction cross section with mass number A for neutrons at 7 MeV calculated using an optical model. Results shown correspond to a square-well potential with model parameters of $V_o = 40$ MeV and $\varsigma = 0.10$ and to a diffuse-boundary potential with the same V_o and $\varsigma = 0.08$. Open circles denote energy averaged experimental data. (b) Same as (a) except the comparison is between volume (dashed curve) and surface (solid curve) absorptions. [Adapted from Emmerich, 1963.]

Spin-orbit coupling

Our discussions thus far have not taken into consideration the fact that the interaction potential is known to be spin-dependent. In Chap. 5 we have seen the importance of spin-orbit coupling in describing the nuclear shell structure. The addition of a spin-orbit interaction term is essential to the understanding of the stability of nuclides with magic number of nucleons. A second aspect is the interaction between the incident neutron, which has a spin, and the target nucleus which also can have a nonzero spin I. The spin-spin coupling effect is particularly important in the case of neutron-proton scattering, as we will see in Chap. 9. When the incident particle has a spin, such as a neutron or a proton, the spin also can couple to the orbital angular momentum of the scattering system, as we have seen in Chap. 5 in the discussion of the nuclear shell model. Thus spin-orbit coupling would exist even if the target nucleus has zero spin I. To include the spin-orbit interaction one can take the interaction potential to be the sum,

$$V = V(r) + V_{SO} \tag{8.68}$$

with

$$V_{SO} = A\left[\frac{1}{r}\frac{dV(r)}{dr}\right]\underline{L}\cdot\underline{s} \tag{8.69}$$

and $V(r)$ could be a diffuse boundary potential such as the Woods-Saxon model, Eq. (8.66). In Eq. (8.69), $\underline{L}$ is the orbital angular momentum of the effective one-body scattering problem, see Sec. 7.3, and $\underline{s}$ is the intrinsic spin of the neutron.

There exist numerical results in the literature that illustrate the effects of the refinements we have just discussed [Emmerich, 1963]. Figure 8.8 shows the use of diffuse boundary potential can indeed give a significantly improved agreement with experiment in the calculation of σ_c. The effects of surface absorption are not as large, but evidence does point to an improvement over volume absorption. The effects of spin-orbit coupling seem to be important only in the angular differential scattering cross section at large scattering angles.

Nonlocal optical model

In determining the various parameters in an optical model one usually adopts those values which have been optimized by fitting to a certain set of experimental data. It has been found these parameters show considerable variation with energy. This is considered an unsatisfactory aspect of the model description since the use of different sets of parameters for different energy ranges implies something fundamental is still missing in the model. Also it casts doubt on the predictive ability of the model description. To avoid this property of the early versions of the optical model, one can formulate a *nonlocal optical model* which does not suffer from having the model parameters vary with energy [Perey, 1962]. It turns out one can also derive an energy-dependent relation between the nonlocal model and an *equivalent local* model, so a set of energy-independent parameters can be determined for the nonlocal model from which energy-dependent parameters for the local model can be obtained without further fitting, and cross section calculation can proceed using the local optical model thereby determined.

We refer the reader to the literature for the development of the nonlocal optical model [Wilmore, 1964; Gersten, 1967]. Figure 8.9 shows a comprehensive test of the nonlocal optical model approach by comparing calculated σ_t and σ_c against experimental data at four energies, 4.1, 7, 14.5, and 24 MeV over the entire range of mass numbers. The parameters used in these calculations have values optimized by fitting the neutron cross sections of lead in the energy range 7 to 14.5 MeV:

$$\begin{aligned} V_o &= 71\,\text{MeV} & a_s &= 0.65 \times 10^{-13}\,\text{cm} \\ W_o &= 15 & a_D &= 0.47 \\ U_{SO} &= 1300 & r_o &= 1.22A^{1/3} \end{aligned}$$

The agreement seen in Fig. 8.9 may be regarded as a demonstration that the nonlocal optical potential approach is capable of predicting fast neutron cross sections up to 30 MeV to within $\sim 10\%$.

It is worth noting the idea of deriving an equivalent local model from the nonlocal model has been studied by requiring the calculated cross sections using the two models to be the same. A nonlinear relation is obtained between the two potentials and analyzed using an iterative procedure [Wilmore, 1964] and an analytical approximation [Gersten, 1967]. Generally speaking, the nonlocal calculations of the total cross section are consistently higher than the equivalent local results by 3–5%. The same behavior also holds for the two contributions to σ_t, shape elastic and compound formation. The equivalent local potential model combines the advantage of robustness of an energy-independent nonlocal potential with the relative simplicity of a local potential calculation.

The nonlocal optical model also gives quite good results for neutron scattering at low energy. This is shown in Fig. 8.10 where we define an effective potential scattering radius by taking the low-energy limit of the shape elastic scattering cross section,

$$4\pi R'^2 = \sigma_{se}(E)|_{E\to 0} = \ell im_{k\to 0}\left[\pi\lambda\!\!\!^{-2}\,|1 - \langle\eta_o\rangle|^2\right] \tag{8.70}$$

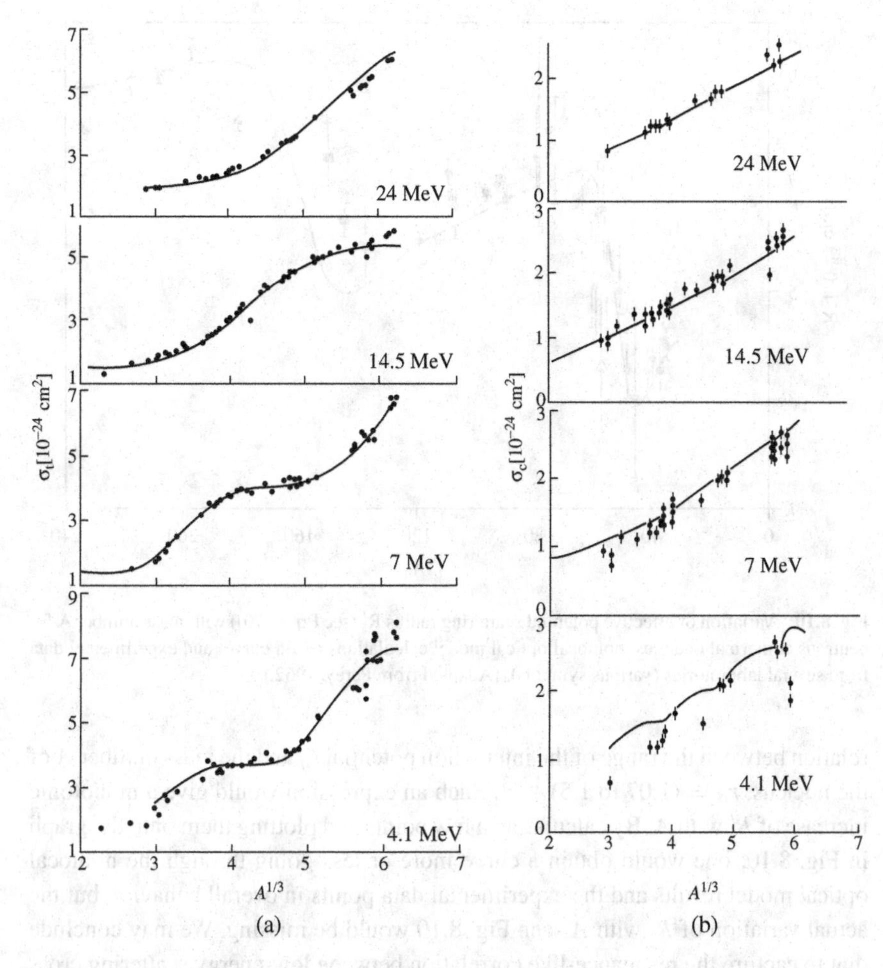

Fig. 8.9 (a) Variations of total cross section with mass number A for neutrons at four energies, $E = 4.1, 7, 14.5, 24$ MeV, calculated using a non-local optical model. Experimental data on energy-averaged cross sections are shown as closed circles. (b) Same as (a) except for reaction cross section. [Adapted from Perey, 1962.]

where $\langle \eta_o \rangle$ is s-wave scattering amplitude calculated using the optical model. The quantity R' plays the role of the scattering length a discussed in Sec. 7.2. It is of interest in the study of thermal neutron scattering which will be discussed in Chaps. 9 and 13. One sees there is considerable variation of elastic scattering with mass number, and the optical model is able to predict quite accurately this behavior across the entire range of A. Notice that in the previous discussions of energy level calculations in Chap. 5 and the cross section calculation in Chap. 7, we used a simple

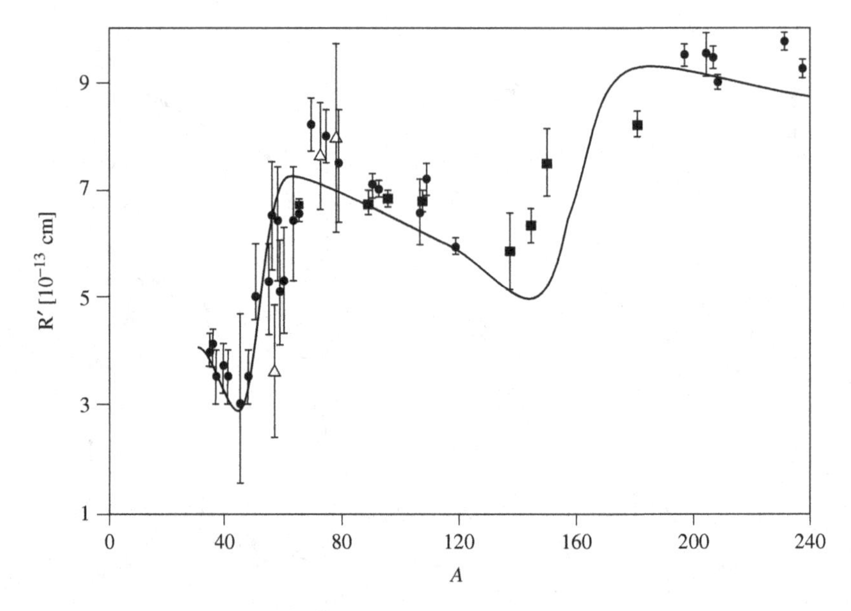

Fig. 8.10 Variation of effective potential scattering radius R′ (see Eq. (8.70)) with mass number A for neutrons at thermal energies, nonlocal optical model calculations (solid curve) and experimental data from several laboratories (various symbols). [Adapted from Perey, 1962.]

relation between the range of the interaction potential r_o and the mass number A of the nucleus, $r_o = (1.07 \text{ to } 1.5)A^{1/3}$. Such an expression would give a monotonic increase of R' with A. By calculating a few points and plotting them onto the graph in Fig. 8.10, one would obtain a curve more or less going through the nonlocal optical model results and the experimental data points in overall behavior, but the actual variation of R' with A seen Fig. 8.10 would be missing. We may conclude that to capture the resonance-like correlation between low-energy scattering cross sections and the mass number, an optical model formulation would be needed.

R-matrix formulation

The general problem of the theory of nuclear reactions is to relate the values of the scattering or collision matrix elements to the dynamics (interactions) of the nuclear structure (nuclear reactions) [Satchler, 1980, see also Preston, 1962; Blatt and Weisskopf, 1952; Lane and Thomas, 1958; Mott and Massey, 1965]. The *R-matrix* or boundary matching theories do not solve this problem. Rather they provide a mathematical reformulation of the problem such that on the one hand they make explicit some of the phenomena (such as resonances) one may expect to observe, and on the other hand they express the measurable quantities (positions and widths of cross section peaks) in terms of parameters whose physical significance can be more readily understood.

These approaches are useful because of the short range of nuclear forces. Two nuclei may be assumed not to be interacting (except for Coulomb forces) if they are separated by a distance greater than $r \sim a$, where a is a characteristic distance expected to be not much larger than the nuclear radius. (Reader should not confuse the present distance a with the scattering length discussed extensively in Chap. 7, Sec. 7.2. It is unfortunate both are distances and denoted by the same symbol in the literature. In the rest of this section there should be no ambiguity since the scattering length will not appear in the following discussion.) We know the form of the exterior wave function is comprised of spherical incoming andoutgoing waves with amplitudes given by the scattering matrix elements as in Eq. (7.40). In the interior region we do not know the wave function *a priori*. What we do know is that its magnitude and derivative must match smoothly onto the outgoing wave across the boundary $r \sim a$. So we define a complete set of states X_α in the interior region with eigenvalues E_α and expand the interior wave in this set (compound states). The expansion coefficients are then related to the observable scattering matrix elements through the interface continuity condition at $r = a$. The hope is that the expansion states may correspond reasonably well to the actual states of the compound nucleus which satisfy the same wave equation but do not have to satisfy the artificial boundary condition at $r = a$.

As an example, consider what happens if the E_α are well separated and the bombarding energy is very close to one of them. In this case the true wave function will be similar to the particular state X_α. Thus, one compound nucleus state is excited very strongly and we have a resonance at $E = E_\alpha$ (see, for example, the lower part of the schematic spectrum in Fig. 5.19). On the other hand, if the energy eigenvalues E_α are very dense (the upper part of the figure) surrounding the energy E, the reaction will excite many such compound states (whether bound or virtual) and we would be better advised to adopt a statistical model which treats the corresponding expansion coefficients as random variables [Satchler, 1980].

Although the mathematical formalism is perfectly general, it is clear this approach is particularly well suited to describing reactions of the compound nucleus type (see Chap. 12). Direct reactions do not involve just a single compound state, nor a random mixture. Rather there is a correlation between the contributions from the various X_α. Boundary-matching theories offer an insight into how this correlation comes about. On the other hand, an approach known as perturbation theory may provide a more convenient framework for describing direct reactions. In the R-matrix formulation the nucleus is treated as basically a black box, with no attempt to account for all the known features of nuclear structure derived from the development of nuclear models. The goal here is to calculate the various cross sections of interest in terms of the collision matrix which we have previously defined in Chap. 7 (see Eq. (7.40)). It turns out that one can express this collision matrix in terms of the quantity that depends on the interior wave function, known as the R-matrix. However, the R-matrix is not sufficient to completely specify the collision matrix, other quantities which pertain to the exterior wave function are also needed. The essential advantage of the R-matrix approach is the appearance of physical parameters that can be determined by fitting the cross sections to experimental data. In other words, while one does not presume to know the logarithmic derivative of the interior wave evaluated at an appropriate interface (surface of the nucleon-nucleus system) at any stage of the formulation, the end result is useful in that this unknown derivative is expressed in terms of energy levels and energy width functions, quantities that can be readily fitted to the measured cross sections. The utility of the theory is therefore to reduce experimental data in terms of a set of (resonance) parameters that have physical meaning and can be conveniently tabulated.

To provide a qualitative understanding of the underlying ideas we give a simplified definition of the R-matrix using the results already discussed in Chaps. 5 and 7. From this definition it will be clear that there is a basic relation between the R-matrix and the collision-matrix η (see Eq. (8.56)). Thus the cross section calculations we have discussed in this section using the η-matrix can just as well be expressed in terms of the R-matrix. We defer to Chap. 12, Sec. 12.2, the discussion of neutron resonance cross sections. There one has examples of how the measured cross sections can be interpreted in terms of the resonance parameters, quantities known as resonance energies and level widths.

For the sake of illustration, we consider a nucleon-nucleus reaction that can be represented by an effective one-body problem (Sec. 7.3), in the form of a one-dimensional time-independent Schrödinger equation, see Eq. (5.1).

$$\frac{d^2 u_E}{dr^2} + (2M/\hbar^2)(E - V)u_E(r) = 0 \quad r < a,\ E > 0 \tag{8.71}$$

Here we take the interaction potential to have a finite range at $r = a$, so $u_E(r)$ is the interior wave function. Since we are treating the nucleon-nucleus system as a black box, we have no information about the interaction V and therefore cannot expect to solve Eq. (8.71) directly. On the other hand, we can manipulate Eq. (8.71) to obtain useful expressions for the logarithmic derivative of $u_E(r)$ at the interface where it has to be equal to the logarithmic derivative of the exterior wavefunction, see for example Eq. (7.31). Suppose we pick two arbitrary energies, E_1 and E_2, and write equations like Eq. (8.71) for the corresponding interior wave functions, $u_1 = u_{E_1}$

and $u_2 = u_{E_2}$. Multiply the first equation by u_2, and the second equation by u_1, integrate the two equations from r = 0 to $r = a$, then subtract the results to obtain

$$\left(u_2\frac{du_1}{dr} - u_1\frac{du_2}{dr}\right)_{r=a} + (2M/\hbar^2)(E_1 - E_2)\int_0^a u_1u_2dr = 0 \quad (8.72)$$

In arriving at the first bracket of terms in Eq. (8.72) we have performed an integration by parts and made use of the property that the wave function must vanish at the origin. Now consider the set of functions $\{X_\lambda\}$ which are the eigenfunctions of the problem. By this we mean X_λ satisfies Eq. (8.71) with boundary condition

$$\left(\frac{dX_\lambda}{dr}\right)_{r=a} = 0 \quad (8.73)$$

We take the eigenfunctions to form an orthonormal set and proceed to expand the interior wave function in this set,

$$u_E(r) = \sum_\lambda A_\lambda X_\lambda(r) \quad (8.74)$$

The expansion coefficients are

$$A_\lambda = \int_0^a u_E X_\lambda dr \quad (8.75)$$

Going back to Eq. (8.72) we now choose E_1 to be E_λ, so u_1 becomes X_λ, and E_2 to be E, so u_2 becomes u_E. Then the integral in Eq. (8.72) is just Eq. (8.75), which means, after applying Eq. (8.73),

$$A_\lambda = \frac{\hbar^2}{2M}\frac{\left(X_\lambda \frac{du_E}{dr}\right)_{r=a}}{E_\lambda - E} \quad (8.76)$$

The expansion Eq. (8.74) now takes the form

$$u_E(r) = G(r, a)\left(a\frac{du_E}{dr}\right)_{r=a} \quad (8.77)$$

with

$$G(r, a) = \frac{\hbar^2}{2Ma}\sum_\lambda \frac{\left(a\frac{du_E}{dr}\right)_{r=a} X_\lambda(a)X_\lambda(r)}{E_\lambda - E} \quad (8.78)$$

The significance of Eq. (8.77) is we can interpret $G(r, a)$as the Green's function that relates the interior wavefunction to it's derivative at the surface. With this Green's function we now define the R-matrix as

$$R = G(a, a)$$

$$= \sum_\lambda \frac{\gamma_\lambda^2}{E_\lambda - E} \quad (8.79)$$

where

$$\gamma_\lambda = \left(\hbar^2/2Ma\right)^{1/2} X_\lambda(a) \quad (8.80)$$

In this illustration R appears as a function. It becomes a matrix when we specify a reaction taking place between specific incoming and outgoing channels. Also we have not been concerned with various angular momenta associated with the channels. It is clear from Eq. (8.79) that R is an energy-dependent function involving two other quantities, E_λ and γ_λ^2. We can anticipate the former represents the energy level of a resonance reaction, while the latter plays the role of a strength function or the level width [Satchler, 1980].

Given the definition of R in Eq. (8.79), the relation Eq. (8.78) shows that R has the interpretation of being the reciprocal of a times the logarithmic derivative of the interior wave function at a. This recognition immediately leads us to a relation between R and the collision function η.

Recalling the wave function Eq. (7.18) from Chap. 7 and keeping only the s-wave ($\ell = 0$) term, we have a simplified form of the radial wave function in the exterior region,

$$u(r) \sim \frac{1}{2k}\left(e^{-ikr} - e^{2i\delta}e^{ikr}\right), \quad r > a \tag{8.81}$$

The first term is clearly the incoming wave component, while the second term is the outgoing wave component modulated by the phase shift. In the present context $e^{2i\delta}$ is the scattering amplitude which should be replaced by the collision function η. Equation (8.81) is a relation between the wave function in the exterior region and the collision function which can be used to evaluate the logarithmic derivative at the surface of the reaction system. A more general form of this relation which can include various angular momentum and spin indices appropriate to channel specifications is

$$u_{ext} \sim (\mathbf{I} - \eta\mathbf{O}) \tag{8.82}$$

where $\mathbf{I}$ and $\mathbf{O}$ represents the incoming and outgoing free-particle wave functions respectively. We can use Eq. (8.82) to evaluate the logarithmic derivative whose reciprocal is the R-function, as indicated in Eq. (8.77) with $r = a$. Thus

$$R = \frac{1}{a}\left[\frac{u_{ext}}{du_{ext}/dr}\right]_{r=a} \tag{8.83}$$

Solving for η one obtains

$$\eta = \frac{I_a}{O_a}\frac{1 - (I_a'/I_a)R}{1 - (O_a'/O_a)R} \tag{8.84}$$

where the subsbscript a denotes the value at $r = a$, and the prime denotes spatial derivative. This is then the fundamental relation between the collision function and the R-function. Notice that knowing R is not sufficient to determine the collision function. This is not surprising since R is purely concerned with the interior wave function, as our brief derivation above indicates. The other parts of Eq. (8.84) therefore represents the free-particle wave in the exterior region evaluated at the interface. In this sense one may note a certain structural similarity between Eqs. (7.84) and (7.40).

REFERENCES

F. Bjorklund and S. Fernbach, *Physical Review* **109**, 1295 (1958).

W. S. Emmerich, in *Fast Neutron Physics*, eds. J. B. Marion and J. L. Fowler (Interscience, New York, 1963), vol I, p. 1072.

R. D. Evans, *The Atomic Nucleus* (McGraw-Hill, New York, 1955), Chap. 12.

E. Fermi, *Nuclear Physics*, Lecture Notes compiled by J. Orear, A. H. Rosenfeld, and R. A. Schluter (University Chicago Press, 1949), revised edition, Chap. VIII.

H. Feshbach, *Ann. Rev. Nuc. Sci.* **8**, 49 (1958).

A. Foderaro, *The Elements of Neutron Interaction Theory* (MIT Press, Cambridge, 1971).

A. Gersten, *Nuc. Phys. A.* **96**, 288 (1967).

P. E. Hodgson, *The Optical Model of Elastic Scattering* (Oxford, London, 1963).

A. M. Lane and R. G. Thomas, R-matrix theory of nuclear reactions, *Rev. Mod. Phys.* **30**, 257 (1958).

J. E. Lynn, *The Theory of Neutron Resonance Reactions* (Oxford, London, 1968).

F. Perey and B. Buck, *Nuc. Phys.* **32**, 353 (1962).

P. Roman, *Advanced Quantum Theory* (Addison-Weley, Reading, 1965), Chap. 3.

G. R. Satchler, *Introduction to Nuclear Reactions* (Halsted, New York, 1980).

A. M. Weinberg and E. P. Wigner, *The Physical Theory of Neutron Chain Reactors* (University of Chicago Press, Chicago, 1958).

D. Wilmore and P. E. Hodgson, *Nuc. Phys.* **55**, 673 (1964).

PART 2

Unit Processes of Nuclear Radiation Interactions

9

Neutron Scattering

Throughout this book, neutron scattering plays a special role — it is the first example of nuclear radiation interaction. We would like to know everything about this interaction, from the nature of the interaction, the kinematics and dynamics, to the various cross sections that are of interest to us, and finally the effects of neutron scattering in the broader study of neutron transport in nuclear systems. In this chapter we confine our discussions to only elastic scattering (no excitation of the target nucleus). Even so there are additional significant aspects of neutron scattering on which the reader can be usefully informed by consulting with the recommended references, not only just those cited in this chapter, but also those cited in Chap. 13.

9.1 NEUTRON-PROTON SCATTERING

We take up the problem of neutron scattering in hydrogen which is well-known for its significance in the science of neutron interactions with matter, as well as in a broad range of nuclear technology applications. The scattering cross section of hydrogen in the form of water has been widely measured to be 20.4 barns over a range of energy from thermal energy up to keV. Our interest here is to apply the method of phase shift analysis discussed in Chap. 7 to this problem. We see very quickly the *s*-wave approximation (the condition of interaction at low energy) is very well justified in the neutron energy range between 1 and 10^3 eV. The scattering-state solution ($E > 0$) gives us the phase shift or equivalently the scattering length. This calculation yields a cross section of only 2.3 barns, which is considerably lower than the experimental value and clearly indicates something important is missing. The reason for the discrepancy lies in the fact we have not taken into account the spin-dependent nature of the *n*-*p* interaction. The neutron and proton spins can form two distinct spin configurations, the two spins being parallel (triplet state) or anti-parallel (singlet), each giving rise to a corresponding scattering length. When this is taken into account, the new estimate is quite close to the experimental value. The conclusion is therefore n-p interaction is spin-dependent and the unusually large value of the well-known neutron scattering cross section for hydrogen is due to this aspect of the nuclear force.

Even before we get into the details it is worthwhile to observe a special connection between the scattering of a neutron by a proton and the bound state of deuteron, as the nucleus composed of a neutron and a proton. The connection is simply both cases involve the same interaction potential. It means we can deduce the basic features of n-p interaction from what is known experimentally about the deuteron nucleus, such as its ground state and the first excited state. One can produce a deuteron by having a neutron incident on a proton to cause an absorption reaction whereby a gamma ray is emitted, a reaction that is known as *radiative capture*.

$$n + H^1 \rightarrow H^2 + \gamma$$

In the terminology of the last chapter, this would be an (n, γ) reaction. Measurements have shown that the γ emitted has an energy of 2.23 MeV, and this constitutes a way to measure the energy of the bound state of the deuteron. One can use this information to constraint the potential parameters like the depth and range of a square-well potential [Meyerhof, 1967].

To discuss the problem of neutron-proton scattering at low energy we begin with the radial wave equation for s-wave, Eq. (7.10) or Eq. (7.14) with $\ell = 0$, using a simple square well potential, with depth V_o and range r_o. Since the wave equation is really simple, the interior and exterior solutions can be written down by inspection,

$$u(r) = B\sin(K'r), \quad r < r_0 \tag{9.1}$$

and

$$u(r) = C\sin(kr + \delta_o), \quad r > r_0 \tag{9.2}$$

where $K' = \sqrt{m(V_o + E)}/\hbar$ and $k = \sqrt{mE}/\hbar$. Applying the interface condition as previously discussed in both Chaps. 5 and 7, we obtain

$$K' \cot(K'r_o) = k\cot(kr_o + \delta_o) \tag{9.3}$$

which is the relation that allows the phase shift δ_0 to be determined in terms of the potential parameters and the incoming energy E. We can simplify the task of estimating the phase shift by recalling the phase shift is simply related to the scattering length by $\delta_o = -ak$ (cf. Eq. (7.28)). Assuming the scattering length a is larger than r_o, we see the right-hand side (RHS) of Eq. (9.3) is approximately $kcot(\delta_0)$. For the LHS, we will ignore E relative to V_o in K', and at the same time ignore E_B relative to V_o in K. Then $K' \sim K$ and the LHS can be set approximately equal to $-\kappa$. Notice that this series of approximations has enabled us to make use of the dispersion relation in the bound-state problem now for the scattering calculation. As a result, Eq. (9.3) becomes

$$k \cot(\delta_o) = -\kappa \tag{9.4}$$

which is a relation between the phase shift and the binding energy.

Once the phase shift δ_o is known, the angular differential scattering cross section is then given by Eq. (7.26),

$$\sigma(\theta) = (1/k^2) \sin^2 \delta_o \tag{9.5}$$

A simple way to make use of Eq. (9.4) is to note the trigonometric relation $\sin^2 x = 1/(1 + \cot^2 x)$,

$$\sin^2 \delta_o = \frac{1}{1 + \cot^2 \delta_o} = \frac{1}{1 + \kappa^2/k^2} \tag{9.6}$$

Thus,

$$\sigma(\theta) \approx \frac{1}{k^2 + \kappa^2} = \frac{\hbar^2}{m} \frac{1}{E + E_B} \approx \frac{\hbar^2}{m E_B} \tag{9.7}$$

The last step follows because we are mostly interested in estimating the scattering cross section in the energy range $1 - 10^3$ eV. Putting in the numerical values of the constants, $\hbar = 1.055 \times 10^{-27}$–erg sec, m$= 1.67 \times 10^{-24}$ g, and and $E_B = 2.23 \times 10^6 \times 1.6 \times 10^{-12}$ ergs, we get

$$\sigma = 4\pi\hbar^2/mE_B \sim 2.3 \text{ barns} \tag{9.8}$$

This value is considerably lower than the experimental value of the scattering cross section of H^1, 20.4 barns, as shown in Fig. 9.1 (the same data are also shown in Fig. 3.6).

The explanation of this discrepancy lies in the neglect of spin-dependent effects. It was suggested by E. P. Wigner in 1933 that neutron-proton scattering should depend on whether the neutron and proton spins are oriented in a parallel configuration (the triplet state, total spin angular momentum equal to $\hbar$) or in an anti-parallel configuration (singlet state, total spin is zero). In the two cases the interaction potentials are different, and therefore the phase shifts also would be different. Following this idea, one can write instead of Eq. (9.7),

$$\sigma(\theta) = \frac{1}{k^2}\left(\frac{1}{4}\sin^2 \delta_{os} + \frac{3}{4}\sin^2 \delta_{ot}\right) \tag{9.9}$$

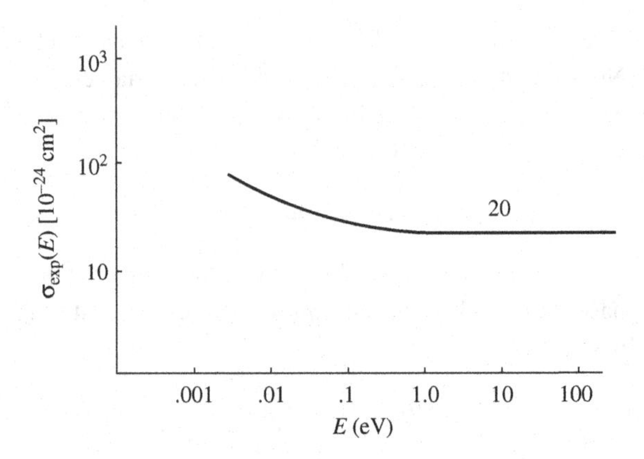

Fig. 9.1 Experimental neutron scattering cross section of *hydrogen*, showing a constant value of 20.4 barns over a wide range of neutron energy. The rise in the cross section at energies below ~0.1 eV can be explained in terms of *molecular binding* effects in the scattering sample.

We have already mentioned that the ground state of the deuteron is a triplet state at $E = -E_B$. If the singlet state produces a virtual state of energy $E = E^*$, then Eq. (9.9) becomes

$$\sigma \approx \frac{\pi\hbar^2}{m}\left(\frac{3}{E_B} + \frac{1}{E^*}\right) \tag{9.10}$$

Taking a value of $E^* \sim 70$ keV, we find from Eq. (9.10) a value of 20.4 barns, thus bringing a relatively straightforward calculation into agreement with experiment.

In summary, experimental measurements have given the following scattering lengths for the two types of $n-p$ interactions, triplet and singlet configurations, and their corresponding potential range and well depth [Preston, 1962].

Interaction	**Scattering length *a* [F]**	**r_o[F]**	**V_o[MeV}**
n-p (*triplet*)	5.4	2	36
n-p (*singlet*)	−23.7	~2.5	18

Notice the scattering length for the triplet state is *positive*, while that for the singlet state is *negative*. This is an illustration of the significance of the sign of the scattering length discussed through Fig. 7.5.

As a final remark, we note that experiments have shown the total angular momentum (nuclear spin) of the deuteron ground state is $I = 1$, where $\underline{I} = \underline{L} + \underline{S}$, with $\underline{L}$ being the orbital angular momentum, and $\underline{S}$ the intrinsic spin, $\underline{S} = \underline{s}_n + \underline{s}_p$.

It is also known that the ground state is mostly 1s ($\ell = 0$), therefore for this state we have $S = 1$ (neutron and proton spins are parallel). We know from experiments the deuteron ground state is *barely bound* at $E_B = 2.23\,\text{MeV}$, so all the higher energy states are not bound. The 1s state with $S = 0$ (neutron and proton spins antiparallel), is therefore a virtual state (see Fig. 5.19); it is unbound by ~60 keV. An important implication here is that nuclear interaction varies with S; in other words, *nuclear forces are spin-dependent*.

Effects of pauli exclusion principle

One might ask if nuclides such as di-neutron and di-proton are stable. The answer lies in the Pauli *Exclusion Principle* which states that no two fermions can occupy the same quantum state. Consider the two electrons in a helium atom. Their wave function may be written as

$$\psi(1,2) = \psi_1(\underline{r}_1)\psi(\underline{r}_2) = A\frac{\sin k_1 r_1}{r_1}\frac{\sin k_2 r_2}{r_2} \tag{9.11}$$

where $\psi_1(\underline{r})$ is the wave functrion of electron 1 at $\underline{r}$. But since we cannot distinguish between electrons 1 and 2, we must get the same probability of finding these electrons if we exchange their positions (or exchange the particles),

$$|\psi(1,2)|^2 = |\psi(2,1)|^2 \Rightarrow \psi(1,2) = \pm\psi(2,1)$$

For fermions (electrons, neutrons, protons) we must choose the (−) sign. Because they obey *Fermi-Dirac statistics* their wave functions must be anti-symmetric under exchange. Thus we should modify Eq. (9.11) and write

$$\psi(1,2) = \psi_1(\underline{r}_1)\psi_2(\underline{r}_2) \mp \psi_2(\underline{r}_1)\psi_1(\underline{r}_2) \equiv \psi_\mp \tag{9.12}$$

If we now include the spin, then an acceptable anti-symmetric wave function is

$$\Psi(1,2) = \psi_-\chi_1(\uparrow)\chi_2(\uparrow) \tag{9.13}$$

so that under an interchange of particles, $1 \leftrightarrow 2$, $\psi(1,2) = -\psi(2,1)$. This corresponds to $S = 1$, symmetric state in spin space. But another acceptable anti-symmetric wave function is

$$\Psi_-(1,2) = \psi_+[\chi_1(\uparrow)\chi_2(\downarrow) - \chi_1(\downarrow)\chi_2(\uparrow)] \tag{9.14}$$

which corresponds to $S = 0$, anti-symmetric state in spin.

For the symmetry of the wave function in configurational space we recall that we have

$$\psi(\underline{r}) \sim \frac{u_\ell(r)}{r} P_\ell^m(\cos\theta) e^{im\varphi} \tag{9.15}$$

which is even (odd) if ℓ is even (odd). Thus, since Ψ has to be anti-symmetric, one can have two possibilities,

$$\ell \textit{ even}, S = 0 \quad (\textit{space symmetric}, \ \textit{spin anti-symmetric})$$

$$\ell \textit{ odd}, S = 1 \quad (\textit{space anti-symmetric}, \ \textit{spin symmetric})$$

These are called $T = 1$ states (T is *isobaric spin*), available to the n-p, n-n, p-p systems. By contrast, states which are symmetric ($T = 0$) are

$$\ell \text{ even}, \quad S = 1$$

$$\ell \text{ odd}, \quad S = 0$$

These are available only to the n-p system for which there is no Pauli Exclusion Principle. The ground state of the deuteron is therefore a $T = 0$ state. The lowest $T = 1$ state is $\ell = 0$, $S = 0$. As mentioned above, this is known to be unbound ($E \sim 60$ KeV). We should therefore expect that the lowest $T = 1$ state in n-n and p-p to be also unbound, i.e., there should be no stable di-neutron or di-proton.

We conclude this section by summarizing several *important features of nuclear forces*, a number of which are relevant to the studies undertaken in this book [Meyerhof, 1967]. There is a dominant short-range part of the nuclear interaction which is central and which combines with a spin-orbit coupling contribution to give the overall shell-model potential (see Eq. (5.45)). There is also a part whose range is much smaller than the nuclear radius, which tends to make the nucleus spherical and to pair up nucleons (see the pairing term in Eq. (4.10)). There is also a part whose range is of the order of the nuclear radius, which tends to distort the nucleus (this is a more sophisticated aspect which we do not consider in this book). There is a spin-spin interaction (an example is the neutron-proton scattering problem just discussed). The force is charge independent if Coulomb interaction is excluded (which we have assumed in Chap. 4). Lastly nuclear force saturates which also has been implicitly assumed in the empirical mass formula, Eq. (4.10).

9.2 NEUTRON ELASTIC SCATTERING

Since a neutron has no charge it can easily penetrate into a nucleus and cause a reaction. Neutrons interact primarily with the nucleus of an atom, except in the special case of magnetic scattering where the interaction involves the neutron spin and the magnetic moment of the atom. Because magnetic scattering is beyond the scope of this book, we will neglect the interaction between neutrons and electrons and think of atoms and nuclei as practically interchangeable. Neutron reactions can take place at any energy, so one has to pay particular attention to the energy variation of the interaction cross section. In a nuclear reactor neutrons can have energies ranging from 10^{-3} eV (1 meV) to 10^{7} eV (10 MeV). This means our study of neutron interactions, in principle, will have to cover an energy range of 10 orders of magnitude. In practice we will limit ourselves to two energy ranges, the slowing down region (eV to keV) and the thermal region (around 0.025 eV).

For a given energy region — thermal, epithermal, resonance, fast — not all the possible reactions are equally important. Which reaction is important depends on the target nucleus and the neutron energy. Generally speaking the important types of interactions, in the order of increasing complexity from the standpoint of basic understanding, can be classified as follows.

(n, n) *elastic scattering*. There are two processes, *potential* scattering which is neutron interaction at the surface of the nucleus (no penetration) as in billiard-ball collisions, and *resonance* scattering which involves the formation and decay of a *compound nucleus*.

(n, γ) *radiative capture*.

(n, n') *inelastic scattering*. This reaction involves the excitation of nuclear levels.

$(n, p), (n, \alpha), \ldots$ emission of charged particle of various types.

(n, f) neutron-induced fission.

If we were interested in fission reactors, the reactions in the order of importance would be fission, capture (in fuel and other reactor materials), scattering (elastic and inelastic), and fission product decay by β-emission. In this chapter, we will study primarily elastic (or potential) scattering. The other reactions involve compound nucleus formation, a process we will discuss in Chap. 12.

Elastic vs. inelastic scattering

Elastic scattering is the simplest process in neutron interactions; it can be analyzed in complete detail. This is also an important process because it is the primary mechanism by which neutrons lose energy in a reactor, from the instant they are emitted as fast neutrons in a fission event to when they appear as thermal neutrons. We can start again with the two-body reaction schematic depicted in Fig. 8.1. For elastic scattering there is no excitation of the nucleus, $Q = 0$. Whatever energy is lost by the neutron is gained by the recoiling target nucleus. In applying the conservation laws we follow the general analysis of Sec. 8.1 by letting $M_1 = M_3 = (M_n)m$, and $M_2 = M_4 = M = Am$. Then Eq. (8.4) becomes

$$E_3\left(1+\frac{1}{A}\right) - E_1\left(1-\frac{1}{A}\right) - \frac{2}{A}\sqrt{E_1 E_3}\cos\theta = 0 \qquad (9.16)$$

Suppose we ask under what condition is $E_3 = E_1$? We see this can occur only when $\theta = 0$, corresponding to forward scattering (no interaction). For all finite θ, E_3 has to be less than E_1, which is reasonable because some energy has to be given to the energy of recoil, E_4. One can readily show that the maximum energy loss by the neutron occurs at $\theta = \pi$, which corresponds to backward scattering,

$$E_3 = \alpha E_1, \quad \alpha = \left(\frac{A-1}{A+1}\right)^2 \qquad (9.17)$$

Equation (9.16) is the starting point for the analysis of neutron moderation (slowing down) in a scattering medium. We will return to it later in this section.

Inelastic scattering is the process by which the incoming neutron excites the target nucleus by causing the nucleus to leave the ground state and go to an excited state with energy E^* above the ground state. Thus $Q = -E^*$ $(E^* > 0)$. We again let the neutron mass be m and the target nucleus mass be M (ground state) or M^* (excited state), with $M^* = M + E^*/c^2$. Since this is a reaction with negative Q, it is an endothermic process requiring energy to be supplied before the reaction can take place. In the case of scattering the only way energy can be supplied is through the kinetic energy of the incoming particle (neutron). To find the minimum energy required for the reaction, we look at the situation where no energy is given to the outgoing particle, $E_3 \sim 0$ and $\theta \sim 0$. Then Eq. (9.16) gives

$$-E^* = -E_{th}\left(\frac{M_4 - M_1}{M_4}\right), \quad \text{or} \quad E_{th} \sim E^*(1 + 1/A) \qquad (9.18)$$

where we have denoted the minimum value of E_1 as E_{th}. Thus we see the minimum kinetic energy required for reaction is always greater than the excitation energy of the nucleus. Where does the difference between E_{th} and E^* go? The answer is it goes into the center-of-mass energy, the fraction of the kinetic energy of the incoming neutron (in the laboratory coordinate) that is not available for reaction (recall the discussions in Sec. 8.1).

Relations between outgoing energy and scattering angle

From the *Q-equation* for elastic scattering, Eq. (9.16), we can obtain a relation between the energy of the outgoing neutron, E_3, and the angle of scattering, θ.

Regarding Eq. (9.16) as a quadratic equation for the variable $\sqrt{E_3}$ as in Chap. 8, we have

$$E_3 - \frac{2}{A+1}\sqrt{E_1 E_3}\cos\theta - \frac{A-1}{A+1}E_1 = 0 \tag{9.19}$$

with solution in the form,

$$\sqrt{E_3} = \frac{1}{A+1}\sqrt{E_1}(\cos\theta + [A^2 - \sin^2\theta]^{1/2}) \tag{9.20}$$

This is a perfectly good relation between E_3 and θ (with E_1 fixed), although it is not a simple one. Nonetheless, it shows a one-to-one correspondence between these two variables. This is what we meant when we said in Sec. 8.1 that elastic scattering is a problem with only one degree of freedom. Whenever we are given either E_3 or θ we can immediately determine the other variable. The reason Eq. (9.20) is not regarded as a simple relation is we can obtain another relation between energy and scattering angle, except in this case the scattering angle is the angle in the center-of-mass coordinate system (CMCS), whereas θ in Eq. (9.20) is the scattering angle in the laboratory coordinate system (LCS). To find the simpler relation we first review the connection between the two coordinate systems.

Relation between LCS and CMCS

Suppose we start with the velocities of the incoming neutron and target nucleus, and those of the outgoing neutron and recoiling nucleus as shown in the Fig. 9.2.

In this diagram, we denote the LCS and CMCS velocities by lower and upper cases respectively, so $\underline{V}_i = \underline{v}_i - \underline{v}_o$, where $\underline{v}_o = [1/(A+1)]\underline{v}_1$ is the velocity of the center-of-mass. Notice the scattering angle in CMCS is labeled θ_c. We see in LCS the center-of-mass moves in the direction of the incoming neutron (with the target nucleus at rest), whereas in CMCS the target nucleus moves toward the center-of-mass which is stationary by definition. One can show (as a homework exercise) in CMCS the post-collision velocities have the same magnitude as the pre-collision velocities, the only effect of the collision is a rotation from $\underline{V}_1$ to $\underline{V}_3$, and $\underline{V}_2$ to $\underline{V}_4$.

Part (c) of Fig. 9.2 is particularly useful for deriving the relations between the LCS and CMCS velocities and angles. Perhaps the most important relation is that between the speed of the outgoing particle v_3 and the scattering angle in CMCS, θ_c.

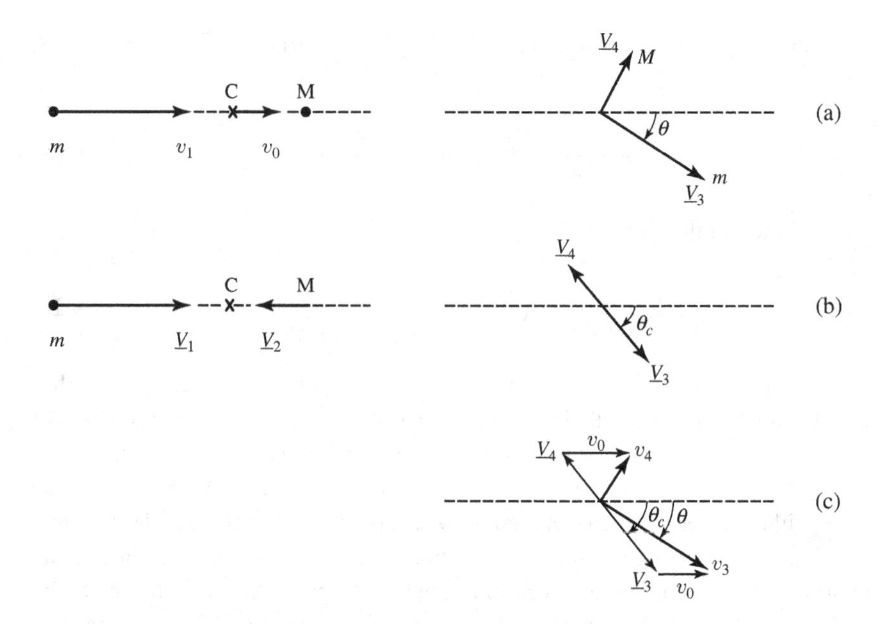

Fig. 9.2 Elastic scattering in LCS (a) and CMCS (b), and the geometric relation between LCS and CMCS post-collision velocity vectors (c). Incident and target particle masses are m and M respectively, and C denotes the center-of-mass.

We can write

$$\begin{aligned}\frac{1}{2}mv_3^2 &= \frac{1}{2}m(\underline{V}_3 + \underline{v}_o)^2 \\ &= \frac{1}{2}m(V_3^2 + v_o^2 + 2V_3 v_o \cos\theta_c) \qquad (9.21)\end{aligned}$$

or

$$E_3 = \frac{1}{2}E_1[(1+\alpha) + (1-\alpha)\cos\theta_c] \qquad (9.22)$$

where $\alpha = [(A-1)/(A+1)]^2$. Compared to Eq. (9.20), Eq. (9.22) is clearly simpler to manipulate. These two relations must be equivalent since no approximations have been made in either derivation. Taking the square of Eq. (9.20) gives

$$E_3 = \frac{1}{(A+1)^2}E_1\left(\cos^2\theta + A^2 - \sin^2\theta + 2\cos\theta\left[A^2 - \sin^2\theta\right]^{1/2}\right) \qquad (9.23)$$

To demonstrate the equivalence of Eqs. (9.22) and (9.23) one needs a relation between the two scattering angles, θ and θ_c. This can be obtained from Fig. 9.2(c) by writing

$$\cos\theta = (v_o + V_3\cos\theta_c)/v_3$$
$$= \frac{1 + A\cos\theta_c}{\sqrt{A^2 + 1 + 2A\cos\theta_c}} \qquad (9.24)$$

The relations Eqs. (9.22), (9.23), and (9.24) all demonstrate a one-to-one correspondence between energy and angle or angle and angle. They can be used to transform distributions from one variable to another, as we will demonstrate next in the discussion of energy and angular distributions of elastically scattered neutrons in the following section.

9.3 ENERGY AND ANGULAR DISTRIBUTIONS, THERMAL MOTIONS

The expressions relating energy and scattering angles just derived can be used to formulate the energy and angular distributions of an elastically scattered neutron. The energy distribution, in particular, is used further in the analysis of neutron energy moderation in nuclear systems, for example, in a nuclear reactor where neutrons are produced at high energies (MeV) which then slow down to thermal energies. In the problem of neutron slowing down the assumption of the target nucleus being initially at rest is well justified. When the neutron energy approaches the thermal region (~0.025 eV), the stationary target assumption is no longer valid. One can relax this assumption by deriving a more general distribution which holds for neutron elastic scattering at any energy. The latter result is appropriate for the analysis of the spectrum (energy distribution) of thermal neutrons, a problem known as *neutron thermalization*. Both neutron slowing and neutron thermalization are essential processes in the study of *neutron transport*, a topic to which Chap. 13 is devoted. It will be seen that transport refers to particle distributions in energy and configuration space. While we do not concern ourselves with spatial transport, here we have an opportunity to begin the discussion of energy transport. A step in this direction is to consider explicitly the energy dependence of the elastic scattering cross section.

From our study of cross section calculation in Chap. 7 we know that for low-energy scattering ($kr_o \ll 1$, which is equivalent to neutron energies below about 10–100 keV), only the *s*-wave contribution to the cross section is important. Moreover, the angular distribution of the scattered neutron is spherically symmetric in CMCS. We will next apply this result to derive the energy distribution of elastically scattered neutrons.

Energy distribution of elastically scattered neutrons

We define $P(\underline{\Omega}_c)$ as the probability the neutron moves in the direction $\underline{\Omega}_c$ (recall this is a unit vector in angular space in CMCS). The physical way of defining P is to let

$P(\underline{\Omega}_c)d\Omega_c =$ the probability that the neutron will be scattered into an element of solid angle $d\Omega_c$ about $\underline{\Omega}_c$

For s-wave scattering the probability is, as just stated, spherically symmetric,

$$P(\underline{\Omega}_c)d\Omega_c = \frac{d\Omega_c}{4\pi} \tag{9.25}$$

Notice $P(\underline{\Omega}_c)$ is a probability distribution in the two angular variables, φ_c and θ_c, and is properly normalized,

$$\int_0^{2\pi} d\varphi_c \int_0^{\pi} \cos\theta_c d\theta_c P(\underline{\Omega}_c) = 1 \tag{9.26}$$

Since there is a one-to-one relation between θ_c and E_3 (cf. Eq. (9.22)), Eq. (9.25) can be transformed to a probability distribution in the outgoing energy, E_3. First we need to reduce Eq. (9.25) from a distribution in two variables to a distribution in the variable θ_c.

Let us define $G(\theta_c)$ as the probability distribution of the scattering angle θ_c. This quantity can be obtained from Eq. (9.25) by simply integrating Eq. (9.25) over all values of the azimuthal angle φ_c,

$$G(\theta_c)d\theta_c = \int_0^{2\pi} d\varphi P(\underline{\Omega}_c) \sin\theta_c d\theta_c \tag{9.27}$$

$$= \frac{1}{2}\sin\theta_c d\theta_c \tag{9.28}$$

Now we write down the transformation from G(θ_c) to the probability distribution in the outgoing energy. For the present discussion of differential cross sections for neutron elastic scattering, our previous notation in labeling particles as 1 through 4 introduced in Chap. 8 is not compatible with the literature on neutron elastic scattering and neutron transport [Lamarsh, 1966; Bell, 1970]. For labeling the energy of the neutron before and after the collision it is conventional to use E and denote either the pre- or post-collision variable by a prime. (Unfortunately there is no convention on which variable should have a prime, the reader therefore should be aware of this ambiguity.) For our own convenience, we will adopt the notation at this point by letting $E_1 = E$ and $E_3 = E'$, and denote the probability distribution for E' as $F(E \to E')$. The transformation between $G(\theta_c)$ and $F(E \to E')$ is the same as that for any distribution function,

$$F(E \to E')dE' = G(\theta_c)d\theta_c \tag{9.29}$$

With $G(\theta_c)$ given by Eq. (9.28) we find

$$F(E \to E') = G(\theta_c)\left|\frac{d\theta_c}{dE'}\right| \tag{9.30}$$

The Jacobian of transformation can be readily evaluated from Eq. (9.22) after relabeling E_1 and E_3 as E and E'. Thus,

$$\begin{aligned} F(E \to E') &= \frac{1}{E(1-\alpha)} \quad \alpha E \le E' \le E \\ &= 0 \quad \text{otherwise} \end{aligned} \tag{9.31}$$

The distribution function, which is sketched in Fig. 9.3, is so simple that one can understand fully all its features. First we see the distribution is *uniform* (a constant) in the interval $(\alpha E, E)$ because the scattering is spherically symmetric. Independence of scattering angle translates into independence of outgoing energy because of the one-to-one correspondence. Secondly, the fact the outgoing energy can lie only in a particular interval follows from the range of scattering angle $(0, \pi)$, from forward to backward scattering. The parameter α depends on the mass of the target, it is zero for hydrogen and approaches unity as $M \gg m$. So the scattered neutron energy lies in the full range *(0, E)* for hydrogen, and for a heavy target, $A \gg 1$, the range shrinks into a narrow region, $(\alpha E, E)$, with α close to unity. Thus the neutron can lose all its energy in one collision with hydrogen, and loses practically no energy if it collides with a heavy target nucleus. Although simple, the distribution, Eq. (9.31), is widely used to analyze neutron energy moderation in

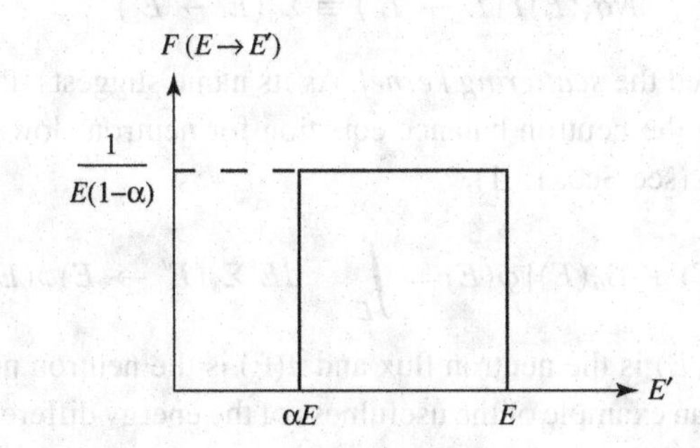

Fig. 9.3 Sketch of $F(E \to E')$, the *conditional* probability that a neutron elastically scattered at initial energy E will have an energy in dE' about E' after the scattering. $F(E \to E')$ is a distribution in E' but not in E.

the slowing down regime. It also serves as a reference behavior for modifications when it is no longer valid to assume the scattering is spherically symmetric in CMCS, or to assume the target nucleus is at rest. We will come back to these two situations later.

Notice that F is a *distribution function*, so its dimension is the reciprocal of its argument, the energy of the scattered neutron. The reader also should keep in mind F is a *conditional probability* specifying the distribution of the energy after scattering *given* that the scattering occurs at the initial energy. It is a distribution in E' but not a distribution in E. F also has been called the *scattering frequency*.

$F(E \to E')$ is properly normalized, meaning its integral over the range of the outgoing energy is necessarily unity as required by particle conservation. This is to be expected since $G(\theta_c)$ is properly normalized. Knowing the probability distribution F one can construct the energy differential cross section

$$\frac{d\sigma_s}{dE'} = \sigma_s(E)F(E \to E') \tag{9.32}$$

such that

$$\int dE' \frac{d\sigma_s}{dE'} = \sigma_s(E) \tag{9.33}$$

which is the "total" (in the sense that it is the integral of a differential) scattering cross section. It is important to keep in mind $\sigma_s(E)$ is a function of the initial (incoming) neutron energy, whereas the integration in Eq. (9.33) is over the final (outgoing) neutron energy. The quantity $F(E \to E')$ is a distribution in the variable E' and a point function of E. We can multiply Eq. (9.31) by the number density of the target nuclei N to obtain

$$N\sigma_s(E)F(E \to E') \equiv \Sigma_s(E \to E') \tag{9.34}$$

which is often called the *scattering kernel*. As its name suggests, this is the quantity that appears in the neutron balance equation for neutron slowing down in an absorbing medium (see Sec. 13.1),

$$[\Sigma_s(E) + \Sigma_a(E)]\,\phi(E) = \int_E^{E/\alpha} dE'\Sigma_s(E' \to E)\phi(E') \tag{9.35}$$

where $\phi(E) = vn(E)$ is the neutron flux and n(E) is the neutron number density. Equation (9.35) is an example of the usefulness of the energy differential scattering cross section Eq. (9.34).

The scattering distribution $F(E \to E')$ can be used to calculate various energy-averaged quantities pertaining to elastic scattering. For example, the average loss for a collision at energy E is

$$\int_{\alpha E}^{E} dE'(E - E')F(E \to E') = \frac{E}{2}(1 - \alpha) \tag{9.36}$$

For hydrogen the energy loss in a collision is one-half its energy before the collision, whereas for a heavy nucleus it is $\sim 2E/A$.

Angular distribution of elastically scattered neutrons

We have already made use of the fact that for s-wave scattering the angular distribution is spherically symmetric in CMCS. This means the angular differential scattering cross section in CMCS is of the form

$$\frac{d\sigma_s}{d\Omega_c} = \sigma_s(\theta_c) \equiv \sigma_s(E)\frac{1}{4\pi} \tag{9.37}$$

One can ask what is the angular differential scattering cross section in LCS? The answer can be obtained by transforming the result Eq. (9.37) from a distribution in the unit vector $\underline{\Omega}_c$ to a distribution in $\underline{\Omega}$. As before (cf. Eq. (9.29)) we write

$$\sigma_s(\theta)d\Omega = \sigma_s(\theta_c)d\Omega_c \tag{9.38}$$

or

$$\sigma_s(\theta) = \sigma_s(\theta_c)\frac{\sin\theta_c}{\sin\theta}\frac{d\theta_c}{d\theta} \tag{9.39}$$

From the relation between $\cos\theta$ and $\cos\theta_c$, Eq. (9.24), we can find

$$\frac{d(\cos\theta)}{d(\cos\theta_c)} = \frac{\sin\theta_c d\theta_c}{\sin\theta d\theta}$$

Thus

$$\sigma_s(\theta) = \frac{\sigma_s(E)}{4\pi}\frac{(\gamma^2 + 2\gamma\cos\theta_c + 1)^{3/2}}{1 + \gamma\cos\theta_c} \tag{9.40}$$

with $\gamma = 1/A$. Since Eq. (9.40) is a function of θ, the factors $\cos\theta_c$ on the right hand side should be expressed in terms of $\cos\theta$ in accordance with Eq. (9.24). The angular distribution in LCS, as given by Eq. (9.40), is somewhat too complicated to sketch simply. From the relation between LCS and CMCS indicated in Fig. 9.2, we can expect that if the distribution is isotropic in CMCS, then the distribution in LCS should be peaked in the forward direction (simply because the scattering angle in LCS is always less than the angle in CMCS). One way to demonstrate this is indeed the case is to calculate the average value of $\mu = \cos\theta$,

$$\bar{\mu} = \frac{\int d\Omega \cos\theta \sigma_s(\theta)}{\int d\Omega \sigma_s(\theta)} = \frac{\int_{-1}^{1} d\mu \mu \sigma_s(\mu)}{\int_{-1}^{1} d\mu \sigma_s(\mu)} = \frac{\int_{-1}^{1} d\mu_c \mu(\mu_c) \sigma_s(\mu_c)}{\int_{-1}^{1} d\mu_c \sigma_s(\mu_c)} = \frac{2}{3A} \tag{9.41}$$

The fact that $\bar{\mu} > 0$ means that the angular distribution is always peaked in the forward direction. However, the bias becomes less pronounced the heavier the target mass; for $A \gg 1$ the distinction between LCS and CMCS vanishes.

Assumptions in deriving $F(E \to E')$

In arriving at the scattering distribution, Eq. (9.31), we have made use of three assumptions, namely,

(i) elastic scattering
(ii) target nucleus at rest
(iii) scattering is isotropic in CMCS (s-wave)

These assumptions imply certain restrictions pertaining to the energy of incoming neutron E and the temperature of the scattering medium. Assumption (i) is valid provided the neutron energy is not high enough to excite the nuclear levels of the compound nucleus formed by the target nucleus plus the incoming neutron. If the neutron energy is sufficient to excite the energy levels of the bound states above the ground state, then inelastic scattering becomes energetically possible. Inelastic scattering is a threshold reaction ($Q < 0$), it can occur in heavy nuclei at $E \sim 0.05 - 0.1$ MeV, or in medium nuclei at $\sim$0.1 – 0.2 MeV. Typically the cross section for inelastic scattering, $\sigma(n, n')$, is of the order of 1 barn or less. In comparison with elastic scattering, which is always present no matter what other reactions can take place and is of order 5–10 barns except in the case of hydrogen where it is 20 barns as we have seen in Sec. 9.1.

Assumption (ii) is valid when the neutron energy is large compared to the kinetic energy of the target nucleus, typically taken to be $k_B T$ assuming the medium is in thermal equilibrium at temperature T. This would be the case for neutron energies $\sim$0.1 eV and above. When the incident neutron energy is comparable to the energy of the target nucleus, the assumption of stationary target is clearly invalid. To take into account the thermal motions of the target, one should know what is the state of the target since the nuclear (atomic) motions in solids are different from those in liquids, such as vibrations in the former versus diffusion in the latter. If we assume the scattering medium can be treated as a gas at temperature T, then the target nucleus moves in a straight line with a speed that is given by the

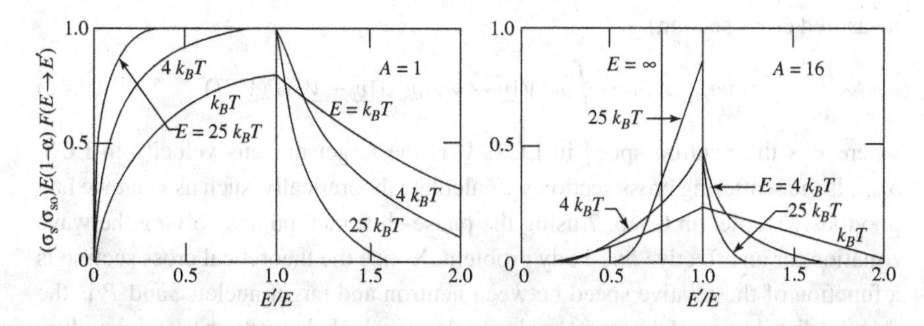

Fig. 9.4 Energy distributions of neutrons elastically scattered at energy E in a gas of nuclei with mass $A = M/m$ at temperature T. Left and right panels are for hydrogen and oxygen respectively. [Adapted from Bell, 1970.]

Maxwellian distribution. In this case one can derive the scattering distribution as an extension of Eq. (9.37) [Bell, 1970]. We do not go into the details here except to show the qualitative behavior of the more general result in Fig. 9.4. From the way the scattering distribution changes with incoming energy E one can get a good intuitive feeling for how the more general $F(E \to E')$ evolves from a relatively spread-out distribution (the curves for $E = k_B T$) to the more restricted form (sharp cutoffs) seen in Fig. 9.3.

Notice for $E \sim k_B T$ there can be appreciable *up-scattering* which is not possible when assumption (ii) is valid. As E becomes larger compared to $k_B T$, up-scattering becomes less important. The condition of stationary nucleus also means that $E \gg k_B T$.

When thermal motions have to be taken into account, the scattering cross section $\sigma_s(E)$ is also changed; it is no longer a constant at $4\pi a^2$, where a is the scattering length. This occurs in the energy region of neutron thermalization; it covers the range (0, 0.1–0.5 eV). We will now discuss the energy dependence of $\sigma_s(E)$. For the case of the scattering medium being a gas of atoms with mass A and at temperature T, it is still relatively straightforward to work out the expression for $\sigma_s(E)$. We will give the essential steps to give the reader a feel for the kind of analysis that one can carry out even for more complicated situations such as neutron elastic scattering in solids and liquids [Parks 1970].

Energy dependence of scattering cross section $\sigma_s(E)$

When the target nucleus is not at rest, one can write down the expression for the elastic scattering cross section measured in the laboratory (we will call it the measured cross section),

$$v\sigma_{meas}(v) = \int d^3V|\underline{v} - \underline{V}|\sigma_{theo}(|\underline{v} - \underline{V}|)P(\underline{V}, T) \tag{9.42}$$

where v is the neutron speed in LCS, $\underline{V}$ is the target nucleus velocity in LCS, σ_{theo} is the scattering cross section we calculate theoretically, such as what we had previously studied in Chap. 7 using the phase-shift method and solving the wave equation for an effective one-body problem. Notice the theoretical cross section is a function of the relative speed between neutron and target nucleus, and P is the thermal distribution of the target nucleus velocity which depends on the temperature of the medium. Equation (9.42) is a general relation between what is calculated theoretically, in solving the effective one-body problem, and what is measured in the laboratory where one necessarily has only an average over all possible target nucleus velocities. By the scattering cross section $\sigma_s(E)$ we mean σ_{meas}. It turns out that we can reduce Eq. (9.42) further by using for P the Maxwellian distribution to obtain the result

$$\sigma_s(v) = \frac{\sigma_{so}}{\beta^2}\left[\left(\beta^2 + \frac{1}{2}\right) erf(\beta) + \frac{1}{\sqrt{\pi}}\beta e^{-\beta^2}\right] \tag{9.43}$$

where $erf(x)$ is the *error function* integral

$$erf(x) = \frac{2}{\sqrt{\pi}}\int_0^x dte^{-t^2} \tag{9.44}$$

and $\beta^2 = AE/k_BT$, and $E = mv^2/2$. Given that the error function has the limiting behavior for small and large arguments,

$$\begin{aligned} erf(x) &\sim \tfrac{2}{\sqrt{\pi}}\left(x - \tfrac{x^3}{3} + \cdots\right) \quad x \ll 1 \\ &\quad 1 - \tfrac{e^{-x^2}}{x\sqrt{\pi}}\left(1 - \tfrac{1}{2x^2} + \cdots\right) \quad x \gg 1 \end{aligned} \tag{9.45}$$

we have the two limiting behavior,

$$\sigma_s(v) \sim \sigma_{so}/v \quad \beta \ll 1 \tag{9.46}$$

$$\sigma_s(v) \sim \quad \sigma_{so} \quad \beta \gg 1 \tag{9.47}$$

Thus the elastic scattering cross section has a $1/v$ variation at low energy (or high temperature) and it approaches a velocity-independent behavior at high energy. The expression Eq. (9.43) is a useful expression describing the energy variation of the scattering cross section over the entire energy range, from thermal to MeV, so far as elastic scattering is concerned. It has been obtained by assuming the target nuclei move as in a gas of non-interacting atoms. This assumption is not realistic

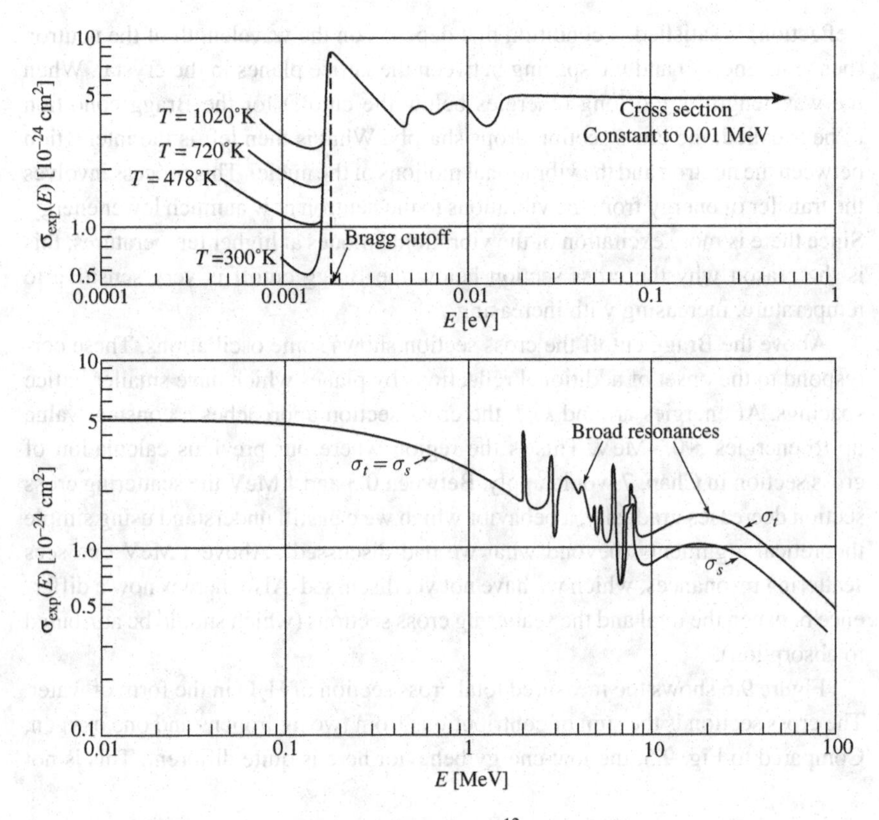

Fig. 9.5 Total and elastic scattering cross sections of C^{12} in the form of graphite. [Adapted from Lamarsh, 1966.]

when the scattering medium is a solid or a liquid. For these situations one can also work out the expressions for the cross section, but the results are more complicated (and beyond the scope of this book). We will therefore settle for a brief, qualitative look at what new features can be seen in the energy dependence of the elastic scattering cross sections of typical solids (crystals) and liquids.

Figure 9.5 shows the total and elastic scattering cross sections of graphite (C^{12}) over the entire energy range of interest to us. At the very low-energy end we see a number of features we have not discussed previously. These all have to do with the fact that the target nucleus (atom) is bound to a crystal lattice and therefore the positions of the nuclei are fixed to well-defined lattice sites and the atomic motions are small-amplitude vibrations about these sites. There is a sharp drop of the cross section below an energy marked Bragg cutoff. Cutoff here refers to Bragg reflection which occurs when the condition for constructive interference (reflection) is satisfied, a condition that depends on the wavelength of the neutron (hence its energy) and the spacing between the lattice planes in the crystal. When the wavelength is too long (energies below the cutoff) for the Bragg condition to be satisfied, the cross section drops sharply. What is then left is the interaction between the neutron and the vibrational motions of the nuclei. This process involves the transfer of energy from the vibrations to the neutron now at much lower energy. Since there is more excitation of the vibrational modes at higher temperatures, this is the reason why the cross section below the Bragg cutoff is very sensitive to temperature, increasing with increasing T.

Above the Bragg cutoff the cross section shows some oscillations. These correspond to the onset of additional reflections by planes which have smaller lattice spacings. At energies around $k_B T$ the cross section approaches a constant value up to energies $\sim$0.3 MeV. This is the region where our previous calculation of cross section in Chap. 7 would apply. Between 0.3 and 1 MeV the scattering cross section decreases gradually, a behavior which we can still understand using simple theoretical arguments (beyond what we had discussed). Above 1 MeV one sees scattering resonances, which we have not yet discussed. Also there is now a difference between the total and the scattering cross sections (which should be attributed to absorption).

Figure 9.6 shows the measured total cross section of H_2O in the form of water. The cross section is the sum of contributions from two hydrogens and one oxygen. Compared to Fig. 9.5, the low-energy behavior here is quite different. This is not

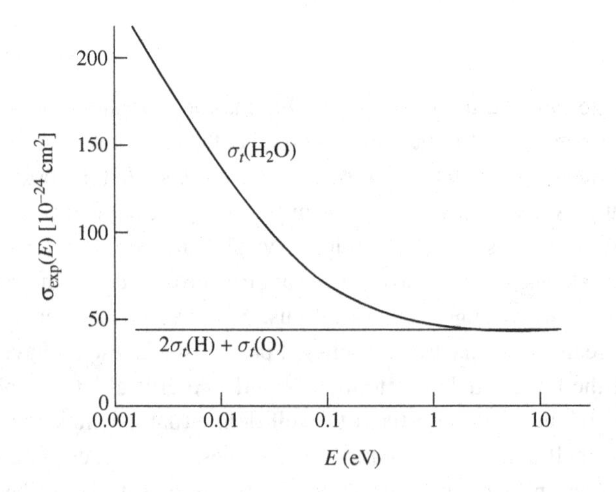

Fig. 9.6 Total cross section of water. [Adapted from Lamarsh, 1966.]

unexpected since a crystal and a liquid are really very different with regard to their atomic structure and atomic motions. In the case of the liquid the cross section rises from a constant value at energies above 1 eV in a manner like the $1/v$ behavior given by Eq. (9.46). Notice the constant value of about 45 barns is just what we know from the hydrogen cross section σ_{so} of 20.4 barns per hydrogen and a cross section of about 5 barns for oxygen.

The importance of hydrogen (water) in neutron scattering has led to another interpretation of the rise of the cross section with decreasing neutron energy, one which focuses on the effect of chemical (molecular) binding. The idea is that at high energies (relative to thermal) the neutron does not see the water molecule. Instead it sees only the individual nuclei as targets which are free-standing and essentially at rest. In this energy range (1 eV and above) the interaction is the same as that between a neutron and free protons and oxygen nuclei. This is why the cross section is just the sum of the individual contributions. When the cross section starts to rise as the energy decreases, this is an indication that the chemical binding of the protons and oxygen in a water molecule starts to have an effect. When the neutron energy is around $k_B T$ the neutron now sees the entire water molecule rather than the individual nuclei. In that case the scattering is effectively between a neutron and a water molecule. What this means is that as the neutron energy decreases the target changes from individual nuclei with their nuclear masses to a water molecule with mass 18. Now one can show the scattering cross section is actually proportional to the square of the reduced mass of the scatterer μ,

$$\sigma_s \propto \mu^2 = \left(\frac{mM}{m+M}\right)^2 = \left(\frac{A}{A+1}\right)^2 \tag{9.48}$$

One can moreover define a *free-atom* cross section appropriate for the energy range where the cross section is a constant, and a *bound-atom* cross section for the energy range where the cross section is rising, with the relation

$$\sigma_{free} = \sigma_{bound}\left(\frac{A}{A+1}\right)^2 \tag{9.49}$$

For hydrogen these two cross sections would have the values of 20 barns and 80 barns respectively.

We close this chapter with a brief consideration of assumption (iii) used in deriving Eq. (9.31). When the neutron energy is in the 10 keV range and higher, the contributions from the higher angular momentum (*p-wave* and above) scattering may become significant. In that case we know the angular distribution will be more forward peaked. This means one should replace Eq. (9.25) by a different form of $P(\underline{\Omega}_c)$ (see the discussions surrounding Eq. (7.47)). Without going through any

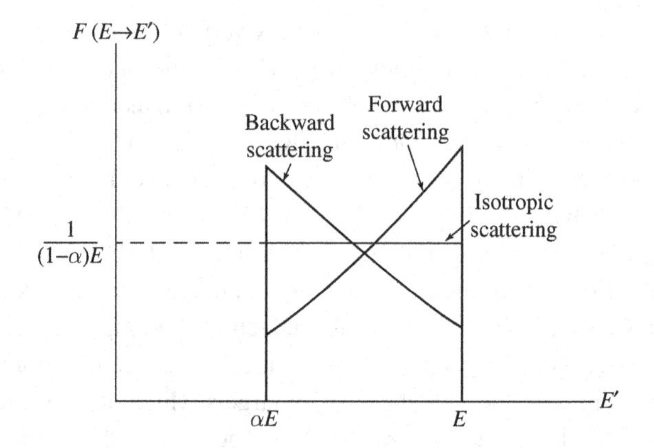

Fig. 9.7 Schematic energy distribution of neutrons elastically scattered by a stationary nucleus showing the trend corresponding to forward or backward scattering in CMCS. The special case of isotropic scattering, shown previously in Fig. 9.3, results in a uniform distribution of the neutron energy E' in the range $(\alpha E, E)$. [Adapted from Lamarsh, 1966.]

more details, we show in Fig. 9.7 the general behavior that one can expect in the scattering distribution $F(E \to E')$ when scattering in CMCS is no longer isotropic.

REFERENCES

J. Byrne, *Neutrons, Nuclei and Matter* (Institute of Physics, 1994).
G. I. Bell and S. Glasstone, *Nuclear Reactor Theory* (Van Nostrand Reinhold, 1970).
R. D. Evans, *Atomic Nucleus* (McGraw-Hill New York, 1955), Chap. 12.
E. Fermi, *Nuclear Physics*, Lecture Notes by J. Orear, A. H. Rosenfeld and R. A. Schluter (University of Chicago Press, 1949).
A. Foderaro, *The Elements of Neutron Interaction Theory* (MIT Press, Cambridge, 1971).
J. R. Lamarsh, *Nuclear Reactor Theory* (Addison-Wesley, Reading, 1966), Chap. 2.
W. E. Meyerhof, *Elements of Nuclear Physics* (McGraw-Hill, New York, 1967), Sec. 3.3.
D. E. Parks, J. R. Beyster, M. S. Nelkin, N. F. Wikner, *Slow Neutron Scattering and Themalization* (W. A. Benjamin, New York, 1970).
M. A. Preston, *Physics of the Nucleus* (Addison-Wesley, Reading, 1962).
E. Segrè, *Nuclei and Particles* (W. A. Benjamin, New York, 1965), Chap. X.
M. M. R. Williams, *Slowing Down and Thermalization of Neutrons* (North-Holland, Amsterdam, 1966).

10

Gamma Scattering and Absorption

We are interested in the interactions of gamma rays, which are electromagnetic radiations produced primarily by nuclear transitions, in an atomic medium. These are typically photons with energies in the range of $\sim 0.1 - 10$ MeV. The attenuation of the intensity of a beam of gamma rays in an absorber follows an exponential variation with the distance of penetration unlike the stopping of charged particles (Chap. 11). Gammas and charged particles are the other two types of radiation interactions we study in this book, along with neutron interactions. It is well to keep in mind that each of these three types of radiation interacts with matter in a different way. A neutron sees the nuclei in the atoms that make up the medium through nuclear forces, which are strong and very short-ranged. For all practical purposes the neutron does not see the electrons, although the topic of neutron-electron interaction is not without fundamental interest in nuclear physics. A gamma, on the other hand, sees mostly the atomic electrons. A charged particle basically interacts through Coulomb forces, therefore it sees both the atomic electrons and the nuclei. This distinction gives rise to very different interaction mechanisms in the broad context of nuclear radiation interactions. as the reader can readily appreciate by comparing the treatments undertaken in the set of three consecutive chapters, Chaps. 9, 10, and 11.

Consider the attenuation of gamma radiation as a result of its interaction with an absorber as shown in Fig. 10.1. It is well established that the intensity, flux of particles, typically in unit of number of photons per cm^2 – sec, varies exponentially with the absorber thickness,

$$I(x) = I_o e^{-\mu x} \tag{10.1}$$

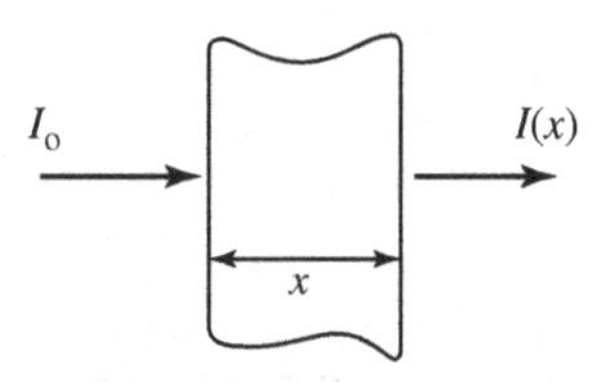

Fig. 10.1 Incident and emerging beams of gamma radiation with intensity I_0 and $I(x)$ respectively after passing through an absorber of thickness x.

where μ is the *linear attenuation coefficient*, a quantity which does not depend on x but will depend on the energy of the incident gamma. By attenuation we mean either scattering or absorption. Since either process will remove the gamma from the beam, the probability of penetrating a distance x is the same as the probability of traveling a distance x without any interaction, $\exp(-\mu x)$. The attenuation coefficient is therefore the *probability per unit path* of interaction; it is essentially what we have called the macroscopic cross section $\Sigma = N\sigma$ in the case of neutron interaction.

There are several different processes of gamma interaction. If each process can be treated as occurring independently of the others, then μ is the sum of the individual contributions. These contributions, of course, are not equally important at any given energy of the gamma radiation. Each process has its own energy variation as well as dependence on the atomic number of the absorber. We will focus on the three most important processes of gamma interactions, *Compton scattering*, *photoelectric effect*, and *pair production*. These can be classified by the target particle with which the photon interacts and the type of process (absorption or scattering), as shown in the matrix below. Photoelectric effect is the absorption of a photon followed by the ejection of an atomic electron. Compton scattering is the inelastic (photon loses energy) relativistic scattering by a free electron. For this process it is implied that the photon energy is at least comparable to the rest mass energy of the electron. If the photon energy is much lower than the rest mass energy, the scattering by a free electron then becomes elastic (no energy loss), a process that we will call *Thomson scattering*. When the photon energy is greater than twice the rest mass energy of electron, the photon can be absorbed and an electron-positron pair is emitted. This is the process of pair production. Other combinations of interaction and process in the matrix (indicated by the entry x) could be discussed, but they are of less interest to us in this book.

Interaction with \	*absorption*	*elastic scattering*	inelastic cattering
atomic electron	*photoelectric effect*	*Thomson*	*Compton*
nucleus	x	x	x
electric field around the nucleus	*pair production*	x	x

For the three processes we have identified, the matrix shows the photon interacts with the atomic electron in *photoelectric effect* and *Compton scattering*, while in *pair production* the photon interacts with the electric field of the nucleus. This is useful to keep in mind as we work our way through the kinematics and the attenuation coefficient in each case. Given what we have just said, the attenuation coefficient becomes

$$\mu = \mu_C + \mu_\tau + \mu_\kappa \tag{10.2}$$

where the subscripts C, τ, and κ will henceforth denote Compton scattering, photoelectric effect, and pair production respectively.

10.1 COMPTON SCATTERING

The treatment of *Compton scattering* is similar to our analysis of neutron scattering in several ways. This analogy should be noted by the reader as the discussion unfolds. The phenomenon is the scattering of a photon with incoming momentum $\hbar\underline{k}$ by a *free, stationary electron*, which is treated *relativistically*. After scattering at anangle θ, the photon has momentum $\hbar\underline{k}'$, while the electron moves off at an angle φ with momentum $\underline{p}$ and kinetic energy T, as shown in Fig. 10.2. (The notation of using a primed variable to denote post-collision energy and momentum is consistent with our practice in Chap. 9 regarding neutron scattering.)

To analyze the kinematics we write the momentum and energy conservation equations,

$$\hbar\underline{k} = \hbar\underline{k}' + \underline{p} \tag{10.3}$$

$$\hbar ck = \hbar ck' + T \tag{10.4}$$

where the relativistic energy-momentum relation for the electron is (see Eqs. (3.3))

$$cp = \sqrt{T\left(T + 2m_e c^2\right)} \tag{10.5}$$

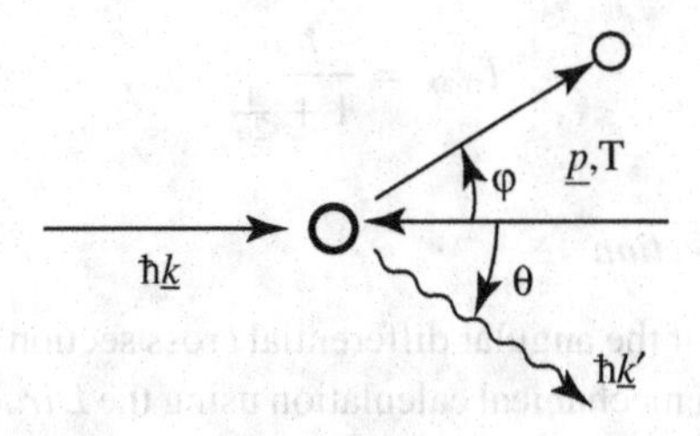

Fig. 10.2 Schematic of *Compton scattering* of a photon with momentum $\hbar\underline{k}$ at an angle θ resulting in momentum $\underline{p}$ and energy T being acquired by the free electron emitted at angle ϕ. The scattered photon has momentum $\hbar k'$.

with c being the speed of light. One should also recall the relations, $\omega = ck$ and $\lambda\nu = c$, with ω and ν being the circular and linear frequencies, respectively ($\omega = 2\pi\nu$), and λ the wavelength of the photon. By algebraic manipulations one can obtain the following results.

$$\lambda' - \lambda = \frac{c}{\nu'} - \frac{c}{\nu} = \frac{h}{m_e c}(1 - \cos\theta) \tag{10.6}$$

$$\frac{\omega'}{\omega} = \frac{1}{1 + \alpha(1 - \cos\theta)} \tag{10.7}$$

$$T = \hbar\omega - \hbar\omega' = \hbar\omega\frac{\alpha(1 - \cos\theta)}{1 + \alpha(1 - \cos\theta)} \tag{10.8}$$

$$\cot\phi = (1 + \alpha)\tan\frac{\theta}{2} \tag{10.9}$$

In Eq. (10.6), the factor $h/m_e c = 2.426 \times 10^{-10}$ cm is called the *Compton wavelength*. The gain in wavelength after scattering at an angle of θ is known as the *Compton shift*. This shift in wavelength is independent of the incoming photon energy, whereas the shift in energy, Eq. (10.8), is dependent on energy. In Eq. (10.7), the parameter $\alpha = \hbar\omega/m_e c^2$ is a measure of the photon energy in units of the electron rest mass energy (0.511 Mev). As $\alpha \to 1$, we see $\omega' \to \omega$ as the process goes from *inelastic* to *elastic*. Low-energy photons are scattered with only a moderate energy change, while high-energy photons can suffer large energy losses. For example, at $\theta = \pi/2$, if $\hbar\omega = 10$ keV, then $\hbar\omega' = 9.8$ keV (a 2% change), but if $\hbar\omega = 10$ MeV, then $\hbar\omega' = 0.49$ MeV (a 20-*fold* change).

Equation (10.8) gives the energy of the recoiling electron, the *Compton electron*, which is of interest because it is often the quantity that is measured in Compton scattering. In the limit of energetic gammas, $\alpha \gg 1$, the scattered gamma energy becomes only a function of the scattering angle. The energy is a *minimum* for backward scattering ($\theta = \pi$), $\hbar\omega' = m_e c^2/2$, while for 90° scattering $\hbar\omega' = m_e c^2$. The maximum energy transfer is given by Eq. (10.8) with $\theta = \pi$,

$$T_{\text{max}} = \frac{\hbar\omega}{1 + \frac{1}{2\alpha}} \tag{10.10}$$

Klein-Nishina cross section

The proper derivation of the angular differential cross section for Compton scattering requires a quantum mechanical calculation using the *Dirac's relativistic theory* of the electron. This work was first published in 1928 by Klein and Nishina [Heitler, 1955]. We will be content here to just quote the formula and discuss some of its

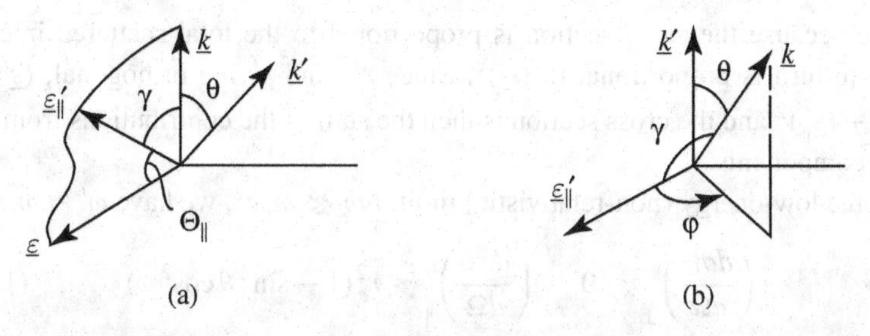

Fig. 10.3 Angular relations among incoming and outgoing wave vectors, $\underline{k}$ and $\underline{k}'$, of the scattered photon, and the electric vectors, $\underline{\varepsilon}$ and $\underline{\varepsilon}'$, which are transverse to the corresponding wave vectors. Subscript $\|$ denotes the electric vector of the scattered photon lying in the plane of the incoming wave vector $\underline{k}$ and its corresponding electric vector ε. [Adapted from Evans, 1955.]

implications. The cross section is

$$\frac{d\sigma_C}{d\Omega} = \frac{r_e^2}{4}\left(\frac{\omega'}{\omega}\right)^2\left[\frac{\omega}{\omega'} + \frac{\omega'}{\omega} - 2 + 4\cos^2\Theta\right] \tag{10.11}$$

where Θ is the angle between the electric vector $\underline{\varepsilon}$ (polarization) of the incident photon and that of the scattered photon, $\underline{\varepsilon}'$. The diagrams shown in Fig. 10.3 are helpful in visualizing the various vectors involved. Recall a photon is an electromagnetic wave characterized by a wave vector $\underline{k}$ and an *electric vector* $\underline{\varepsilon}$ which is perpendicular to $\underline{k}$. For a given incident photon with $(\underline{k}, \underline{\varepsilon})$ we can decompose the scattered photon electric vector $\underline{\varepsilon}'$ into a component $\underline{\varepsilon}'_\perp$ perpendicular to the plane containing $\underline{k}$ and $\underline{\varepsilon}$, and a parallel component $\varepsilon_\|$ which lies in this plane. For the perpendicular component, $\cos\Theta_\perp = 0$, and for the parallel component we notice

$$\cos\gamma = \cos\left(\frac{\pi}{2} - \Theta\right) = \sin\Theta = \sin\theta\cos\varphi \tag{10.12}$$

Therefore,

$$\cos^2\Theta = 1 - \sin^2\theta\cos^2\varphi \tag{10.13}$$

The decomposition of the scattered photon electric vector means that the angular differential cross section can be written as a sum of two contributions,

$$\begin{aligned}\frac{d\sigma_C}{d\Omega} &= \left(\frac{d\sigma_C}{d\Omega}\right)_\perp + \left(\frac{d\sigma_C}{d\Omega}\right)_\| \\ &= \frac{r_e^2}{2}\left(\frac{\omega'}{\omega}\right)^2\left[\frac{\omega}{\omega'} + \frac{\omega'}{\omega} - 2\sin^2\theta\cos^2\varphi\right]\end{aligned} \tag{10.14}$$

This is because the cross section is proportional to the total scattered intensity which in turn is proportional to $(\underline{\varepsilon}')^2$. Since $\underline{\varepsilon}'_{\perp}$ and $\underline{\varepsilon}'_{\parallel}$ are orthogonal, $(\underline{\varepsilon}')^2 = (\underline{\varepsilon}'_{\perp})^2 + (\underline{\varepsilon}'_{\parallel})^2$ and the cross section is then the sum of the contributions from each of the components.

In the low-energy (non-relativistic) limit, $\hbar\omega \ll m_e c^2$, we have $\omega' \approx \omega$, then

$$\left(\frac{d\sigma_C}{d\Omega}\right)_{\perp} \sim 0, \quad \left(\frac{d\sigma_C}{d\Omega}\right)_{\parallel} \sim r_e^2(1 - \sin^2\theta\cos^2\varphi) \tag{10.15}$$

Equation (10.15) means that if the incident radiation is polarized (photons have a specific polarization vector), then the scattered radiation is also polarized. But if the incident radiation is unplolarized, then we have to average $d\sigma_C/d\Omega$ over all the allowed directions of $\underline{\varepsilon}$ (remembering $\underline{\varepsilon}$ is perpendicular to $\underline{k}$). Since $d\sigma_C/d\Omega$ depends only on angles θ and φ, the result for *unpolarized* radiation is given by averaging over φ. Thus,

$$\left(\frac{d\sigma_C}{d\Omega}\right)_{unpol} = \frac{1}{2\pi}\int_0^{2\pi} d\varphi \left(\frac{d\sigma_C}{d\Omega}\right)_{\parallel}$$
$$= \frac{r_e^2}{2}(1 + \cos^2\theta) \tag{10.16}$$

where $r_e \equiv e^2/m_e c^2 = 2.818 \times 10^{-13}$ cm is called the *classical radius of the electron*. This is a well-known expression for the angular differential cross section for *Thomson scattering*. Integrating this over all solid angles gives

$$\sigma_C = \int d\Omega \left(\frac{d\sigma_C}{d\Omega}\right)_{unpol} = \frac{8\pi}{3} r_e^2 \equiv \sigma^o \tag{10.17}$$

which is the *Thomson cross section*.

Returning to the general result Eq. (10.14) we have in the case of un-polarized radiation,

$$\frac{d\sigma_C}{d\Omega} = \frac{r_e^2}{2}\left(\frac{\omega'}{\omega}\right)^2 \left(\frac{\omega}{\omega'} + \frac{\omega'}{\omega} - \sin^2\theta\right) \tag{10.18}$$

We can rewrite this result in terms of α and cosθ by using Eq. (10.7),

$$\frac{d\sigma_C}{d\Omega} = \frac{r_e^2}{2}(1 + \cos^2\theta)\left(\frac{1}{1 + \alpha(1 - \cos\theta)}\right)^2$$
$$\times \left[1 + \frac{\alpha^2(1 - \cos\theta)^2}{(1 + \cos^2\theta)\left[1 + \alpha(1 - \cos\theta)\right]}\right] \tag{10.19}$$

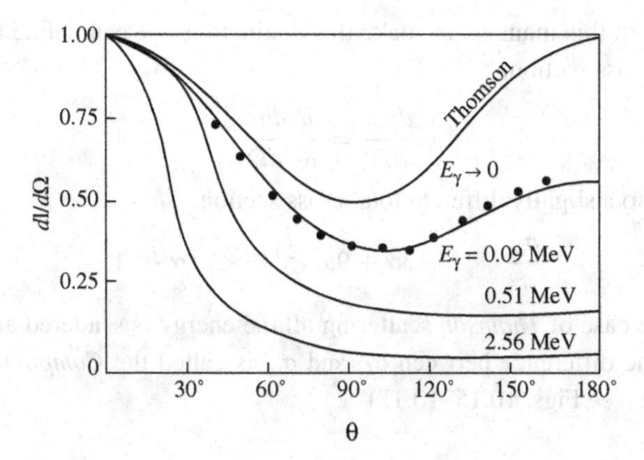

Fig. 10.4 Angular distributions of Compton scattering at various incident energies E_γ. All curves are normalized at 0°. Note the low-energy limit of *Thomson* scattering. [Adapted from Heitler, 1955.]

The behavior of $d\sigma_C/d\Omega$ is shown in Fig. 10.4. Notice at any given α (recall Eq. (10.7)) the angular distribution is peaked in the forward direction. As α increases, the forward peaking becomes more pronounced. The deviation from *Thomson* scattering is largest at large scattering angles; even at $\hbar\omega \sim 0.1\,\text{MeV}$ the assumption of *Thomson* scattering is not valid. In practice the *Klein-Nishina* cross section has been found to be in excellent agreement with experiments at least out to $\hbar\omega = 10m_ec^2 \sim 5\,\text{MeV}$.

To find the total cross section per electron for Compton scattering, one can integrate Eq. (10.19) over solid angles. The analytical result can be found in Evans [Evans, 1955, p. 684]. We note the two limiting cases,

$$\begin{aligned} \frac{\sigma_C}{\sigma^o} &= 1 - 2\alpha + \frac{26}{5}\alpha^2 - \cdots \qquad \alpha \ll 1 \\ &= \frac{3}{8}\frac{m_ec^2}{\hbar\omega}\left[\ell n\frac{2\hbar\omega}{m_ec^2} + \frac{1}{2}\right] \qquad \alpha \gg 1 \end{aligned} \tag{10.20}$$

We see that at high energies ($\geq 1\,\text{MeV}$) the Compton cross section decreases with energy like $1/\hbar\omega$.

Collision, scattering and absorption cross sections

In discussing the *Compton effect* a distinction should be made between *collision* and *scattering*. Here collision refers to ordinary scattering in the sense of removal of the photon from the beam. This is what we have been discussing above. Since the electron recoils, not all the original energy $\hbar\omega$ is scattered, only a fraction ω'/ω

is dissipated in this manner. To make this distinction one can define a *scattering* differential cross section,

$$\frac{d\sigma_{sc}}{d\Omega} = \frac{\omega'}{\omega}\frac{d\sigma_C}{d\Omega} \tag{10.21}$$

which leads to a slightly different total cross section,

$$\frac{\sigma_{sc}}{\sigma^o} \sim 1 - 3\alpha + 9.4\alpha^2 - \cdots \quad \alpha \ll 1 \tag{10.22}$$

Notice in the case of *Thomson* scattering all the energy is scattered and none are absorbed. The difference between σ_C and σ_{sc} is called the *Compton absorption cross section* (see Figs. 10.15–10.17).

Energy distribution of Compton electrons and photons

We have been discussing the angular distribution of the Compton scattered photons in terms of the differential cross section $\frac{d\sigma_C}{d\Omega}$. To transform the angular distribution to an energy distribution we need first to reduce the angular distribution of two angle variables, θ and φ, to a distribution in θ, in the same way as we had treated neutron scattering in Chap. 9. We therefore define

$$\frac{d\sigma_C}{d\theta} = \int_0^{2\pi} d\varphi \frac{d\sigma_C}{d\Omega}\sin\theta = \frac{d\sigma_C}{d\Omega}2\pi\sin\theta \tag{10.23}$$

and write

$$\frac{d\sigma_C}{d\omega'} = \frac{d\sigma_C}{d\theta}\left|\frac{d\theta}{d\omega'}\right| \tag{10.24}$$

with ω' and θ being related through Eq. (10.7). Since we can also relate the scattering angle θ to the angle of electron recoil ϕ through Eq. (10.9), we obtain the distribution of electron energy by performing two transformations, from θ to ϕ first, then from ϕ to T by using the relation

$$T = \hbar\omega\frac{2\alpha\cos^2\phi}{(1+\alpha)^2 - \alpha^2\cos^2\phi} \tag{10.24}$$

as found by combining Eqs. (10.8) and (10.9). Thus,

$$\frac{d\sigma_C}{d\Omega_e}2\pi\sin\phi d\phi = \frac{d\sigma_C}{d\Omega}2\pi\sin\theta d\theta \tag{10.25}$$

$$\frac{d\sigma_C}{d\phi} = \frac{d\sigma_C}{d\Omega_e}2\pi\sin\phi \tag{10.26}$$

$$\frac{d\sigma_C}{dT} = \frac{d\sigma_C}{d\phi}\left|\frac{d\phi}{dT}\right| \tag{10.27}$$

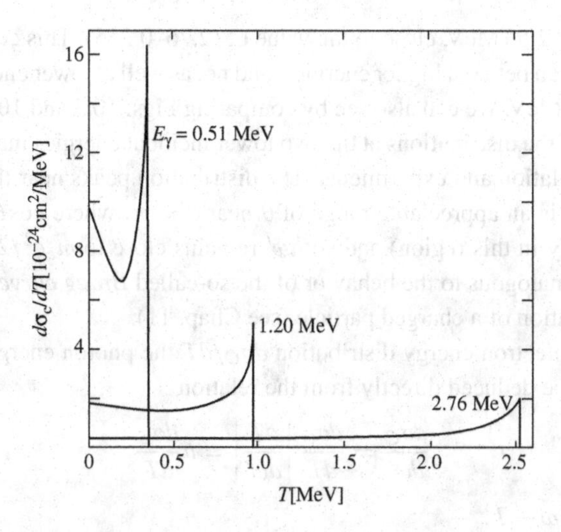

Fig. 10.5 Energy distributions of *Compton electrons* for several incident gamma-ray energies. [Adapted from Meyerhof, 1967.]

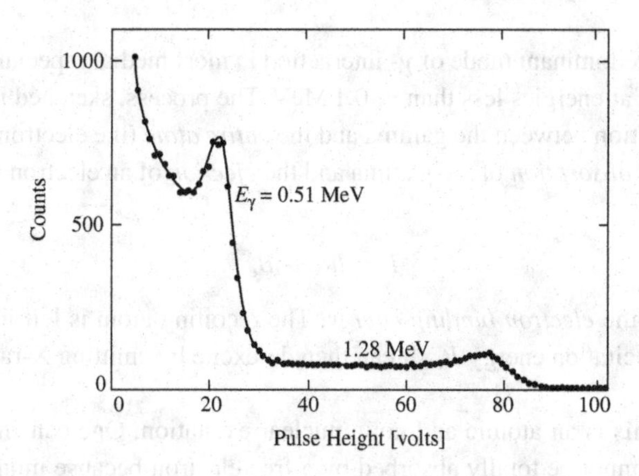

Fig. 10.6 Pulse-height spectra of *Compton electrons* produced by 0.51- and 1.28-MeV gamma rays. [Adapted from Meyerhof, 1967.]

These results show all the distributions are related to one another. In Fig. 10.5, we see several calculated electron recoil energy distributions which can be compared with the experimental data given in Fig. 10.6.

For a given incident gamma energy the recoil energy is maximum at $\theta = \pi$, where $\hbar\omega'$ is smallest. Previously we had found if the photon energy is high enough, the outgoing photon energy is a constant at ~ 0.255 MeV. In Fig. 10.5, we see for incident photon energy of 2.76 MeV the maximum electron recoil energy is approximately 2.53 MeV, close to the value of (2.76–0.255). This correspondence should hold even better at higher energies, and not as well at lower energies, such as 1.20 and 0.51 MeV. We can also see by comparing Figs. 10.5 and 10.6 the relative magnitudes of the distributions at the two lower incident energies match quite well between calculation and experiments. The distribution peaks near the *cutoff* T_{max} because there is an appreciable range of θ near $\theta = \pi$, where $\cos\theta \sim 1$ (cosine changes slowly in this region) and so $\hbar\omega'$ remains close to $m_e c^2/2$. This feature of pile-up is analogous to the behavior of the so-called *Bragg curve* depicting the specific ionization of a charged particle (see Chap. 11).

From the electron energy distribution $d\sigma_C/dT$ the photon energy distribution $d\sigma_C/d\omega'$ can be deduced directly from the relation

$$\frac{d\sigma_C}{d\omega'} = \frac{d\sigma_C}{dT}\left|\frac{dT}{d\omega'}\right| = \hbar\frac{d\sigma_C}{dT} \tag{10.28}$$

since $\hbar\omega' = \hbar\omega - T$.

10.2 PHOTOELECTRIC EFFECT

This is the predominant mode of γ- interaction in most media, especially the high-Z absorbers, at energies less than ~ 0.1 MeV. The process, sketched in Fig. 10.7, is the interaction between the gamma and the *entire atom* (the electron cloud) that results in the *absorption* of the gamma and the *ejection* of an electron with kinetic energy

$$T \sim \hbar\omega - B_e \tag{10.29}$$

where B_e is the *electron binding energy*. The recoiling atom is left in an excited state at an excitation energy B_e. It can then de-excite by emitting X-rays or Auger electrons.

Notice this is an atomic and not a nuclear excitation. One can show that the incident γ cannot be totally absorbed by a free electron because momentum and energy conservations cannot be satisfied simultaneously, but total absorption can occur if the electron is initially bound in an atom. The most tightly bound electrons

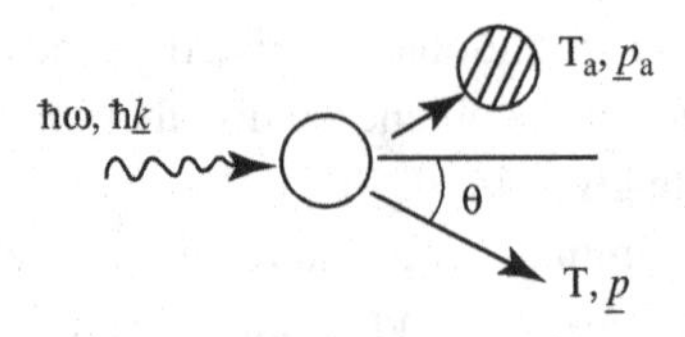

Fig. 10.7 Schematic of the *photoelectric effect*, absorption of an incident gamma ray causing the ejection of an atomic electron and the excitation of the recoiling atom.

have the greatest probability of absorbing the γ. It is known theoretically and experimentally that $\sim$ 80% of the absorption occurs in the K (innermost) shell, so long as $\hbar\omega > B_e(K - shell)$. Typical ionization potentials of K electrons are 2.3 keV (*Al*), 10 keV (Cu), and $\sim$ 100 keV (Pb).

The conservations equations for the photoelectric effect are

$$\hbar\underline{k} = \underline{p} + \underline{p}_a \tag{10.30}$$

$$\hbar\omega = T + T_a + B_e \tag{10.31}$$

where $\underline{p}_a$ is the momentum of the recoiling atom whose kinetic energy is T_a. Since $T_\alpha \sim T(m_e/M)$, where M is the mass of the atom, one can usually ignore T_a.

The theory describing the photoelectric effect is essentially a first-order perturbation theory calculation [Heitler, 1955, Sec. 21]. In this case the transition taking place is between an initial state consisting of two particles, a bound electron, wave function $\sim e^{-r/a}$, with $a = a_o/Z$, a_o being the *Bohr radius* $(= \hbar^2/m_e e^2 = 0.529 \times 10^{-8}$ cm), and an incident photon, wave function $e^{i\underline{k}\cdot\underline{r}}$, and a final state consisting of a free electron whose wave function is $e^{i\underline{p}\cdot\underline{r}/\hbar}$. The interaction potential is of the form $\underline{A} \cdot \underline{p}$, where A is the vector potential of the electromagnetic radiation and $\underline{p}$ is the electron momentum. The result of this calculation is

$$\frac{d\sigma_\tau}{d\Omega} = 4\sqrt{2}\frac{r_e^2 Z^5}{(137)^4}\left(\frac{m_e c^2}{\hbar\omega}\right)^{7/2}\frac{\sin^2\theta\cos^2\varphi}{\left(1 - \frac{v}{c}\cos\theta\right)^4} \tag{10.32}$$

where v is the photoelectron speed. The polar and azimuthal angles specifying the direction of the photoelectron are defined in Fig. 10.8. One should pay particular

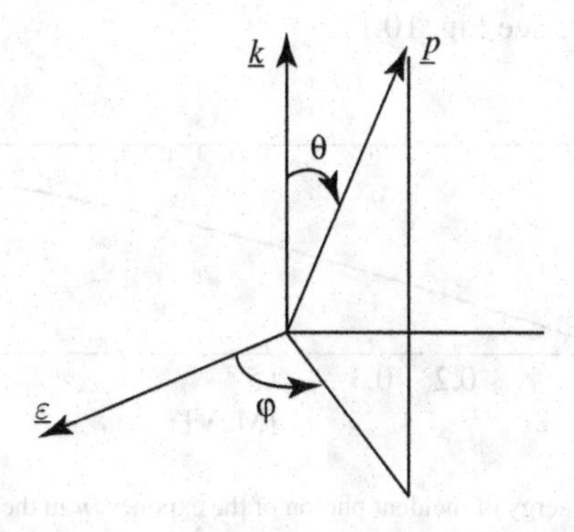

Fig. 10.8 Spherical coordinates defining the photoelectron direction of emission relative to the incoming photon wave vector and its polarization vector.

notice to the Z^5 and $(\hbar\omega)^{-7/2}$ variation. As for the angular dependence, the numerator in Eq. (10.32) suggests an origin in $(\underline{p} \cdot \underline{\varepsilon})^2$, while the denominator suggests $\underline{p} \cdot \underline{k}$. Integration of Eq. (18.4) gives the total cross section

$$\sigma_\tau = \int d\Omega \frac{d\sigma_\tau}{d\Omega} = 4\sqrt{2}\sigma^o \frac{Z^5}{(137)^4}\left(\frac{m_e c^2}{\hbar\omega}\right)^{7/2} \tag{10.33}$$

where one has ignored the angular dependence in the denominator and has multiplied the result by a factor of 2 to account for 2 electrons in the *K-shell* [Heitler, 1955, p. 207].

In practice, it has been found the charge and energy dependence behave more like

$$\sigma_\tau \propto Z^n/(\hbar\omega)^3 \tag{10.34}$$

with n varying from 4 to 4.6 as $\hbar\omega$ varies from ~ 0.1 to 3 MeV, and with the energy exponent decreasing from 3 to 1 when $\hbar\omega \geq m_e c^2$. The qualitative behavior of the photoelectric cross section is illustrated in Figs.10.9 and 10.10. The former shows a power law variation with Z that increases gradually with increasing photon energy, whereas the latter shows the Z dependence along with an approximate power law variation with energy.

Edge absorption

As the photon energy increases the cross section σ_τ can show discontinuous jumps known as *edges*. These correspond to the onset of additional contributions when the energy is sufficient to eject an inner shell electron. The effect is more pronounced in the high-Z material. See Fig. 10.17.

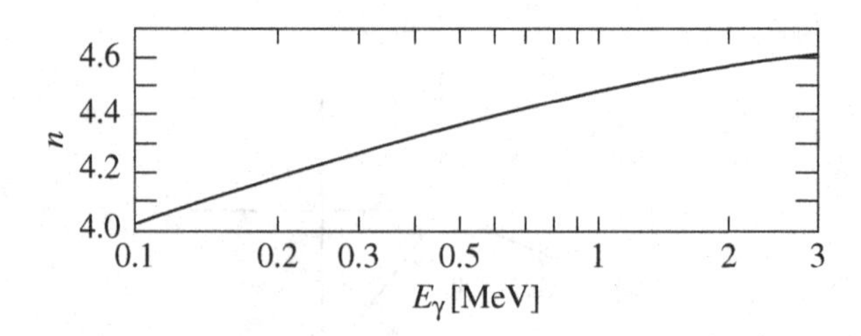

Fig. 10.9 Variation with energy of incident photon of the exponent n in the total cross section for the photoelectric effect. [Adapted from Evans, 1955].

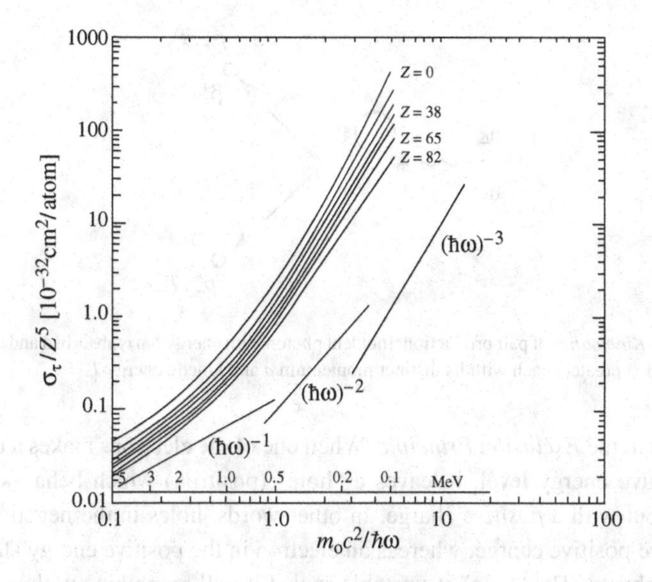

Fig. 10.10 Photoelectric cross sections showing approximate inverse power-law behavior which varies with the energy of the incident photon. [Adapted from Evans, 1955.]

10.3 PAIR PRODUCTION

Pair production is a process, which can occur only when the energy of the incident γ exceeds 1.02 MeV, twice the rest mass of an electron. The photon is absorbed in the vicinity of the nucleus, a positron-electron pair is produced, and the atom is left in an excited state, as indicated in Fig. 10.11. The conservation equations are therefore

$$\hbar \underline{k} = \underline{p}_+ + \underline{p}_- \tag{10.35}$$

$$\hbar\omega = (T_+ - m_e c^2) + (T_- + m_e c^2). \tag{10.36}$$

One finds these conditions cannot be satisfied simultaneously without the presence of an atomic nucleus. Since the nucleus takes up some momentum, it also takes up some energy in the form of recoil.

The existence of *positron* is a consequence of the *Dirac's relativistic theory* of the electron which allows for negative energy states,

$$E = \pm(c^2 p^2 + m_e^2 c^4)^{1/2}. \tag{10.37}$$

One assumes all the negative states are filled, representing a "sea" of electrons which are generally not observable because no transitions into these states can

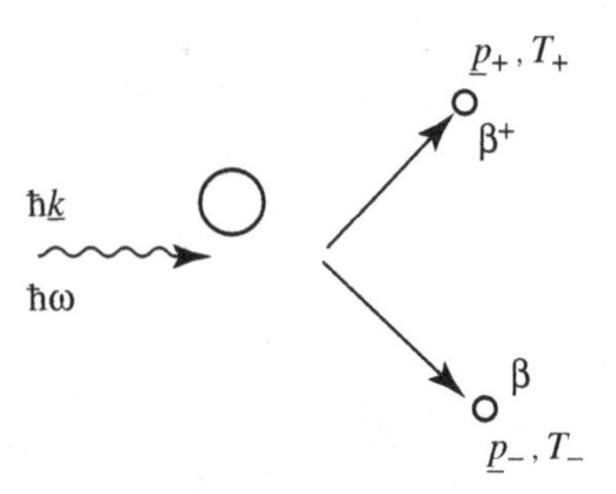

Fig. 10.11 *Kinematics* of pair production. Incident photon with energy $\hbar\omega$ is absorbed and an electron-positron pair is created, each with its distinct momentum $\underline{p}$ and kinetic energy T.

occur due to the *Exclusion Principle*. When one of the electrons makes a transition to a positive energy level, it leaves a "hole" (positron) which behaves likes an electron but with a *positive* charge. In other words, holes in the negative energy states have positive charge, whereas an electron in the positive energy states have negative charge. The "hole" is unstable in that it will recombine with an electron when it loses most of its kinetic energy (thermalized). This recombination process is called *pair annihilation*. It usually produces two γ (*annihilation radiation*), each of energy 0.511 Mev, emitted back-to-back. A positron and an electron can form an atom called the *positronium*. Its life time is $\sim 10^{-7}$ to 10^{-9} sec depending on the relative spin orientation, the shorter lifetime corresponding to antiparallel orientation.

Pair production is intimately related to the process of *Bremsstrahlung* in which an electron undergoes a transition from one positive energy state to another while a photon is emitted. One can take over directly the theory of *Bremsstrahlung* for the transition probability with the incident particle being a photon instead of an electron, and using the appropriate density of states for the emission of the positron-electron pair [Heitler, 1955, Sec. 26]. For the positron the energy differential cross section is

$$\frac{d\sigma_\kappa}{dT_+} = 4\sigma_o Z^2 \frac{T_+^2 + T_-^2 - \frac{2}{3}T_+T_-}{(\hbar\omega)^3}\left[\ell n\left(\frac{2T_+T_-}{\hbar\omega m_e c^2}\right) - \frac{1}{2}\right] \tag{10.38}$$

where $\sigma_o = r_e^2/137 = 5.8 \times 10^{-4}$ barns. This result holds under the condition of the Born approximation, which is a high-energy condition ($Ze^2/\hbar v_\pm \ll 1$), and no screening, which requires $2T_+T_-/\hbar\omega m_e c^2 \ll 137/Z^{1/3}$. Here screening means the partial reduction of the nuclear charge by the potential of the inner-shell electrons. As a result of the Born approximation, the cross section is symmetric in T_+ and T_-. Screening effects will lead to a lower cross section.

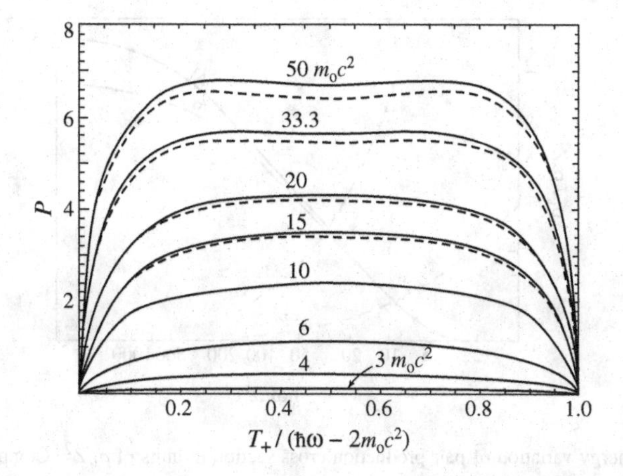

Fig. 10.12 Calculated energy distribution of positron emitted in pair production, as expressed in Eq. (10.39), with correction for screening (dashed curves) for photon energies above $10m_ec^2$ [Adapted from Evans, 1955].

To see the energy distribution we rewrite Eq. (10.38) as

$$\frac{d\sigma_\kappa}{dT_+} = \frac{\sigma_o Z^2}{\hbar\omega - 2m_ec^2} P \tag{10.39}$$

where the dimensionless factor P is a rather complicated function of $\hbar\omega$ and Z; its behavior is depicted in Fig. 10.12. We see the cross section increases with increasing energy of the incident photon. Since the value is slightly higher for Al than for Pb, it means the cross section varies with Z somewhat weaker than Z^2. At lower energies the cross section favors equal distribution of energy between the positron and the electron, while at high energies a slight tendency toward unequal distribution could be noted. Intuitively we expect the energy distribution to be biased toward more energy for the positron than for the electron simply because of Coulomb repulsion of the positron and attraction of the electron by the nucleus.

The total cross section for pair production is obtained by integrating Eq. (10.39). Analytical integration is possible only for extremely relativistic cases [Evans, 1955, p. 705],

$$\sigma_\kappa = \int dT_+ \left(\frac{d\sigma_\kappa}{dT_+}\right) = \sigma_o Z^2 \left[\frac{28}{9} \ell n \left(\frac{2\hbar\omega}{m_ec^2}\right) - \frac{218}{27}\right], \quad \textit{no screening} \tag{10.40}$$

$$= \sigma_o Z^2 \left[\frac{28}{9} \ell n \left(\frac{183}{Z^{1/3}}\right) - \frac{2}{27}\right], \quad \textit{complete screening} \tag{10.41}$$

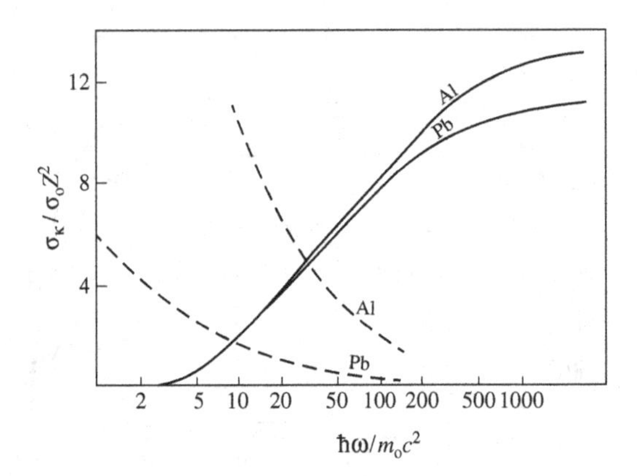

Fig. 10.13 Energy variation of pair production cross section in units of $\sigma_o Z^2$. Compton scattering results are shown as dashed line.

Moreover, Eqs. (10.40) and (10.41) are valid only for $m_e c^2 \ll \hbar\omega \ll 137\, m_e c^2 Z^{-1/2}$ and $\hbar\omega \gg 137 m_e c^2 Z^{-1/2}$ respectively. The behavior of this cross section is shown in Fig. 10.13, along with the cross section for Compton scattering. Notice screening leads to a saturation effect (energy independence); however, it is not significant below $\hbar\omega \sim 10\,\text{MeV}$

Mass attenuation coefficients

So long as the different processes of photon interaction are not correlated, the total linear attenuation coefficient μ for γ-interaction can be taken to be the sum of contributions from Compton scattering, photoelectric effect, and pair production, with $\mu = N\sigma$ and N being the *number of atoms per* cm^3. Since $N = N_o \rho / A$, where N_o is *Avogardro's number* (see Table 3.4) and ρ the mass density of the absorber, it is useful to express the interaction in terms of the *mass attenuation coefficient* μ/ρ. This quantity is known to be more or less independent of the density and physical state of the absorber. It will be seen in Chap. 11 for charged particle interactions Z/A is approximately constant for all elements for the same reasons. Here we see in the product $N\sigma$ we get a factor of $1/A$ from N, and we can get a factor of Z from any of the three cross sections σ, so we can also take advantage of Z/A being roughly constant in describing photon interaction (the same argument does not hold for neutrons even though it is useful to think of neutron interactions in terms of the macroscopic cross section Σ). We will therefore write

$$\mu = \mu_C + \mu_\tau + \mu_\kappa \tag{10.42}$$

$$\mu_C / \rho = (N_o / A)\, Z\sigma_C, \quad \sigma_C \sim 1/\hbar\omega \quad \textit{per electron} \tag{10.43}$$

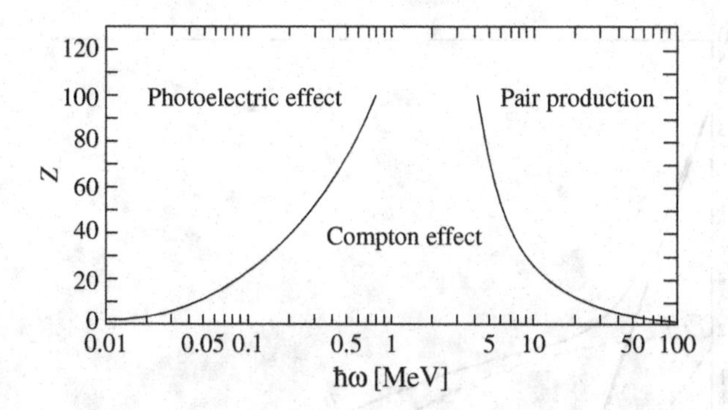

Fig. 10.14 Regions where one of the three γ-interactions is dominant over the other two. Solid curves are boundaris where the two competing effects are approximately equally important. [Adapted from Evans, 1955.]

$$\mu_\tau/\rho = (N_o/A)\,\sigma_\tau, \quad \sigma_\tau \sim Z^5/(\hbar\omega)^{7/2} \quad per\ atom \tag{10.44}$$

$$\mu_\kappa/\rho = (N_o/A)\,\sigma_\kappa, \quad \sigma_\kappa \sim Z^2 \ell n(2\hbar\omega/m_e c^2) \quad per\ atom \tag{10.45}$$

It should be clear by now the three processes we are discussing are not equally important for a given region of Z and $\hbar\omega$. Generally speaking, photoelectric effect is important at low energies and high Z, Compton scattering is important at intermediate energies ($\sim 1-5$ MeV) and all Z, and pair production becomes dominant at higher energies and high Z. This is illustrated in Fig. 10.14.

We also show several mass attenuation coefficients of specific absorber media which are commonly encountered, air, water, and lead (for shielding). These should be examined with reference to the qualitative region of dominance shown in Fig. 10.14, as well as relative to each other in Figs. 10.15–10.17. One should make note of the magnitude of the attenuation coefficients, their energy dependence, and the contribution associated with each process. As a summary statement concerning how theory compares with measurements one may say the agreement is good to about 3% for all elements at $\hbar\omega < 10 m_e c^2$. At higher energies disagreement sets in at high Z (can reach $\sim$ 10% for Pb), which is due to the use of Born approximation in calculating σ_κ. If one corrects for this, then agreement to within $\sim$ 1% is obtained out to energies $\sim 600\, m_e c^2$.

10.4 RADIATION DETECTION

We have just concluded the study of gamma radiation interactions with matter in which Compton scattering, photoelectric effect, and pair production were discussed

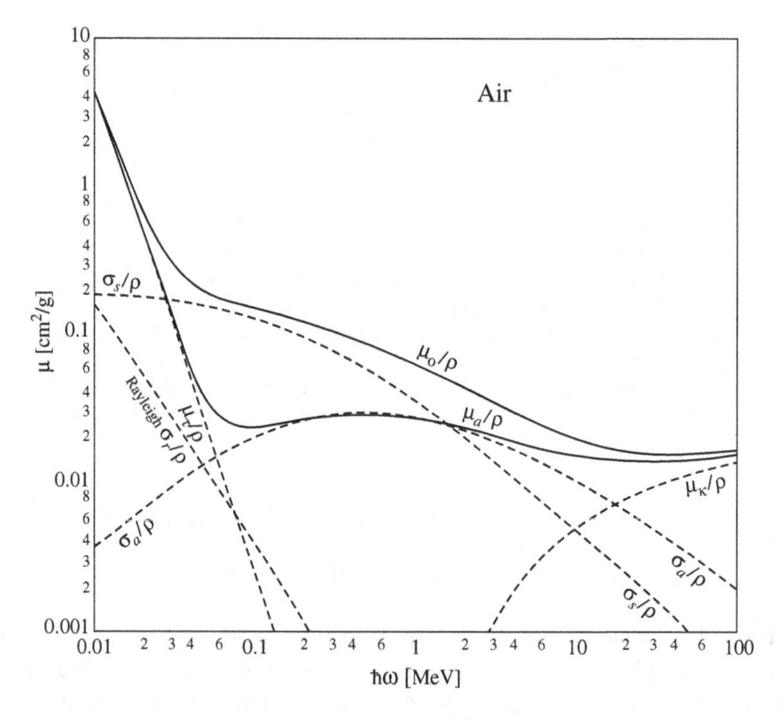

Fig. 10.15 Mass attenuation coefficient for photons in *air* computed from tables of atomic cross sections. [Adapted from Evans, 1955.]

separately. A topic which makes integrated use of this information is the general problem of detection of nuclear radiation. Although this subject properly belongs to a discussion of experimental aspects of nuclear physics, it is nevertheless appropriate to make contact with it at this point in our considerations of gamma interactions. One can think of two reasons for this. First, radiation detection is a central part of the foundational knowledge for all students in the department of Nuclear Science and Engineering [Knoll, 1979]. Even though we cannot do justice to it due to the limited scope of this book, it is certainly worthwhile to bring it into the discussion when an opportunity presents itself. Secondly, an analysis of the features observed experimentally in pulse-height spectra of gamma radiation is a practical application of what we have just learned about the γ-interaction processes of Compton scattering, photoelectric effect and pair production.

We start by noting that regardless of the type of nuclear radiation, the interactions taking place in a material medium invariably result in ionizations and excitations which then can be detected. Heavy charged particles and electrons produce ion pairs in ionization chambers, or light emission (excitation of atoms) in scintillation

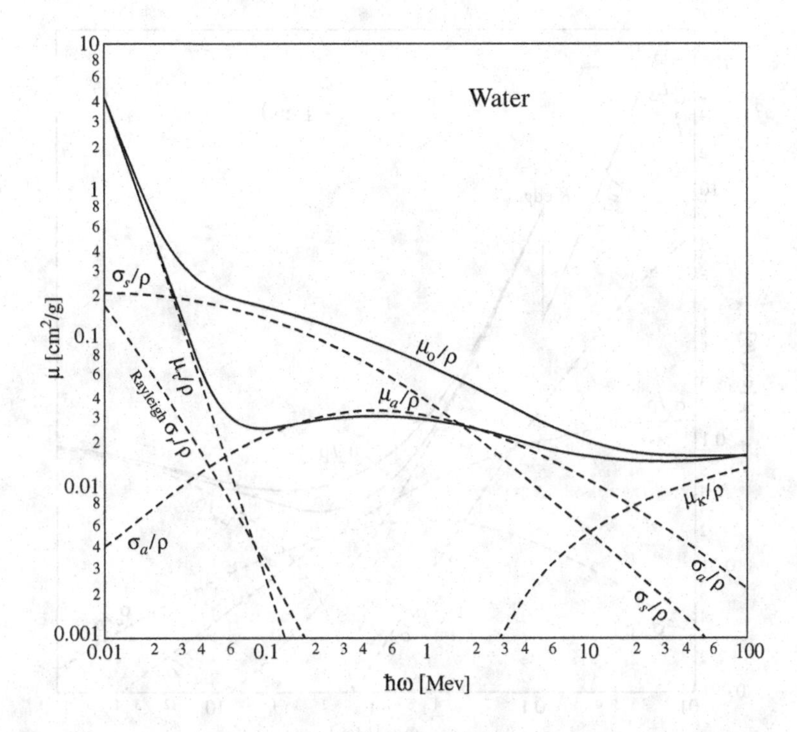

Fig. 10.16 Mass attenuation coefficients for photons in *water*. [Adapted from Evans, 1955.]

counters, or electron-hole pairs in semiconductor detectors. Neutrons collide with protons which recoil and produce ionization or excitation. In the case of gammas, all three processes we have just discussed give rise to energetic electrons which in turn cause ionization or excitation. Thus the basic mechanisms of nuclear radiation detection involve measuring the ionization or excitation occurring in the detector in a way to allow one to deduce the energy of the incoming radiation. A useful summary of the different types of detectors and methods of detection is can be found in Meyerhof [Meyerhof, 1967, p. 107]. Let us now focus on the detection of γ radiation. Suppose we are concerned with the problem of detecting two γ rays, at energies 1.37 MeV and 2.75 MeV, emitted from the radioactive nuclide Na^{24}. Measurements have been made in the form of *pulse-height spectrum*, number of counts per channel in a multichannel analyzer plotted against the pulse height, as shown in Fig. 10.18. These results were obtained using a Na-I scintillation detector, which is relatively large in size but not very good in energy resolution. The spectra are seen to have two sets of features, one for each incident γ. By a set, we

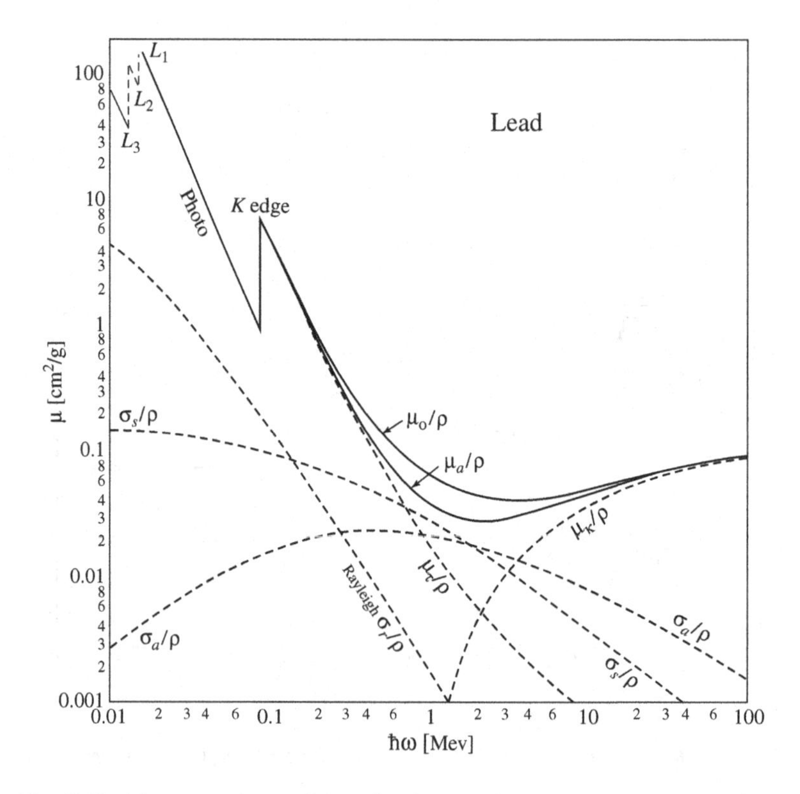

Fig. 10.17 Mass attenuation coefficients for photons in *lead*. [Adapted from Evans, 1955.]

mean a *photopeak* at the incident energy, a *Compton edge* at an energy approximately 0.25 MeV ($m_e c^2/2$) less than the incident energy, and two so-called *escape peaks* denoted as $P1$ and $P2$. The escape peaks refer to pair production processes where either one or both annihilation photons escape detection by leaving the counter.

Thus, $P1$ should be 0.511 MeV below the incident energy, and $P2$ should be 0.511 MeV below $P1$. The other features that can be seen in Fig. 10.18 are a peak at 0.511 MeV, clearly to be identified as the annihilation photon, a backscattered peak associated with Compton scattering at $\theta = \pi$ which should be positioned at $m_e c^2/2$, and finally an unidentified peak which we may attribute to x-rays emitted from excited atoms.

One can notice in Fig. 10.18 the various peaks are quite broad. This is a known characteristic of the scintillation detector, namely, relatively poor energy resolution. In contrast, a semiconductor detector, such as *Li-drifted Ge*, would have much higher energy resolution, as can be seen in Fig. 10.19. In addition to the sharper

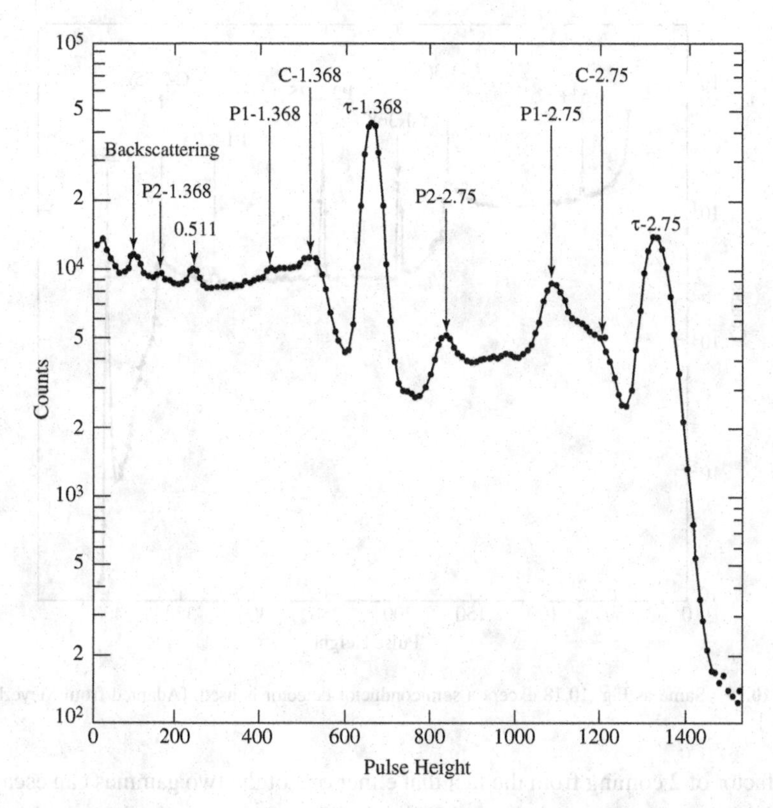

Fig. 10.18 Pulse-height spectra of 1.37 MeV and 2.75 MeV γ obtained using a Na-*I* detector. [Adapted from Meyerhof, 1967.]

lines, one should notice the peaks measured using the semiconductor detector have *different* relative intensities compared to the peaks obtained using the scintillation detector. In particular, looking at the relative intensities of $P1$ and $P2$, we see that $P1 > P2$ in Fig. 10.18, whereas the opposite holds, $P2 > P1$, in Fig. 10.19.

This difference can be explained by noting the scintillation detector is larger in physical size than the semiconductor detector. In this particular case the former is a cylinder 7.6 cm in diameter and 7.6 cm in length, whereas the latter is 1.9 cm in diameter and 0.5 cm in length. Thus one can expect that the probability that a photon will escape from the detector can be quite different in these two cases.

To follow up on this distinction, let us define P as the probability of escape. In a one-dimensional configuration (for example, Fig. 10.1), $P \sim e^{-\mu x}$, where μ is the linear attenuation coefficient and x is the dimension of the detector. Now the probability that one of the two annihilation gammas will escape is $P1 = 2P(1-P)$,

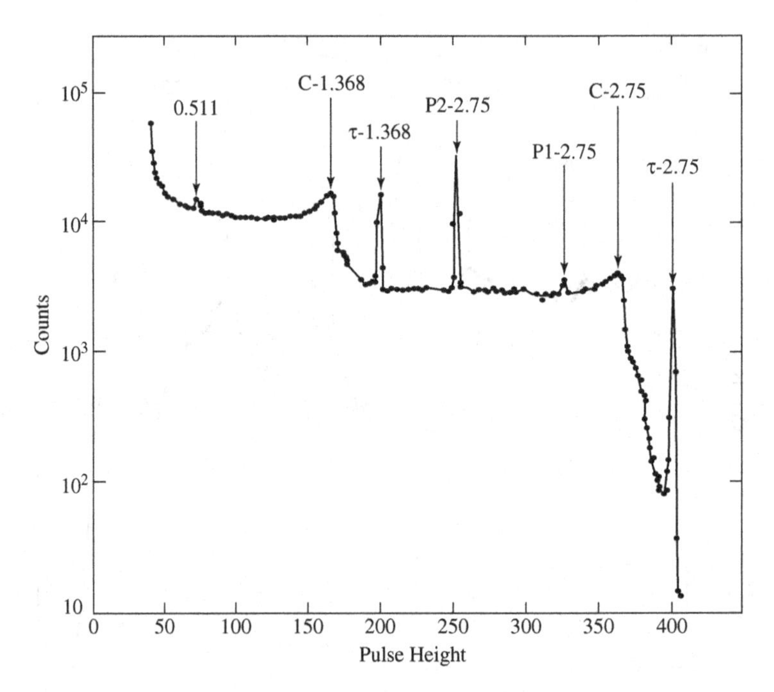

Fig. 10.19 Same as Fig. 10.18 except a semiconductor detector is used. [Adapted from Meyerhof, 1967.]

the factor of 2 coming from the fact that either one of the two gammas can escape. For both gammas to escape the probability is $P2 = P^2$. So we see that whether $P1$ is larger or smaller than $P2$ depends on the magnitude of P. If P is small, $P1 > P2$, but if P is close to unity, then $P2 > P1$. For the two detectors in question, it is to be expected that P is larger for the semiconductor detector. Without putting in actual numbers we can infer from an inspection of Figs. 10.18 and 10.19 that P is small enough in the case of the scintillation detector for $P1$ to be larger than $P2$, and also P is close enough to unity in the case of the semiconductor detector for $P2$ to be larger than $P1$.

REFERENCES

R. D. Evans, *Atomic Nucleus* (McGraw-Hill New York, 1955), Chaps. 23–25.

W. Heitler, *Quantum Theory of Radiation* (Oxford, 1955), Sec. 26.

G. F. Knoll, *Radiation Detection and Measurement* (Wiley, New York, 1979).

W. E. Meyerhof, *Elements of Nuclear Physics* (McGraw-Hill, New York, 1967), Secs. 3–6.

11

Charged Particle Stopping

When a swift charged particle enters a materials medium it interacts with the electrons and nuclei in the medium and begins to lose energy as it penetrates into the material. The interactions can be thought of as individual *collisions* between the charged particle and the atomic electrons surrounding a nucleus or the nucleus itself (considered separately). The energy given off during these collisions will result in *ionization*, the production of ion-electron pairs, in the medium. Also it can appear in the form of *electromagnetic radiation*, a process known as *bremsstrahlung* (braking radiation). We are interested in describing the energy loss per unit distance traveled by the charged particle, and the range of the particle in various materials. The range is defined as the distance traveled from the point of entry to the point of essentially coming to rest.

A charged particle is called "heavy" if its rest mass is large compared to the rest mass of the electron. Thus, mesons, protons, α-particles, and of course fission fragments, are all heavy charged particles. By the same token, electrons and positrons are "light" particles.

If we ignore nuclear forces and consider the interactions arising only from Coulomb forces, we can distinguish four types of charged-particle interactions:

(a) *Inelastic Collision with Atomic Electrons*. This is the principal process of energy transfer, particularly if the particle velocity is below the level where bremsstrahlung is significant. It leads to the excitation of the atomic electrons (still bound to the nucleus) and to ionization (electron stripped off the nucleus). Inelastic here refers to the excitation of electronic levels.

(b) *Inelastic Collision with a Nucleus*. This process can leave the nucleus in an excited state or the particle can radiate (bremsstrahlung).

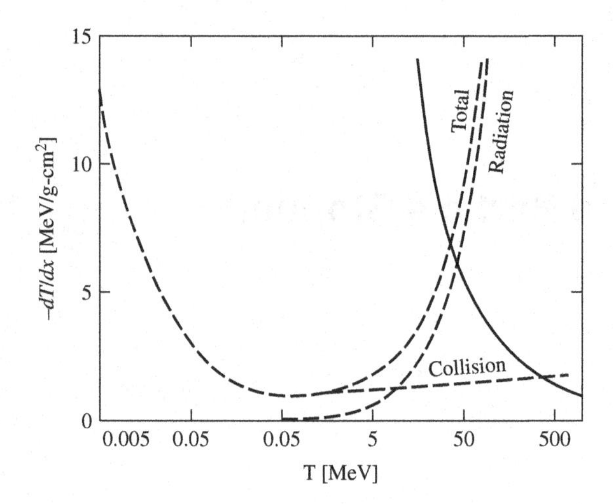

Fig. 11.1 Energy variations of the stopping power, $-dT/dx$, of electron (dashed curve) and proton (solid curve) in Pb in units of MeV/g-cm^2. For electron the contributions from *collisions* and *radiation* are shown separately. For proton radiation loss occurs at energies higher than the range indicated. [Adapted from Meyerhof, 1967.]

(c) *Elastic Collision with a Nucleus*. This process is known as *Rutherford scattering*. There is no excitation of the nucleus, nor emission of radiation. The particle loses energy only through the recoil of the nucleus.

(d) *Elastic Collision with Atomic Electrons.* The process is elastic deflection which results in a small amount of energy transfer. It is significant only for charged particles that are low-energy electrons.

In general, interaction of type (a), which is sometimes simply called collision, is the dominant process of energy loss, unless the charged particle has a kinetic energy exceeding its rest mass energy, in which case the radiation process, type (b), becomes important. For heavy particles, radiation occurs only at such kinetic energies, $\sim 10^3$ MeV, that it is of little interest to nuclear engineers. The characteristic behavior of electron and proton energy loss in a high-Z medium like *lead* is shown in Fig. 11.1 [Meyerhof, 1967]. This is an important illustrative result to guide the discussions to follow.

11.1 STOPPING POWER, COLLISIONS AND IONIZATION

The kinetic energy loss per unit distance suffered by a charged particle, to be denoted as $-dT/dx$, is conventionally known as the *stopping power*. It is a positive quantity since dT/dx is always < 0. There are quantum mechanical as well as classical theories for calculating this basic quantity. One would like to express

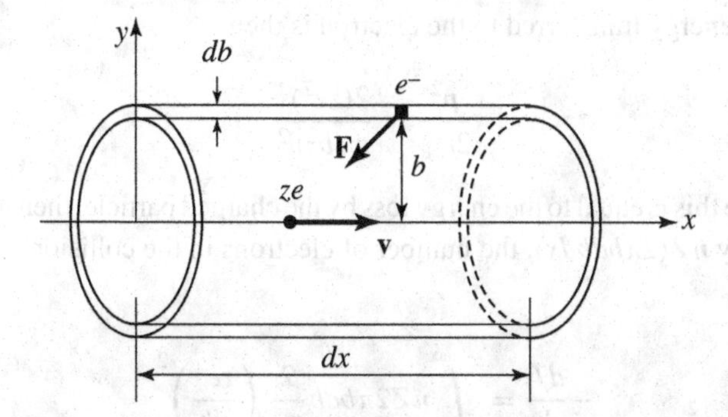

Fig. 11.2 Collision-cylinder configuration for estimating the energy loss to an atomic electron, at impact parameter b, by an incident charged particle, charge ze, mass M, moving with speed v along the x-direction. The interaction force **F** is acting along the line joining the charged particle and the electron.

$-dT/dx$ in terms of the properties specifying the incident charged particle, such as its velocity v and charge ze, and the properties pertaining to the atomic medium, the charge of the atomic nucleus Ze, the density of atoms n, and the average ionization potential $\bar{I}$.

We consider a crude derivation of a formula for $-dT/dx$. We begin with an estimate of the energy loss suffered by an incident charged particle when it interacts with a free and initially stationary electron. For this purpose a convenient configuration to use is a collision cylinder whose radius is the impact parameter b and whose length is the small distance traveled dx shown in Fig. 11.2. In this set-up, we ask for the net momentum transferred to the electron as the incident charged particle moves from one end of the cylinder to the other end. We expect the momentum transfer to be mostly directed in the perpendicular direction because the force in the direction of travel, F_x, changes sign when the charged particles passes by the atomic electron. So the net momentum along the horizontal direction vanishes, only a force along the negative vertical ($-y$) direction remains. So we write

$$\int dt F_x(t) \approx 0 \tag{11.1}$$

$$\begin{aligned} p_e &= \int dt F_y(t) \\ &= \int \frac{ze^2}{x^2+b^2} \frac{b}{(x^2+b^2)^{1/2}} \frac{dx}{v} \\ &\cong \frac{ze^2 b}{v} \int_{-\infty}^{\infty} \frac{dx}{(x^2+b^2)^{3/2}} = \frac{2ze^2}{vb} \end{aligned} \tag{11.2}$$

The kinetic energy transferred to the electron is then

$$\frac{p_e^2}{2m_e} = \frac{2(ze^2)^2}{m_e b^2 v^2} \tag{11.3}$$

If we assume this is equal to the energy loss by the charged particle, then multiplying Eq. (11.3) by $nZ(2\pi bdbdx)$, the number of electrons in the collision cylinder, we obtain

$$\begin{aligned} -\frac{dT}{dx} &= \int_{b_{\min}}^{b_{\max}} nZ2\pi bdb \frac{2}{m_e}\left(\frac{ze^2}{vb}\right)^2 \\ &= \frac{4\pi(ze^2)^2 nZ}{m_e v^2} \ell n \left(\frac{b_{\max}}{b_{\min}}\right) \end{aligned} \tag{11.4}$$

where $b_{\max}$ and $b_{\min}$ are the maximum and minimum impact parameters which one should specify according to the physical description he wishes to treat.

In reality, the atomic electrons are of course not free electrons, so the charged particle must transfer at least an amount of energy equal to the first excited state of the atom. If we take the time interval of energy transfer to be $\Delta t \approx b/v$, then $(\Delta t)_{\max} \sim 1/\nu$, where $h\nu \approx \bar{I}$ is the mean ionization potential. This gives an estimate of the maximum impact parameter,

$$b_{\max} \approx hv/\bar{I} \tag{11.5}$$

An empirical expression for $\bar{I}$ is$\bar{I} \approx kZ$, with $k \sim 19\,\text{eV}$ for H and $\sim 10\,\text{eV}$ for Pb. Next we estimate $b_{\min}$ by invoking the uncertainty principle to say the electron position cannot be specified more precisely than its de Broglie wavelength (in the relative coordinate system of the electron and the charged particle). Since the electron momentum in the relative coordinate system is $m_e v$, we have

$$b_{\min} \approx h/m_e v \tag{11.6}$$

Combining these two estimates we arrive at

$$-\frac{dT}{dx} = \frac{4\pi z^2 e^4 nZ}{m_e v^2} \ell n \left(\frac{2m_e v^2}{\bar{I}}\right) \tag{11.7}$$

In Eq. (11.7), we have inserted a factor of 2 in the argument of the logarithm, this is to make our formula agree with the result of a quantum mechanical calculation first carried out by *Bethe* using the *Born approximation*.

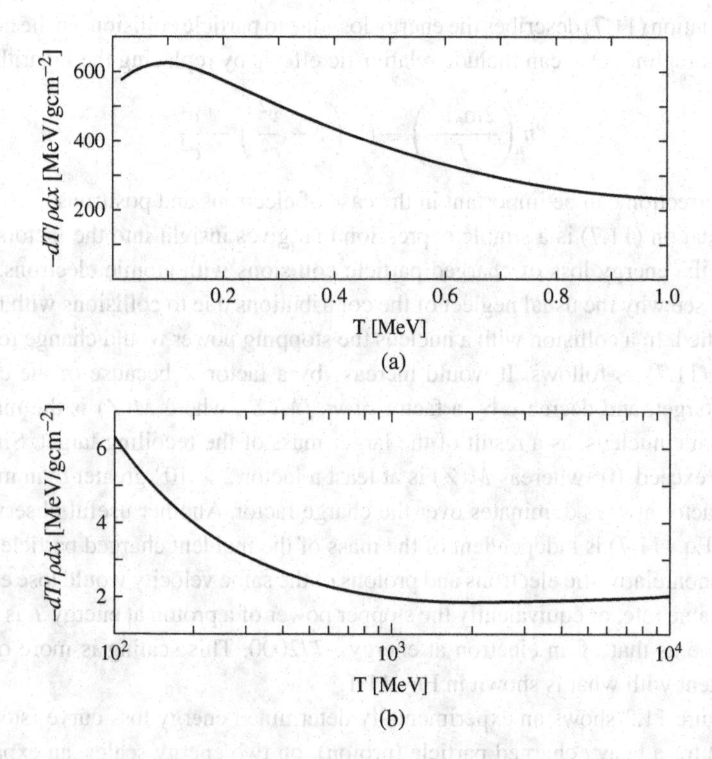

Fig. 11.3 The experimentally determined stopping power, $(-dT/\rho dx)$, for protons in air, (a) low-energy region where the *Bethe formula* applies down to $T \sim 0.3\,\text{MeV}$ with $\bar{I} \sim 80\,\text{eV}$. Below this range charge loss due to electron capture causes the stopper power to reach a peak and start to decrease (see also Fig. 11.4 below), (b) high-energy region where a broad minimum occurs at $T \sim 1500\,\text{MeV}$. [Adapted from Meyerhof, 1967.]

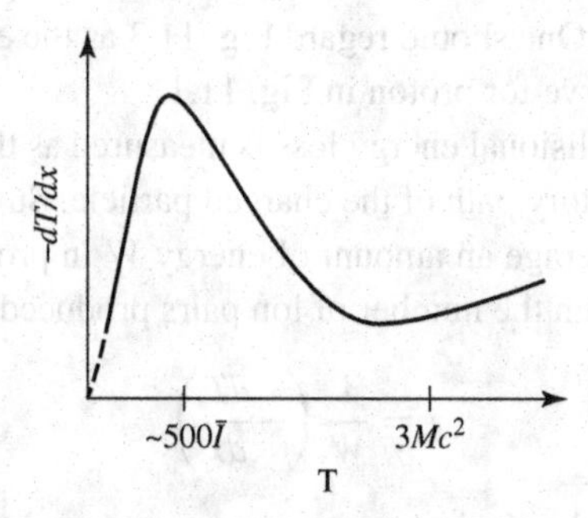

Fig. 11.4 Schematic of overall variation of stopping power $-dT/dx$ of a charged particle with kinetic energy T and rest mass M. $\bar{I}$ is the mean ionization potential of the absorber.

Equation (11.7) describes the energy loss due to particle collisions in the nonrelativistic regime. One can include relativistic effects by replacing the logarithm by

$$\ell n\left(\frac{2\mathrm{m_e}v^2}{\bar{I}}\right) - \ell n\left(1-\frac{v^2}{c^2}\right) - \frac{v^2}{c^2}$$

This correction can be important in the case of electrons and positrons.

Equation (11.7) is a simple expression that gives insight into the factors governing the energy loss of charged particle collisions with atomic electrons. First we can see why the usual neglect of the contributions due to collisions with nuclei is justified. In a collision with a nucleus the stopping power would change relative to Eq. (11.7) as follows. It would increase by a factor Z because of the charge of the target, and decrease by a factor of $m_e/M(Z)$, where $M(Z)$ is the mass of the atomic nucleus, as a result of the larger mass of the recoiling target. Since Z cannot exceed 10^2 whereas $M(Z)$ is at least a factor 2×10^3 greater than $\mathrm{m_e}$, the mass factor always dominates over the charge factor. Another useful observation is that Eq. (11.7) is independent of the mass of the incident charged particle. This means nonrelativistic electrons and protons of the same velocity would lose energy at the same rate, or equivalently the stopper power of a proton at energy T is about the same as that of an electron at energy ~$T/2000$. This scaling is more or less consistent with what is shown in Fig. 11.1.

Figure 11.3 shows an experimentally determined energy loss curve (stopping power) for a heavy charged particle (proton), on two energy scales, an expanded low-energy region where the stopping power decreases smoothly with increasing kinetic energy of the charged particle T below a certain peak centered about 0.1 MeV, and a more compressed high-energy region where the stopping power reaches a broad minimum around 10^3 MeV. Notice also a slight upturn as one goes to higher energies past the broad minimum which we expect is associated with relativistic corrections. One should regard Fig. 11.3 as the extension, at both ends of the energy, of the curve for proton in Fig. 11.1.

Experimentally, collisional energy loss is measured as the number of ion pairs formed along the trajectory path of the charged particle. Suppose a heavy charged particle loses on the average an amount of energy W in producing an ion pair, an electron and an ion. Then the number of ion pairs produced per unit path is

$$i = \frac{1}{W}\left(-\frac{dT}{dx}\right) \tag{11.8}$$

Equation (11.7) is valid only in a certain energy range because of the assumptions we have made in its derivation. We have seen from Figs. 11.1 and 11.3 that the atomic stopping power varies with energy in the manner sketched below in

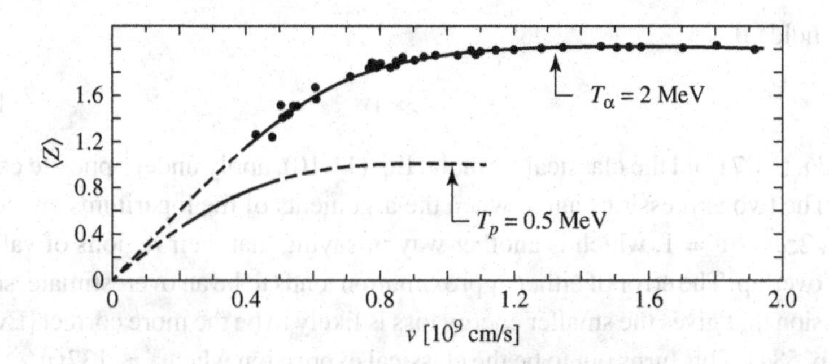

Fig. 11.5 Loss of charge with speed as a charged particle slows down. Results shown are the average charge for α-particles and protons with initial energies as indicated. [Adapted from Meyerhof, 1967.]

Fig. 11.4. In the intermediate energy region, $500\bar{I} < T \leq Mc^2$, where M is the mass of the charged particle, the stopping power behaves like $1/T$, which is roughly what is predicted by Eq. (11.7). In this region the relativistic correction is small and the logarithm factor varies slowly. At higher energies the logarithm factor along with the relativistic correction terms give rise to a gradual increase so that a broad minimum is set up in the neighborhood of $\sim 3\ Mc^2$. At energies below the maximum in the stopping power, $T < 500\bar{I}$, Eq. (11.7) is not valid because the charged particle is moving slowly enough to capture electrons and begin to lose its charge. Figure 11.5 shows the correlation between the mean charge of a charged particle and its velocity. This is a difficult region to analyze theoretically. For α-particles and protons the range begins at $\sim$1 MeV and 0.1 MeV respectively [Bethe, 1953].

Equation (11.7) is generally known as the Bethe formula. It is a quantum mechanical result derived on the basis of the Born approximation which is essentially an assumption of weak scattering, namely the amplitude of the electron wave scattered by the atomic electron field should be small compared to the amplitude of the incident wave [Williams, 1945]. The approximation is valid if

$$\frac{ze^2}{\hbar v} = \left(\frac{e^2}{\hbar c}\right)\frac{z}{v/c} = \frac{z}{137(v/c)} \ll 1 \tag{11.9}$$

where ze is the charge of the incident particle and v its velocity. On the other hand, *Bohr* has used classical theory to derive a somewhat similar expression for the stopping power,

$$-\left(\frac{dT}{dx}\right)_{class} = \frac{4\pi z^2 e^4 nZ}{m_e v^2}\ell n\left[\frac{M\hbar v}{2ze^2(m_e + M)}\frac{2m_e v^2}{\bar{I}}\right] \tag{11.10}$$

which holds if

$$\frac{ze^2}{\hbar v} \gg 1 \tag{11.11}$$

Thus Eq. (11.7) and the classical formula, Eq. (11.10), apply under opposite conditions. The two expressions agree when the arguments of the logarithms are equal, that is, $2ze^2/\hbar v = 1$, which is another way of saying that their regions of validity do not overlap. The error of either approximation tends to be an overestimate, so the expression that gives the smaller energy loss is likely to be the more correct [Evans, 1955, p. 584]. This turns out to be the classical expression when $z > 137(v/c)$, and Eq. (11.7) when $2z < 137(v/c)$. Knowing the charge of the incident particle and its velocity, one can use this criterion to choose the appropriate stopper power formula. In the case of fission fragments (*high Z* nuclides) the classical result should be used. Also, it should be noted a quantum mechanical theory has been developed by *Bloch* that gives the Bohr and Bethe results as limiting cases.

For heavy particles the condition Eq. (11.9) can be interpreted as the necessary requirement for the particle to conserve its full charge (i.e. no electron pickup), however for electrons it simply establishes a threshold under which the Born approximation is no longer valid and the amplitude of the scattered wave is comparable to that of the incoming wave. Physically this implies that a strong potential is the cause of the scattering (and as such the scattered field is not negligible) and a perturbative analysis (Born approximation) is no longer valid.

Atomic electrons do not generate such large potentials. This means the primary interactions at low energies are not the inelastic electron-electron collisions, but collisions between the incident electron and the atoms of the material which have a larger electric field. The behavior of those interactions governed by the *Rutherford scattering formula* is thus responsible for the drop in the stopping power as the incident particle energy goes to 0. This is physically intuitive because we expect the electron, being substantially less massive than the atom, will lose very little energy in those collisions. In fact the energy loss will decrease with decreasing incoming energy as in any other elastic collision. Other contributions involving different mechanisms also may have to be considered at those low energies. Experimentally the drop in the stopping power has been observed, suggesting elastic collisions as the dominant mechanism. This is consistent with the theoretical arguments. The Bethe formula, Eq. (11.7), is most appropriate for heavy charged particles. For fast electrons (relativistic) one should use

$$-\frac{dT}{dx} = \frac{2\pi e^4 nZ}{m_e v^2}\left[\ell n\left(\frac{m_e v^2 T}{\bar{I}^2(1-\beta^2)}\right) - \beta^2\right] \tag{11.12}$$

where $\beta = v/c$.

11.2 RADIATION LOSS

The sudden deflection of an electron by the Coulomb field of the nuclei can cause the electron to radiate, producing a continuous spectrum of X-rays called *bremsstrahlung*. The fraction of electron energy converted into bremsstrahlung increases with increasing electron energy and is greater for media of high atomic number. (This process is important in the production of X-rays in conventional X-ray tubes.)

According to the classical theory of electrodynamics [Jackson, 1962, p. 509], the acceleration produced by a nucleus of charge Ze on an incident particle of charge ze and mass M is proportional to Zze^2/M. The intensity of radiation emitted is proportional to $(ze \times \text{acceleration})^2 \sim (Zz^2e^3/M)^2$. Notice the $(Z/M)^2$ dependence; this shows *bremsstrahlung* is more important in a high-Z medium. Also it is more important for electrons and positrons than for protons and α-particles. Another way to understand the $(Z/M)^2$ dependence is to recall the derivation of stopping power in Sec. 11.1 where the momentum change due to a collision between the incident particle and a target nucleus is $(2ze^2/vb) \times Z$. The factor Z represents the Coulomb field of the nucleus. This was unity since we had an atomic electron as the target. The recoil velocity of the target nucleus is therefore proportional to Z/M, and the recoil energy, which is the intensity of the radiation emitted, is then proportional to $(Z/M)^2$.

In an individual deflection by a nucleus, the electron can radiate any amount of energy up to its kinetic energy T. The spectrum of *bremsstrahlung* wavelength for a thick target is of the form $I(\lambda) = \text{const}/\lambda^2$, with a cutoff at $\lambda_{\text{min}} = hc/T$. This converts to a frequency spectrum which is a constant up to the maximum frequency of $\nu_{\text{max}} = T/h$. The shape of the spectrum is independent of Z, and the intensity varies with electron energy like $1/T$.

In the quantum mechanical theory of *bremsstrahlung* a plane wave representing the electron enters the nuclear field and is scattered. There is a small but finite chance that a photon will be emitted in the process. The theory is intimately related to the theory of pair production where an electron-positron pair is produced by a photon in the field of a nucleus. Because a radiative process involves the coupling of the electron with the electromagnetic field of the emitted photon, the cross sections for radiation are of the order of the fine-structure constant [Dicke, 1960, p. 11], $e^2/\hbar c(= 1/137)$, times the cross section for elastic scattering. This means that most of the deflections of electrons by atomic nuclei results in elastic scattering, only in a small number of instances is a photon emitted. Since the classical theory of *bremsstrahlung* predicts the emission of radiation in every collision in which the electron is deflected, it is incorrect. However, when averaged over all collisions the classical and quantum mechanical cross sections are of the same order of magnitude,

$$\sigma_{rad} \sim \frac{Z^2}{137}\left(\frac{e^2}{m_e c^2}\right)^2 \text{cm}^2/nucleus \tag{11.13}$$

where $e^2/m_e c^2 = r_e = 2.818 \times 10^{-13}$ cm is the *classical radius of electron.* In the few collisions where photons are emitted a relatively large amount of energy is radiated. In this way the quantum theory replaces the multitude of small-energy losses predicted by the classical theory by a much smaller number of larger-energy losses. The spectral distributions are therefore different in the two theories, with the quantum description being in better agreement with experiments.

Given a nucleus of charge Ze and an incident electron of kinetic energy T, the quantum mechanical differential cross section for the emission of a photon with energy in $d(h\nu)$ about $h\nu$ is

$$\left[\frac{d\sigma}{d(h\nu)}\right]_{rad} = \sigma_o B Z^2 \frac{T + m_e c^2}{T}\frac{1}{h\nu} \tag{11.14}$$

where $\sigma_o = \left(e^2/m_e c^2\right)^2/137 = 0.580 \times 10^{-3}$ *barns* and $B \sim 10$ is a very slowly varying dimensionless function of Z and T. A general relation between the energy differential cross section and the energy loss per unit path length is

$$-\frac{dT}{dx} = n\int_0^T dE E\frac{d\sigma}{dE} \tag{11.15}$$

where $d\sigma/dE$ is the differential cross section for energy loss E (recall Sec. 9.2). Applying this to Eq. (11.14) we have

$$-\left(\frac{dT}{dx}\right)_{rad} = n\int_0^T d(h\nu)h\nu\left[\frac{d\sigma}{d(h\nu)}\right]_{rad}$$

$$= n(T + m_e c^2)\sigma_{rad} \quad \text{ergs/cm} \tag{11.16}$$

where

$$\sigma_{rad} = \sigma_o Z^2 \int_0^1 d\left(\frac{h\nu}{T}\right) B \equiv \sigma_o Z^2 \overline{B} \tag{11.17}$$

is the total *bremsstrahlung* cross section. The variation of $\overline{B}$, the *bremmstrahlung* cross section in units of $\sigma_o Z^2$, with the kinetic energy of an incident electron is shown in Fig. 10.6 for media of various Z [Evans, 1967, p. 605].

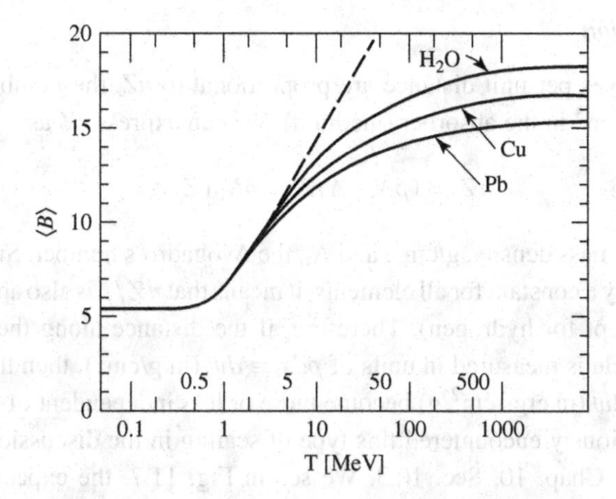

Fig. 11.6 Variation of the constant $\overline{B}$ in Eq. (11.17) with the kinetic energy of electron. [Adapted from Evans, 1955, p. 605]. Solid curves include screening corrections, while in the dashed cuve the corrections are neglected. Note the dependence on the medium in which radiative loss is occurring.

Comparison of various cross sections

It is instructive to compare the cross sections describing the interactions that we have considered between an incident electron and the atoms in the medium. For nonrelativistic electrons, $T \leq 0.1$ MeV and $\beta = v/c \leq 0.5$, we have the following cross sections (all in barns/atom),

$$\sigma_{ion} = \frac{2\alpha Z}{\beta^4}\ell n\left(\frac{\sqrt{2}T}{\bar{I}}\right) \quad ionization \tag{11.18}$$

$$\sigma_{nuc} = \frac{\alpha Z^2}{4\beta^4} \quad backscattering\ by\ nuclei \tag{11.19}$$

$$\sigma_{el} = \frac{2\alpha Z}{\beta^4} \quad elastic\ scattering\ by\ atomic\ electrons \tag{11.20}$$

$$\sigma'_{rad} = \frac{8\alpha}{3\pi}\frac{1}{137}\frac{Z^2}{\beta^2} \quad bremsstrahlung \tag{11.21}$$

where $\alpha = 4\pi(e^2/m_ec^2)^2 = 1.00$ barn. The difference between σ_{rad} and σ'_{rad} is the former corresponds to fractional loss of *total* energy, $dT/(T+m_ec^2)$, while the latter corresponds to fractional loss of *kinetic* energy, *dT/T*.

Mass absorption

Ionization losses per unit distance are proportional to nZ, the number of atomic electrons per cm^3 in the absorber (medium). We can express nZ as

$$nZ = (\rho N_o/A)Z = \rho N_o(Z/A) \tag{11.22}$$

whereρ is the mass density, g/cm^3, and N_0 the Avogadro's number. Since the ratio (Z/A) is nearly a constant for all elements, it means that nZ/ρ is also approximately constant (except for hydrogen). Therefore, if the distance along the path of the charged particle is measured in units of $\rho dx \equiv dw$ (in g/cm^2), then the ionization losses, $-dT/dw$ (in ergs cm^2/g) become more or less independent of the material. We have previously encountered this type of scaling in the discussion of gamma attenuation in Chap. 10, Sec. 10.3. We see in Fig. 11.7, the expected behavior of energy loss being material independent holds only approximately, as $-dT/dw$ actually decreases with increasing Z. Two reasons may be given for this behavior, Z/A is not really constant but decreasing slightly as Z increases, and $\bar{I}$ is known to increase linearly with Z.

We have seen ionization losses per path length vary mainly as $1/v^2$ while radiative losses increase with increasing energy. The two losses become roughly comparable when $T \gg Mc^2$, or $T \gg m_ec^2$ in the case of electrons. The ratio can

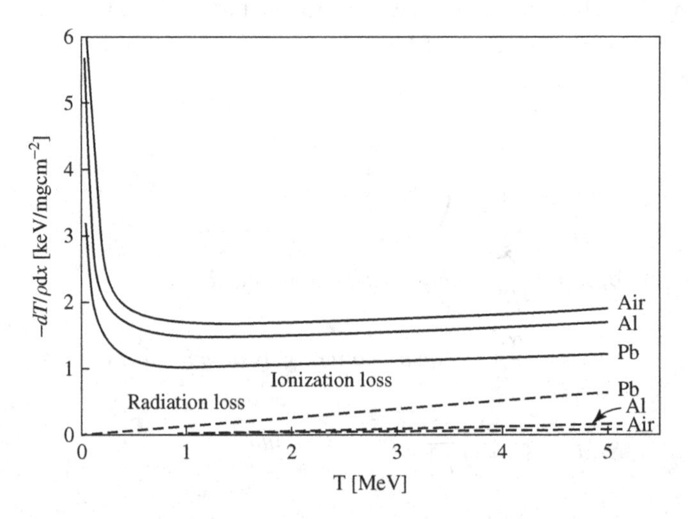

Fig. 11.7 Mass absorption energy losses, $-dT/dw$, for electrons in three absorbers, *air*, *Al*, and *Pb* [Adapted from Evans, 1955, p. 609]. Ionization losses are denoted by the solid curves, whereas radiation losses (*bremsstrahlung*) are indicted by the dashed lines. Note the ordering by absorber is opposite in the two cases.

be approximately expressed as

$$\frac{(dT/dx)_{rad}}{(dT/dx)_{ion}} \approx Z\left(\frac{m_e}{M}\right)^2\left(\frac{T}{1400 m_e c^2}\right) \tag{11.23}$$

where for electrons, $M \to m_e$. The two losses are therefore equal in the case of electrons for $T = 18\, m_e c^2 = 9\,\text{MeV}$ in *Pb*, and $T \sim 100\,\text{MeV}$ in water or air.

11.3 RANGE-ENERGY RELATIONS

Upon entering an absorber a charged particle immediately starts to interact with the many electrons in the medium. For a heavy charged particle the deflection from any individual encounter is small, so the track of the heavy charged particle tends to be quite straight, except at the very end of its travel when it has lost practically all its kinetic energy. In this case one could calculate the range of the particle, the distance beyond which it cannot penetrate, by integrating the stopping power,

$$R = \int_0^R dx = \int_{T_o}^0 \left(\frac{dx}{dT}\right) dT = \int_0^{T_o} \left(-\frac{dT}{dx}\right)^{-1} dT \tag{11.24}$$

where T_o is the initial kinetic energy of the particle. An estimate of R is given by taking the Bethe formula, Eq. (11.7), for the stopping power and ignoring the v-dependence in the logarithm. One finds simply

$$R \propto \int_0^{T_o} T dT = T_o^2 \tag{11.25}$$

This is an example of a *range-energy relation*. Given what we have said about the region of applicability of Eq. (11.7) one might expect this behavior to hold at low energies. At high energies it is more reasonable to take the stopping power to be a constant, in which case,

$$R \propto \int_0^{T_o} dT = T_o \tag{11.26}$$

We will see below both limiting behavior are seen in measurements (Figs. 11.10 and 11.11).

Experimentally one finds the energy loss by measuring the number of ion pairs produced from an ionization event. The amount of energy W required for a particle of certain energy to produce an ion pair is generally known. The number of ion pairs, i, produced per unit path length (*specific ionization*) of the charged particle is then

$$i = \frac{1}{W}\left(-\frac{dT}{dx}\right) \tag{11.27}$$

The quantity W depends on complicated processes such as atomic excitation and secondary ionization in addition to primary ionization. On the other hand, for a given material it has been found to be approximately independent of the nature of the particle or its kinetic energy. For example, in air the values of W are 35.0, 35.2, and 33.3 eV for 5 keV electrons, 5.3 MeV alphas, and 340 MeV protons, respectively.

The specific ionization is an appropriate measure of the ionization processes taking place along the path (track length) of the charged particle. It is useful to regard Eq. (11.27) as a function of the distance traveled by the particle. Such results are shown in Fig. 11.8, where one sees a characteristic shape of the ionization curve for a heavy charged particle. Ionization is constant or increasing slowly during the early-to-mid stages of the total travel, then it rises appreciably, reaching a peak value before dropping sharply to zero.

We have already mentioned that as the charged particle loses energy and slows down, the probability of capturing electron increases. So the mean charge of a beam of particles will decrease with the decrease in their speed (cf. Fig. 11.3). This is the reason why the specific ionization shows a sharp drop. The value of $-dT/dx$ along a particle track is also called specific energy loss. A plot of $-dT/dx$ along the track

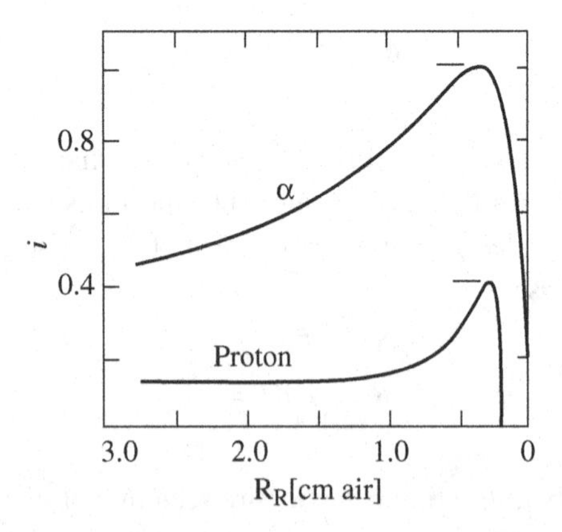

Fig. 11.8 Specific ionization of heavy particles in air, normalized to the peak value (indicated by a horizontal bar) for α-particles at 6600 *ion pairs/mm* [Adapted from Meyerhof, 1967, p. 80]. For protons the peak value is 2750 *ion pairs/mm*. Residual range R_R is the distance still to travel before coming to rest. Proton range is 0.2 cm shorter than that of α-particle.

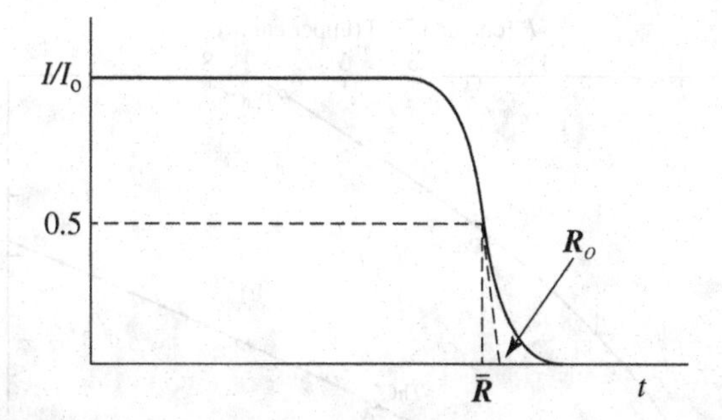

Fig. 11.9 Schematic of the determination of range by transmission experiment. [Adapted from Knoll, 1979.] $I(t)$ is the detected number of heavy charged particles passing through an absorber of thickness t, and I_o is the number detected without the absorber. $\bar{R}$ and R_o are the mean and extrapolated ranges, respectively. End of the curve shows the effects of straggling.

of a charged particle is known as a *Bragg* curve. It should be emphasized the Bragg curve differs from a plot of $-dT/dx$ for an individual particle in that the former is an average over a large number of particles. Hence the Bragg curve includes the effects of *straggling* (statistical distribution of range values for particles having the same initial velocity) and has a pronounced tail beyond the extrapolated range.

A typical result of measuring the range of charged particles in a transmission experiment is shown in Fig. 11.9. The mean range $\bar{R}$ is defined as the absorber thickness at which the intensity is reduced to one-half of the initial value. The extrapolated range R_o is obtained by linear extrapolation at the inflection point of the transmission curve. This is an example that $I/I_o$is not always an exponential, in contrast to gamma attenuation, Fig. 10.1. In charged particle interactions the distinction between heavy and light charged particles makes it necessary that we think beyond $I/I_o \sim e^{-\mu x}$, One should be concerned also with the range R.

In practice one uses range-energy relations that are mostly empirically determined. For a rough estimate of the range one can turn to the *Bragg-Kleeman* rule,

$$\frac{R}{R_1} = \frac{\rho_1 \sqrt{A}}{\rho \sqrt{A_1}} \tag{11.28}$$

where the subscript 1 denotes the reference medium which is conventionally taken to be air at 15°C, 760 mm Hg ($\sqrt{A_1} = 3.81$, $\rho_1 = 1.226 \times 10^{-3}$ g/cm^3). Then

$$R = 3.2 \times 10^{-4} \frac{\sqrt{A}}{\rho} \times R_{\text{air}} \tag{11.29}$$

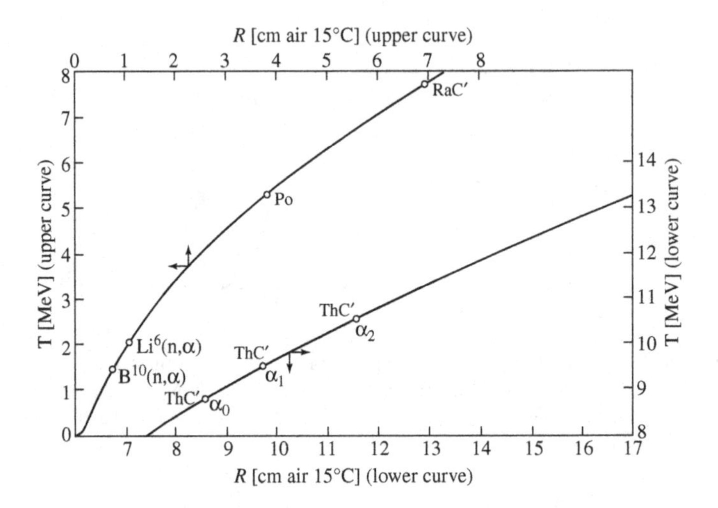

Fig. 11.10 Range-energy relations of α-particles (from the indicated nuclear reactions or alpha emitters) in air [Evans, 1955, p. 650].

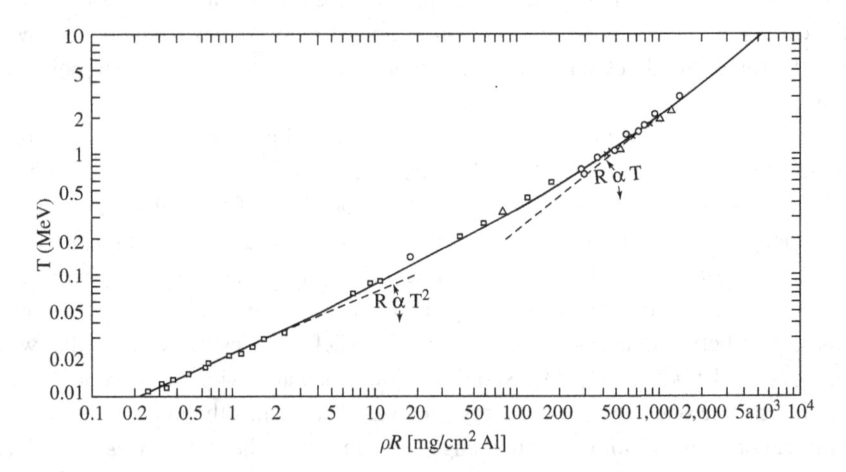

Fig. 11.11 Range-energy relation for electrons in aluminum. Various measurements are denoted by the symbols, while the solid curve is an empirical relation fitted to the data [Evans, 1955, p. 624]. At low and high energies the limiting behavior given by Eqs. (11.25) and (11.26) are validated.

with ρ in g/cm^3. In general such an estimate is good to within about $\pm 15\%$. Figure 11.10 shows the range-energy relations for α-particles in air. Notice at low energy the variation is quadratic, as predicted by Eq. (11.25), and at high energy the relation is more or less linear, as given by Eq. (11.26). The same qualitative behavior, seen in the range-energy relation for electrons, is evident in Fig. 11.11.

We have mentioned heavy charged particles traverse essentially in a straight line until reaching the end of its range where straggling effects manifest. In contrast, for electrons large deflections are quite likely during their traversal, so the trajectory of electron in a thick absorber is a series of zigzag paths. While one can still speak of the range R, the concept of path length is now of little value. The total path length S is appreciably greater than the range R.

The transmission curve I/I_o for a heavy charged particle was shown in Fig. 11.9. The curve has a different characteristic shape for monoenergetic electrons and a still different shape for β-rays (electrons with a distribution of energies). Although the curve for monoenergetic electrons depends to some extent on experimental arrangements, one may regard it as roughly a linear variation which is characteristic of single interaction event in removing the electron. That is, the fraction of electrons getting through is proportional to $1 - P$, where P is the interaction probability which is in turn proportional to the thickness. For the β-ray transmission curve which essentially has the form of an exponential, the shape is an accidental consequence of the β-ray spectrum and of the differences between the scattering and absorption of electrons which have various initial energies [cf. Evans, 1955, p. 625].

Cerenkov radiation

Electromagnetic radiation is emitted when a charged particle passes through a medium under the condition

$$v_{group} \equiv \beta c > v_{phase} \equiv c/n \tag{11.30}$$

where n is the index of refraction of the medium. When $\beta n > 1$, there is an angle (a direction) where constructive interference occurs. This type of radiation is a particular form of energy loss, due to soft collisions, and is not an additional amount of energy loss. Soft collisions involve small energy transfers from charged particles to distant atoms which become excited and subsequently emit coherent radiation (see Evans, 1955, p. 589).

REFERENCES

H. A. Bethe and J. Ashkin, Passage of Radiation Through Matter, in *Experimental Nuclear Physics*, E. Segrè, ed. (Wiley, New York, 1953), Vol. I, p. 166.

R. H. Dicke and J. P. Wittke, *Introduction to Quantum Mechanics* (Addison-Wesley, Reading, 1960).

R. D. Evans, *The Atomic Nucleus* (McGraw-Hill, New York, 1955), Chaps. 18–22.

W. E. Meyerhof, *Elements of Nuclear Physics* (McGraw-Hill, New York, 1967).
J. D. Jackson, *Classical Electrodynamics* (Wiley, New York, 1962), p. 509.
E. Segrè, ed., *Experimental Nuclear Physics*, Vol. I (Wiley, New York, 1953).
E. J. Williams, *Rev. Mod. Phys.* **17**, 217 (1945).
G. F. Knoll, *Radiation Detection and Measurement* (Wiley, New York, 1979).

12

Neutron Reactions

Among the many models of nuclear reactions we have encountered two opposing types. One is the compound-nucleus type of models first proposed by Bohr (1936), in which the incident particle (a neutron) interacts strongly with the entire target nucleus, and the decay of the resulting compound nucleus is *independent* of the mode of formation, The other is the independent-particle type of models in which the incident particle interacts with the nucleus through an effective averaged potential. A well-known example of the first type is the *liquid drop model*, and three examples of the second type are a model proposed by Bethe (1940), the nuclear shell model with spin-orbit coupling (cf. Chap. 5), and a model with a complex potential, the optical model, proposed by Feshbach, Porter and Weisskopf (1949) discussed Sec. 8.3. Each of these models describes well some aspects of what we now know about nuclear structure and reactions, and not so well some of the other aspects. Since we have already examined the nuclear shell model in some detail, we will focus in the brief discussion here on the compound nucleus model, which in some sense may be considered to be in the same class as the liquid drop model. As we will see, this approach is well suited for describing reactions which show single resonance behavior, a sharp peak in the energy variation of the cross section. In contrast, the optical model, which we have discussed in Chap. 8, is good for the *gross* behavior of the cross section in the sense of averaging over an energy interval.

Neutron cross sections of heavy nuclides are central to the functional performance of nuclear systems, the interactions with U^{235} and U^{238} being the cases of great interest. While this is an advanced topic which lies mostly beyond the scope of this book, it is nevertheless appropriate that we provide an introduction to the understanding of neutron resonance cross sections through the concept of formation and decay of a compound nucleus. This discussion is to be contrasted with the optical model calculations discussed in Sec. 8.3, where one is interested in the overall behavior in the sense of an energy average. We will consider the interpretation of neutron resonance cross sections for radiative capture and elastic scattering in terms of the *Breit-Wigner formulas*. Even though we do not go into any analysis of the fission cross section, we will summarize a number of significant features of the fission process in Sec. 12.3, as well as to follow up on the differences between U^{235} and U^{238} previously noted in Sec. 3.4.

12.1 REACTION ENERGETICS

We recall the *Q-equation* introduced in Chap. 8 for a general reaction depicted in Fig. 8.1, and write it as

$$Q = T_3\left(1 + \frac{M_3}{M_4}\right) - T_1\left(1 - \frac{M_1}{M_4}\right) - \frac{2}{M_4}(M_1 M_3 T_1 T_3)^{1/2}\cos\theta \qquad (12.1)$$

Since $Q = T_3 + T_4 - T_1$, the reaction can take place only if M_3 and M_4 emerge with positive kinetic energies (all kinetic energies are LCS unless specified otherwise),

$$T_3 + T_4 \geq 0, \quad \text{or} \quad Q + T_1 \geq 0. \qquad (12.2)$$

We will see that this condition, although quite reasonable from an intuitive standpoint, is *necessary but not sufficient* for the reaction to occur.

We have previously emphasized in the discussion of neutron interaction in Chap. 9 that a fraction of the kinetic energy brought in by the incident particle (a neutron) M_1 goes into the motion of the center-of-mass and is therefore not available for reaction. To see what is the energy available for reaction we look to the kinetic energies of the reacting particles in CMCS. First, the kinetic energy of the center-of-mass, in the case where the target nucleus is at rest, is

$$T_o = \frac{1}{2}(M_1 + M_2)v_o^2 \qquad (12.3)$$

where the center-of-mass speed is $v_o = [M_1/(M_1 + M_2)]v_1$, v_1 being the speed of the incident particle. The kinetic energy available for reaction is the kinetic energy of the incident particle T_1 minus the kinetic energy of the center-of-mass given by Eq. (12.3). Denoting this energy as T_i, we have

$$\begin{aligned} T_i = T_1 - T_o &= \frac{M_2}{M_1 + M_2} T_1 \\ &= \frac{1}{2}M_1 V_1^2 + \frac{1}{2}M_2 v_o^2. \end{aligned} \qquad (12.4)$$

The second line in Eq. (12.4) shows that T_i is also the sum of the kinetic energies of particles 1 and 2 in CMCS (we follow the same notation as before in using capital letters to denote the velocities in CMCS). In addition to the kinetic energy available for reaction, there is also the rest-mass energy available for reaction, as represented by the *Q-value*. Thus the total energy available for reaction is the sum of T_i and Q. A *necessary and sufficient* condition for reaction is therefore

$$E_{avail} = Q + T_i \geq 0. \tag{12.5}$$

We can rewrite Eq. (12.5) as

$$T_1 \geq -Q\frac{M_1 + M_2}{M_2}. \tag{12.6}$$

If $Q > 0$, Eq. (12.6) is always satisfied, which is expected since the reaction is exothermic. For $Q < 0$, Eq. (12.6) shows the requirement that the threshold energy, the minimum value of the incident particle kinetic energy for reaction, must be greater or equal to the rest-mass deficit. The reason for needing more energy than the rest-mass deficit is again that energy is needed for the kinetic energy of the center-of-mass.

At threshold, $Q + T_i = 0$. So M_3 and M_4 both move in LCS with speed v_o ($V_3 =$ and $V_4 = 0$). At this condition the total kinetic energies of the reaction products is

$$(T_3 + T_4)_{thres} = \frac{1}{2}(M_3 + M_4)v_o^2. \tag{12.7}$$

Since we have $M_3V_3 = M_4V_4$ from momentum conservation, we can say in general

$$Q + T_i = \frac{1}{2}M_3V_3^2 + \frac{1}{2}\frac{(M_3V_3)^2}{M_4}. \tag{12.8}$$

With Q and T_1 specified, we can find V_3 from Eq. (12.8) but not the direction of $\underline{V}_3$. It turns out that for T_1 just above threshold of an endothermic reaction, an interesting situation exists where at a certain scattering angle in LCS one can have *two different* kinetic energies in LCS (see the discussion in Sec. 8.1 concerning Fig. 8.3).

Energy-level diagrams for nuclear reactions

In Chap. 6, we have discussed how the various energies involved in nuclear decay can be appropriately displayed in an energy-level diagram. The same argument applies to nuclear reactions. Figure 12.1 shows the energies involved in an endothermic reaction. In this case the reaction can end up in two different states, depending on whether the product nucleus M_4 is in the ground state or in an excited state (*).

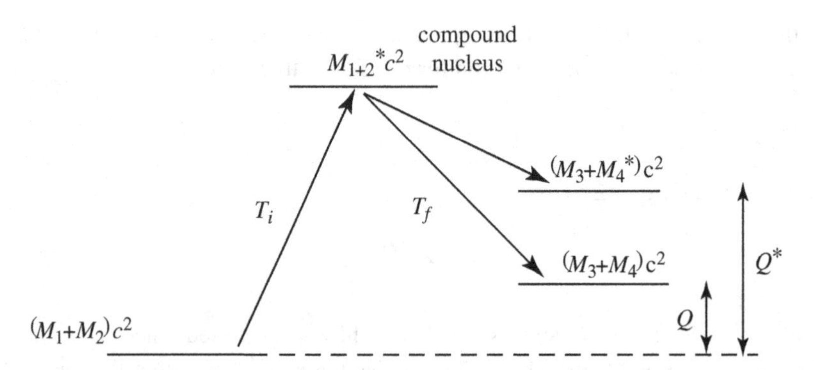

Fig. 12.1 Energy-level diagram for an endothermic reaction in which the product nucleus (particle 4) is left at the the ground state (rest mass M_4) or at an excited state (mass M_4^* and excitation Q^*). T_f is the kinetic energy of the reaction products.

T_f denotes the kinetic energy of the reaction products in CMCS, which one can write as

$$\begin{aligned} T_f &= Q + T_i \\ &= \frac{1}{2}M_3V_3^2 + \frac{1}{2}M_4V_4^2. \end{aligned} \tag{12.9}$$

Since both T_i and T_f can be considered kinetic energies in CMCS, one can say the kinetic energies appearing in the energy-level diagrams generally should be in CMCS.

12.2 COMPOUND NUCLEUS REACTIONS

The concept of compound nucleus model for nuclear reactions is schematically depicted in Fig. 12.2. The idea is that an incident particle reacts with the target nucleus in two ways, a scattering that takes place at the surface of the nucleus (stage (I) in Fig. 12.2) which is, properly speaking, not a reaction, and a reaction that takes place after the incident particle has entered into the nucleus, stages (II) and (III). Recall the discussions in Chaps. 7 and 8 concerning the decomposition of the scattering amplitude into a surface component and an interior component. The former is what we have been studying as elastic scattering, it is also known as *shape elastic* or *potential scattering*. This process is always present in that it is always allowed. The interactions which takes place after the particle has penetrated into the target nucleus, stage (II), can be considered an *absorption* process, leading to the formation of a compound nucleus (this need not be the only process possible,

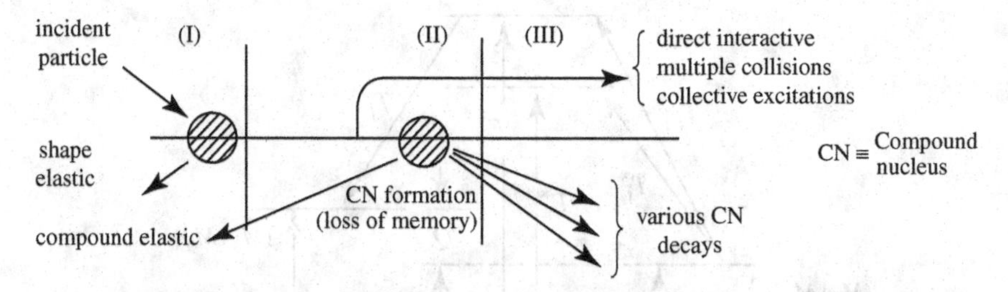

Fig. 12.2 Schematic of three stages of the compound nucleus model of nuclear reaction, shape elastic scattering a the nuclear surface (I), formation of compound nucleus CN with a characteristic lifetime (II), and subsequent decay of CN (III).

the others can be direct interaction, multiple collisions, and collective excitations, etc.). This is the part that we will want to discuss further.

In neutron reactions the formation of *compound nucleus* (CN) is quite likely at incident energies of $\sim 0.1 - 1$ MeV. Physically it corresponds to a large reflection coefficient in the inside edge of the potential well. Once CN is formed it is assumed that it will decay in a manner *independent* of the mode in which it was formed (complete loss of memory). This is the basic assumption of the model, namely, the formation and decay of CN, stage (III), can be treated as two *separate* processes. The approximation of two processes being independent of each other is expressed by writing the interaction as a two-stage reaction,

$$a + X \rightarrow C^* \rightarrow b + Y$$

the asterisk indicating that the CN is in an excited state. Notice we have changed the labeling of reactants and reaction products once again, in order to conform with the conventional notation, from the discussion of kinematics to the discussion of reactions. By now the reader can readily follow the translation from particles (1, 2, 3, 4) to (*a*, *X*, *b*, *Y*). The first arrow denotes the formation stage and the second the decay stage. For this reaction the cross section $\sigma(a, b)$ may be written as

$$\sigma(a, b) = \sigma_C(T_i) P_b(E) \tag{12.10}$$

where $\sigma_C(T_i)$ is the cross section for the *CN* formation at kinetic energy T_i, the available kinetic energy for reaction as discussed above, and $P_b(E)$ is the probability that the *CN* at energy level E will decay by emission of particle b. It is implied that σ_C and P_b are to be evaluated separately since the formation and decay processes are assumed to be decoupled. The energy-level diagram for this reaction is shown in Fig. 12.3 for an endothermic reaction ($Q < 0$). Notice that E is the *CN* excitation and it is measured relative to the rest-mass energy of the nucleus $(a + X)$. If this nucleus should have an excited state (a virtual level) at E^* which is close to

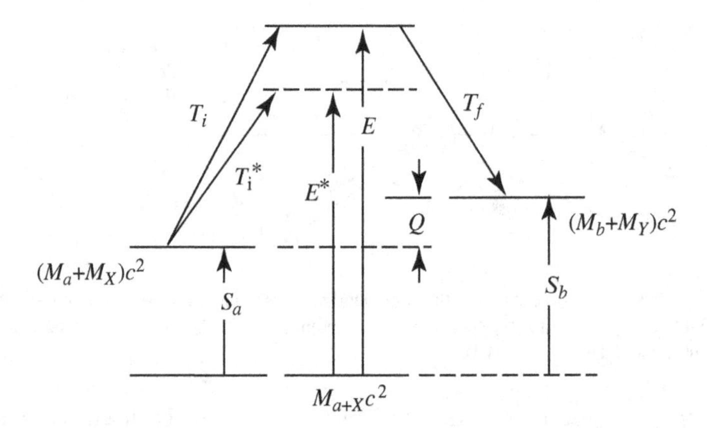

Fig. 12.3 Energy-level diagram for the reaction $a + X \to b + Y$ via CN formation and decay. Two scenarios are depicted, a general case of excitation of the compound nucles $(a + X)$ to energy E above its ground state with corresponding incoming and outgoing kinetic energies T_i and T_f , respectively, and a special case of excitation to energy E^*, which is a virtual level of the compound nucleus, with corresponding specific value T_i^*. The latter excitation is a resonance condition, for which the reaction cross section will show a peak in the vicinity of $T_i{\sim}T_i^*$.

E, then one can have a resonance condition. If the incoming particle a should have a kinetic energy such that the kinetic energy available for reaction has the value T_i^*, then the CN excitation energy matches an excited level of the nucleus $(a + X)$, $E \sim E^*$. Therefore the CN formation cross section $\sigma_C(T_i)$ will show a peak in its variation with T_i as an indication of a resonance reaction.

The condition for a resonance reaction is thus a relation between the incoming kinetic energy and the rest-mass energies of the reactants. Figure 12.3 shows that this relation can be stated as $T_i \sim T_i^*$, or

$$(M_a + M_X)c^2 + T_i^* = M_{a+X}c^2 + E^* \tag{12.11}$$

Each virtual level E^* has a certain *energy width*, denoted as Γ, which corresponds to a finite lifetime of the state (level), $\tau = \hbar/\Gamma$. A sharp resonance peak means a virtual level with long lifetime.

The cross section for CN formation has to be calculated quantum mechanically [Burcham, 1963, p. 532, Blatt, 1952, pp. 398]. One finds

$$\sigma_C(T_i) = \pi \bar{\lambda}^2 g_J \frac{\Gamma_a \Gamma}{(T_i - T_i^*)^2 + \Gamma^2/4} \tag{12.12}$$

where $g_J = \frac{2J+1}{(2I_a+1)(2I_X+1)}$ and $\underline{J} = \underline{I}_a + \underline{I}_X + \underline{L}_a$. In this expression $\bar{\lambda}$ is the reduced wavelength (wavelength/2π) of particle a in CMCS, $\underline{J}$ is the total angular momentum, the sum of the spins of particles a and X and the orbital angular momentum associated with particle a (recall particle X is stationary), Γ_a is the energy width (partial width) for the incoming channel ($a+X$), and Γ (without any index) is the *total* decay width, the sum of all partial widths (a channel is the system of the reactants or the products in a reaction). The idea here is CN formation can result from a number of channels, each with its own partial width. In our case the channel is the reaction with particle a, and the partial width Γ_a is a measure of the strength of this channel. Given our relation Eq. (12.11) we can also regard the CN formation cross section to be a function of the excitation energy E, in which case $\sigma_C(E)$ is given by Eq. (12.12) with $(E-E^*)^2$ replacing the factor $(T_i - T_i^*)^2$ in the denominator.

To complete the cross section expression Eq. (12.10) we need to specify the probability for the decay of the compound nucleus. This is a matter that involves the excitation energy E and the decay channel where particle b is emitted. Treating this process like radioactive decay, we can say

$$P_b(E) = \Gamma_b(E)/\Gamma(E) \tag{12.13}$$

where $\Gamma(E) = \Gamma_a(E) + \Gamma_b(E) +$ width of any other decay channel allowed by the energetics and selective rules. Typically one includes a radiation partial width Γ_γ since gamma emission is usually an allowed process. Combining Eqs. (12.12) and (12.13) we have the cross section for a resonance reaction. In neutron reaction theory this result is known as the *Breit-Wigner* formula for a single resonance. There are two cross sections of interest to us, one for neutron absorption and another for neutron elastic scattering. They are usually written as

$$\sigma(n,\gamma) = \pi\bar{\lambda}^2 g_J \frac{\Gamma_n\Gamma_\gamma}{(T_i - T_i^*)^2 + \Gamma^2/4} \tag{12.14}$$

$$\sigma(n,n) = 4\pi a^2 + \pi\bar{\lambda}^2 g_J \frac{\Gamma_n^2}{(T_i - T_i^*)^2 + \Gamma^2/4} + 4\pi\bar{\lambda} g_J a\Gamma_n \frac{(T_i - T_i^*)}{(T_i - T_i^*)^2 + \Gamma^2/4} \tag{12.15}$$

In $\sigma(n,n)$ the first term is the potential scattering contribution, what we have previously called the *s-wave* part of elastic scattering, with a being the scattering length. The second term in Eq. (12.15) is the compound elastic scattering contribution. It is the term responsible for the peak behavior of the cross section. The last term represents the *interference* between potential scattering and resonant scattering. Notice the interference is *destructive* at energy below the resonance energy and *constructive* above the resonance. We will soon see these are the characteristic

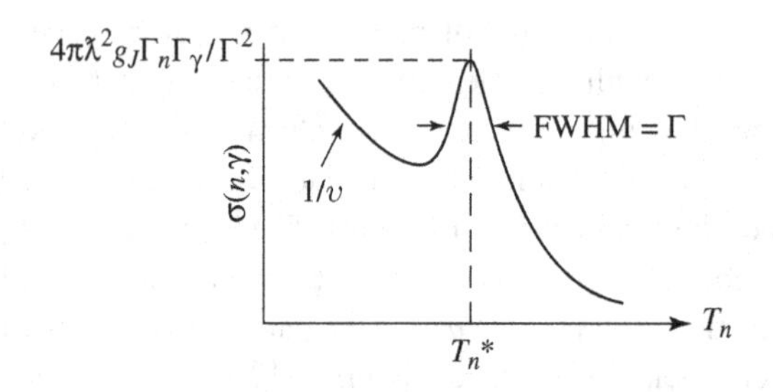

Fig. 12.4 Schematic of *Breit-Wigner* resonance behavior for neutron absorption displayed by the cross section for radiative capture $\sigma(n, \gamma)$. T_n is the kinetic energy of the neutron in LCS, which should be very close to T_i because the target is usually a heavy nuclide, the mass of which the difference between the two kinetic energies negligible.

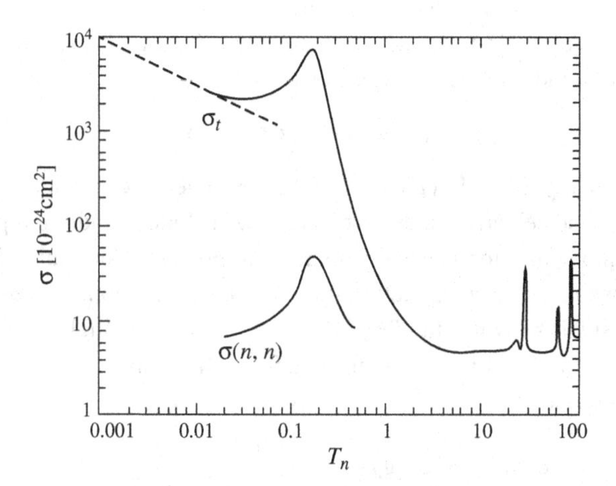

Fig. 12.5 Total and elastic neutron scattering cross sections of *Cd* showing a resonant absorption peak and a resonant scattering peak, respectively, at very different magnitudes. [Adapted from Larmarsh, 1966.]

signatures of the presence of interference in a resonance reaction. In Fig. 12.4, we show schematically the energy behavior of the absorption cross section in the form of a resonance peak. Below the resonance peak the cross section varies like $1/v$ as one can deduce from Eq. (12.14) by noting the energy dependence of the various factors, along with $\Gamma_n \sim \sqrt{T}$, and $\Gamma_\gamma \sim$ *constant*. Notice also the full width at half maximum is governed by the total decay width Γ. Figure 12.5 shows a well-known absorption peak in *Cd* which is widely used as an absorber of low-energy neutrons. One can see the resonance behavior in both the total cross section, which is dominated by absorption, and the elastic scattering cross section in the inset.

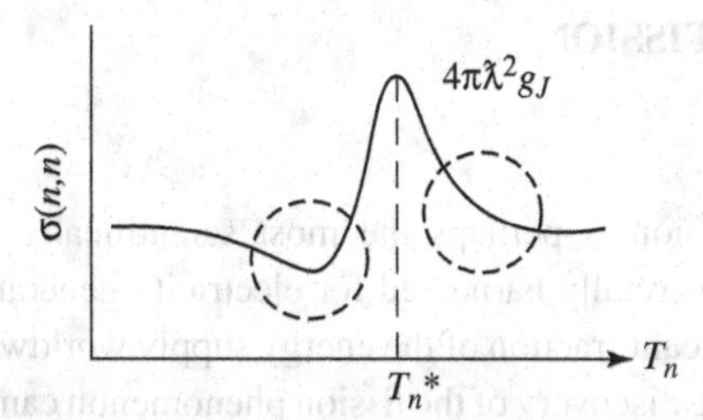

Fig. 12.6 Interference effects in elastic neutron scattering, below and above the resonance, showing destructive and constructive interference, respectively.

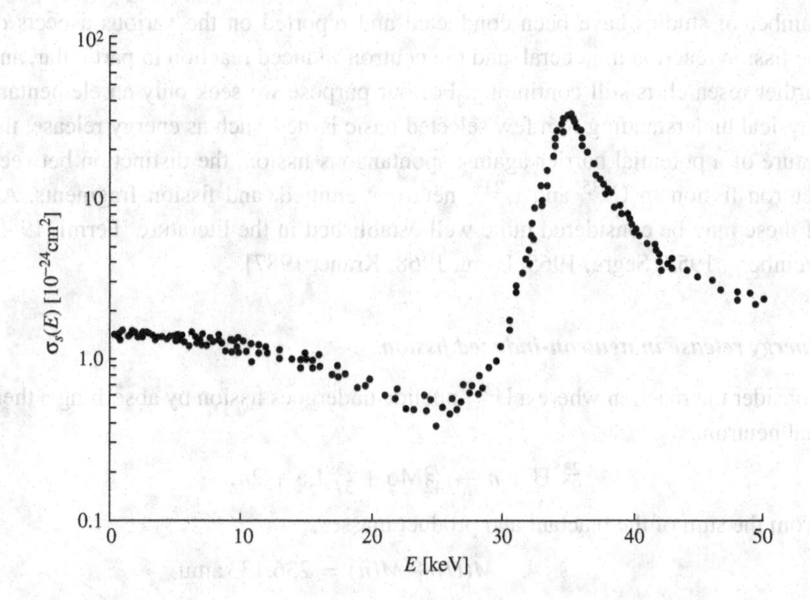

Fig. 12.7 Experimental scattering cross section of Al^{27} showing the interference effects between potential and resonance scattering, and an asymptotically constant value (potential scattering) sufficiently far away from the resonance in the vicinity of about 36 keV. [Adapted from Lynn, 1968.]

We conclude our discussion of compound nucleus reactions by returning to the feature of constructive and destructive interference between potential scattering and resonance scattering in the elastic scattering cross section. Figure 12.6 shows this behavior schematically, and Fig. 12.7 shows that such effects are indeed observed [Lynn, 1968]. Admittedly this feature is not always seen in the data. While the present example is carefully chosen and should not be taken as being a typical situation, the characteristic of destructive and constructive interference can be seen quite generally among the neutron resonant cross sections.

12.3 NEUTRON FISSION

The fission reaction

Neutron-induced fission is perhaps the most scientifically spectacular nuclear reaction being commercially harnessed for electricity generation. Nuclear power accounts for a significant fraction of the energy supply worldwide. As we have discussed in Chap. 1, the discovery of the fission phenomenon came as quite a surprise to the scientific community. Now that we have essentially covered the fundamentals of nuclear radiation interactions in the preceding chapters, we may ask how much do we really understand about this reaction. One can be assured exhaustive number of studies have been conducted and reported on the various aspects of the fission reaction in general, and the neutron-induced reaction in particular, and further research is still continuing. For our purpose we seek only an elementary physical understanding of a few selected basic issues, such as energy release, the nature of a potential barrier against spontaneous fission, the distinction between neutron fission in U^{235} and U^{238}, neutrons emitted, and fission fragments. All of these may be considered quite well-established in the literature [Fermi, 1949; Weinberg, 1958; Segrè, 1965; Lynn, 1968; Krane, 1987].

Energy release in neutron-induced fission

Consider the reaction where a U^{235} nuclide undergoes fission by absorbing a thermal neutron,

$$^{235}_{92}\mathrm{U} + n \rightarrow {}^{95}_{42}\mathrm{Mo} + {}^{139}_{57}\mathrm{La} + 2n.$$

From the sum of the reactant and product masses,

$$M(U) + M(n) = 236.133\,\text{amu}$$

$$M(Mo) + M(La) + 2M(n) = 235.918$$

we obtain the Q-value, or the mass difference Δ, to be 0.215 amu, or 200 MeV. While this is the total energy released, not all of it is given off instantaneously. The two fission fragments, being in highly excited states and neutron rich, will immediately begin to emit particles through various decay modes. The typical distribution of energy release (in MeV) can be separated into prompt release, consisting of the kinetic energies of the fission fragments (165 ± 5), kinetic energies of fission neutrons (5), prompt γ (6 ± 1), and subsequent release due to fission product decays, in the form of β (8 ± 1.5), antineutrinos (12 ± 2.5) and γ (6 ± 1). All the energies except those of the neutrinos are recoverable in the form of heat. In addition, each parasitic absorption (capture) of a neutron leads to capture γ with energies ~12 MeV which are also recoverable. The total energy per fission given off as heat is therefore ~202 MeV.

Effective fission barrier

We have just seen the Q-value for the n-fission reaction is positive and large. With the reaction being exothermic, one can ask why does it not occur spontaneously? An answer one can give is experimentally spontaneous fission of uranium is not observed, thus implying the existence of a threshold energy. Another way to answer the question is to consider the stability limits of heavy nuclides against spontaneous fission using our knowledge of nuclear masses. Recall from Chap. 4 the semi-empirical mass expression contains a contribution from the electrostatic repulsion between the protons and another contribution due to surface tension. The *Liquid Drop* model is useful for calculating the potential energy change when the nuclear drop suffers an ellipsoidal deformation from spherical shape [Fermi, 1949, p. 164]. This involves considering the two competing effects of electrostatic repulsion and capillary cohesion. The result is a spherical nuclear drop would be stable if $Z^2/A < 45$. This is an interesting prediction that can be readily tested against known data. Fig. 12.8 shows how close to this stability limit are the heavy

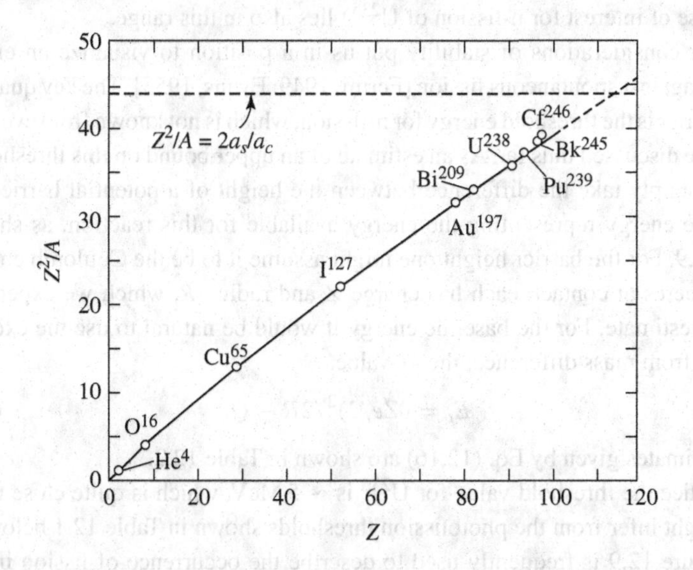

Fig. 12.8 Variation of Z^2/A with Z for selected nuclides covering the full range of Z showing the proximity of the heavy nuclides to the stability limit estimated according to the *Liquid Drop* model. Parameters a_s and a_c are the coefficients in the semi-empirical mass formula, Eq. (4.10) for the surface and Coulomb terms respectively. [Adapted from Evans, 1955, p. 390.]

Table 12.1 Photofission thresholds of several heavy nuclides [Weinberg, 1958, p. 107].

Th^{232}	U^{233}	U^{235}	U^{238}	Pu^{239}
5.9	5.5	5.75	5.85	5.5

nuclides of interest [Evans, 1949, p. 390]. Notice the order of decreasing Z^2/A from *Californium* to *Plutonium* to *Uranium* is in accord with what we know about the relative stability of these nuclides.

Another way to address the question of stability is to consider the photofission reaction (γ, f). It is known experimentally that photofission is a threshold event. The minimum excitation energy required (in MeV) is shown in Table 12.1.

The threshold for n-fission is less than the γ-fission threshold of the compound nucleus by the binding energy of the neutron to the target nucleus, As a result some of the n-fission thresholds are "negative" which means fission can be induced by thermal neutron absorption. Concerning the stability of uranium nuclides against spontaneous fission we thus have an indication from photofission data that the fission threshold of uranium nuclides U^{233}, U^{235}, U^{238} are in the range 5.5 to 5.85 MeV. From this we may reasonably assume the γ-fission threshold for U^{236} (the case of interest for n-fission of U^{235}) lies also in this range.

Our considerations of stability put us in a position to visualize an effective barrier against spontaneous fission [Fermi, 1949; Evans, 1955]. The key quantity in this barrier is the threshold energy for n-fission, which is not known from everything we have discussed thus far. As an estimate of an upper bound on this threshold one might simply take the difference between the height of a potential barrier and a baseline energy representing the energy available for this reaction, as shown in Fig. 12.9. For the barrier height one might assume it to be the Coulomb energy of two spheres at contact, each has charge Z and radius R, which we expect to be an overestimate. For the baseline energy it would be natural to use the excitation energy from mass difference, the Q-value.

$$E_t = (Ze/2)^2/2R - Q. \tag{12.16}$$

The estimates given by Eq. (12.16) are shown in Table 12.2.

Notice the threshold value for U^{236} is ~ 5 MeV, which is quite close to what one might infer from the photofission thresholds shown in Table 12.1 below.

Figure 12.9 is frequently used to describe the occurrence of fission in terms of a potential barrier. It portrays schematically the mechanism of fission as an activation process. The situation is analogous to the model of α-decay in Sec. 6.2, see Fig. 6.11. Starting with the system at an initial energy level, an amount of

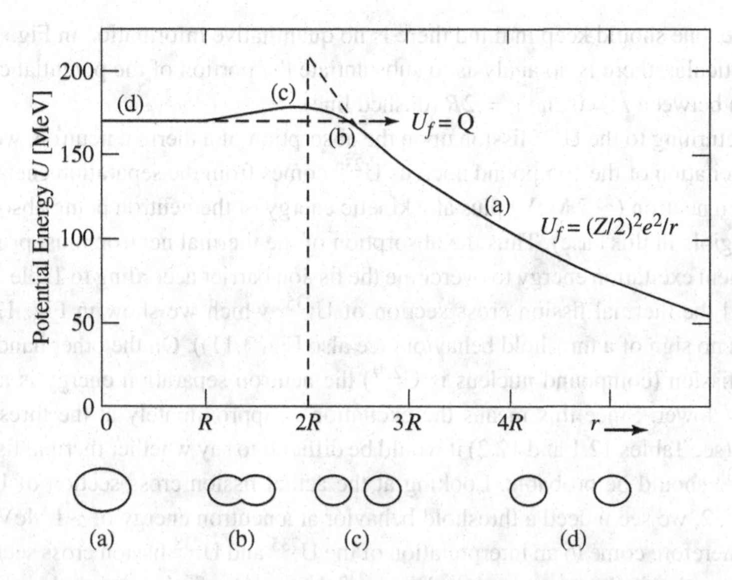

Fig. 12.9 Schematic of variation of an *effective* potential barrier for fission, U_f, with a coordinate r representing the separation distance of two equal fission fragments [Evans, 1955, p. 387]. Sketches at the bottom of the figure suggest the evolution from initial deformation of a liquid drop (stage a) to beginning of break-up (b), to point of separation (c), and to fully separated equal droplets (d).

Table 12.2 n-fission threshold energies (in MeV) estimated using Eq. (12.16). [Weinberg, 1958, p. 107.]

A	16	60	100	140	200	236
$(Ze/2)^2/2R$	4	32	62	110	175	210
$-Q$	14.5	16	-15	-48	-135	-205
E_t	18.5	48	47	62	40	~ 5

energy is required for the system to go over (tunnel through) a potential barrier at a height greater than the initial energy. In the approximation underlying Eq. (12.16), the barrier height and the initial energy level and the barrier height are given by the two terms in Eq. (12.16) respectively. Moreover, if one assumes the activation to correspond to exciting a liquid drop and causing it to break up into two droplets (the Liquid Drop model), then the coordinate r in Fig. 12.9 can be interpreted as the separation between the two droplets. In this scenario the system configuration is visualized as starting out as a deforming ellipsoid at $r = 0$ (stage a), evolving toward a break-up (b), reaching the point of break-up as two equal spheres in contact (c), and fully separating as r increases further. Although this is an intuitively appealing picture, one should keep in mind there is no quantitative information in Fig. 12.9. In particular, there is no analysis to substantiate the portion of the potential curve drawn between $r = 0$ and $r = 2R$ (dashed line).

Returning to the U^{235} fission upon the absorption of a thermal neutron, we see the excitation of the compound nucleus U^{236} comes from the separation energy of an even neutron (~ 7 MeV) plus any kinetic energy of the neutron being absorbed (negligible in this case). Thus the absorption of the thermal neutron must provide sufficient excitation energy to overcome the fission barrier according to Table 12.2. Indeed the thermal fission cross section of U^{235}, which we show in Fig. 12.10, shows no sign of a threshold behavior (see also Fig. 3.11)). On the other hand, for U^{238} fission (compound nucleus is U^{239}) the neutron separation energy is about 1 MeV lower. Since this means the excitation is approximately at the threshold value (see Tables 12.1 and 12.2) it would be difficult to say whether thermal fission in U^{238} should be probable. Looking at the actual fission cross section of U^{238}, Fig. 3.12, we see indeed a threshold behavior at a neutron energy of $\sim$1 MeV. We may therefore come to an interpretation of the U^{235} and U^{238} fission cross sections by noting that in the compound nucleus U^{236} the actual n-fission threshold is likely

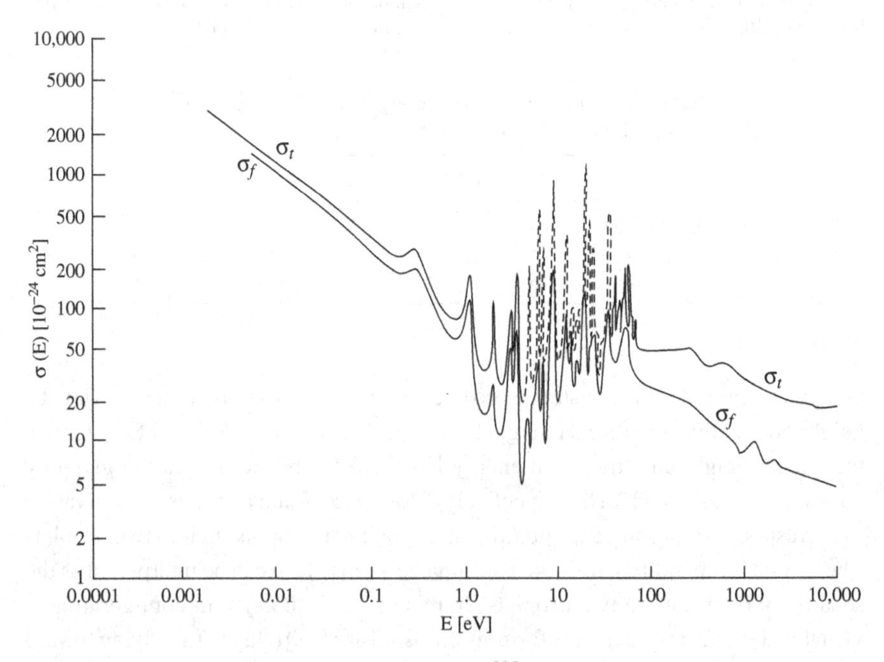

Fig. 12.10 Total and fission neutron cross sections of U^{235} in the thermal and low-energy resonance regions. [Adapted from Weinberg, 1958, p. 118.]

to be just about equal to the neutron separation. Assuming the n-fission threshold is the same between U^{236} and U^{239}, the fact that the neutron separation energy is 1 MeV lower in the latter means that this deficit needs to be made up by the kinetic energy of the neutron that is being absorbed, thus explaining the existence of a threshold energy for fission in U^{238}.

There are many fascinating details about the fission process for the interested reader to pursue. An excellent account can be found in the monograph by Weinberg and Wigner [Weinberg, 1958, Chap. V]. A way to visualize the overall event is to construct a time sequence of the various physical phenomena in the form of a time evolution diagram. One can start with the excitation of the compound nucleus (time 0), followed by the distortion response of the nucleus leading to its fragmentation (time 1), and continue with the two fragments moving apart at great speed (time 2). As the highly excited fragments move in the medium each will emit particles, neutrons first (time 3) and then γ rays (time 4), until the fragments come to a stop (time 5), at which point the fragments may be regarded as fission products. Such a kinetics evolution is depicted in Fig. 12.11. With this diagram as a guide one can go into further details of how the fission process evolves at each stage. We have already commented on the nature of the excitation of the compound nucleus, the energy being provided by the neutron separation energy plus whatever kinetic energy the neutron has. This excitation causes the nucleus

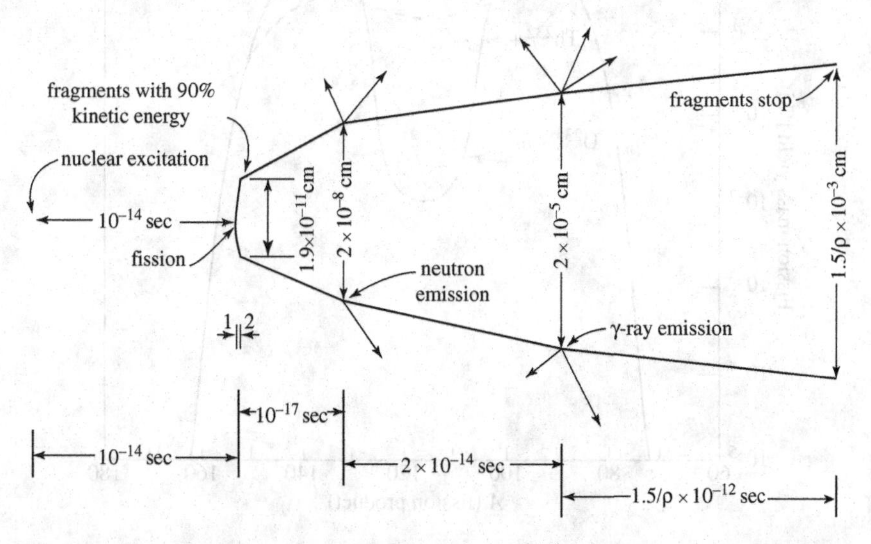

Fig. 12.11 Schematic of the kinetics of evolution of a fission process depicting the nuclear excitation leading to nuclear fragmentation and subsequent nuclear particle emissions in space and time. [Adapted from Weingberg, 1958, p. 115.]

to undergo distortion from spherical to an ellipsoidal shape toward an instability in about 10^{-14} sec (lifetime of the *CN* derived from measurement of the fission width Γ_f), at a separation between the two fragments of $\sim 2 \times 10^{-11}$ cm (range of nuclear forces). The fragments are highly charged and once they are beyond the range of nuclear forces they will be strongly accelerated by the Coulomb repulsion to speeds of the order of 10^9 cm/sec. The fragments have been measured to have a *bimodal* mass distribution with peaks at around 94 and 142, as shown in Fig. 12.12. While a simple explanation of the asymmetry is lacking, we do know the fission yield becomes much more symmetric at significantly higher excitation energy. See,for example, the data on fission of Th^{232} by the bombardment of 37 MeV α-particles in Fig. 12.12. Recall the fragments are middleweight nuclei which are "neutron rich" since their N/Z ratio, being that for stable heavy nuclei, is greater than the N/Z ratio for stable middleweight nuclei. Thus one expects neutrons to be emitted at the instant of fission (these are the 2.5 ± 0.1 prompt neutrons). Both the number of neutrons emitted and their energy spectrum are relevant in the present discussion for their role in sustaining the chain reactions in a nuclear reactor. Table 12.3 [Segrè 1965, p. 494] shows the probability of emitting

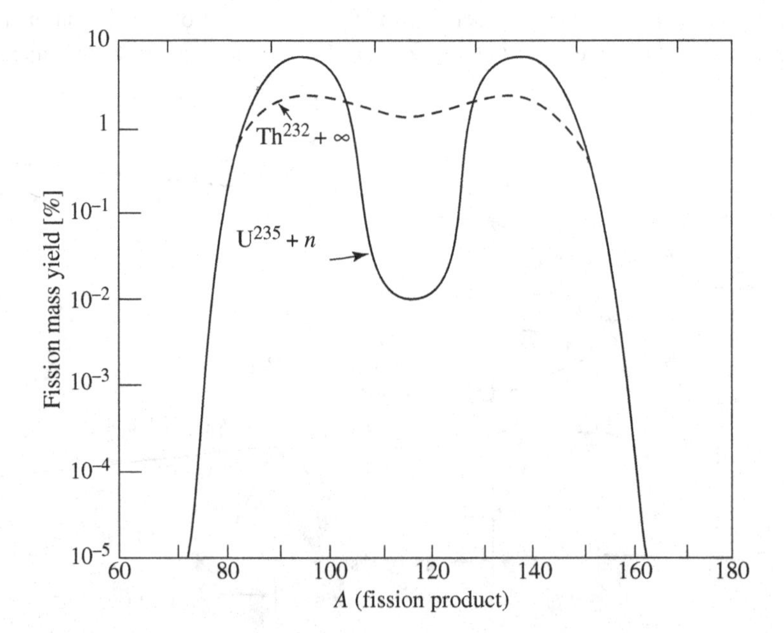

Fig. 12.12 Distribution of mass (atomic number) of fission products from neutron-induced fission of U^{235} at thermal energies showing a bi-modal form (solid curve). Also shown are the results of fission of Th^{232} induced by a-particles at 37 MeV (dashed curve). [Adapted from Evans, 1955, p. 393.]

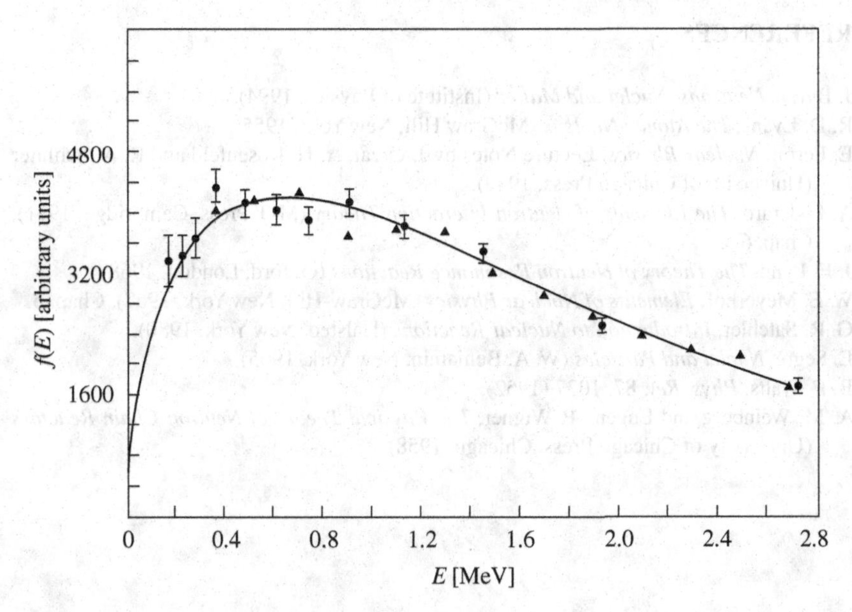

Fig. 12.13 The energy distribution of neutrons emitted from thermal fission of U^{235}, measurements are shown by the symbols and the solid line is given by Eq. (12.17). [Adapted from Weingberg, 1958, p. 113.]

a certain number of neutrons in bombarding $\underline{U}^{235}$ by neutrons at two different energies. The average number of neutrons is seen to increase slightly from 2.45 at 80 keV to 2.65 at 1.25 MeV. As for the energy distribution of the fission neutrons, experimental data for U^{235} fission are shown in Fig. 12.13. An analytic expression, known as the *Watts spectrum*, represents the data well [B. E. Watts, *Phys. Rev.* **87**, 1037 (1952)],

$$f(E) = A(E)^{1/2} \exp(-0.775E) \tag{12.17}$$

with $A = 2 \times 0.775(0.775/\pi)^{1/2}$. The spectrum peaks at about 0.7 MeV with an average energy at ~1.9 MeV. The average number of neutrons emitted per fission event (ν) varies somewhat from one fissile nucleus to another. It is mildly energy dependent. For U^{235}, $\nu = \nu(thermal) + constant \times E$ (MeV), with $\nu(thermal) =$ 2.43 and $constant = 0.1346\,\text{MeV}^{-1}$. Some of the neutrons emitted from fission appear with significant time delays, up to seconds because β decay processes can be quite slow. The role of delayed neutrons is critical in reactor control under normal operations; without delayed neutrons the population of neutrons from one generation to the next will change exponentially (see Eq. (13.126)).

REFERENCES

J. Byrne, *Neutrons, Nuclei and Matter* (Institute of Physics, 1994).

R. D. Evans, *The Atomic Nucleus* (McGraw Hill, New York, 1955).

E. Fermi, *Nuclear Physics*, Lecture Notes by J. Orear, A. H. Rosenfeld and R. A. Schluter (University of Chicago Press, 1949).

A. Foderaro, *The Elements of Neutron Interaction Theory* (MIT Press, Cambridge, 1971), Chap. 6.

J. E. Lynn, *The Theory of Neutron Resonance Reactions* (Oxford, London, 1968).

W. E. Meyerhof, *Elements of Nuclear Physics* (McGraw-Hill, New York, 1967), Chap. 5.

G. R. Satchler, *Introduction to Nuclear Reactions* (Halsted, New York, 1980).

E. Segrè, *Nuclei and Particles* (W. A. Benjamin, New York, 1965).

B. E. Watts, *Phys. Rev.* **87**, 1037 (1952).

A. M. Weinberg and Eugene P. Wigner, *The Physical Theory of Neutron Chain Reactors* (University of Chicago Press, Chicago, 1958).

PART 3

Cumulative Effects of Nuclear Radiation Interactions

13

Neutron Transport

We have emphasized throughout this book the relation between *Nuclear Radiation Interactions* and *Radiation Transport*, as the two foundational areas of *Nuclear Science and Engineering*. In the process we have made clear the connection lies most directly in the cross sections, the basic quantities that characterize our study of interactions. It will be seen in this chapter these cross sections are the physical input to the study of neutron transport, thus demonstrating the inter-relation between *interactions*, studied as a set of individual processes, and *transport*, described as the cumulative behavior evolving over a series of interactions.

Particle Transport is the broad subject of movements of particles under gradients and driving forces that exist within the system, as well as conditions externally imposed on the system. These movements are described by a governing equation usually formulated under a set of assumptions concerning the particle interactions. To predict the behavior of nuclear systems such as a nuclear reactor, one needs to know, in principle, the time-dependent distribution of neutrons in space and energy, and direction of travel. This information can be obtained from the neutron transport equation which we will introduce in this chapter. It turns out the problem can be generally divided into two parts, one pertaining to the moderation of neutrons from high energy, where they were emitted by a fission reaction, to around thermal energy $k_B T$, where T is the temperature of the reactor core, and another part pertaining to the diffusion of the neutrons that have slowed down to the thermal energy range. Not surprisingly, the two problems are known as *neutron slowing down* and *thermalization* respectively. We will see neutron cross sections of various types appear directly in the equations governing neutron transport, slowing down, and thermalization. It follows therefore the predicted behavior of the nuclear system will depend directly on the cross section variations. This is the most direct way for the reader of this book way to appreciate the link between the interaction processes studied in this book and the transport applications encountered in nuclear science and engineering.

Each of the sections treated in this chapter is a subject practically worthy of a book by itself. Indeed special treatises and monographs on neutron diffusion, slowing down, and thermalization all exist in the literature [Case, 1953, Lamarsh, 1966; Bell, 1970; Williams, 1966; Parks, 1970; Duderstadt, 1979]. Since the focus of this book is nuclear interactions, we will provide only an introduction to neutron transport at a level comparable to the preceding chapters, favoring basic definitions and simple derivations to give physical understanding of the fundamental issues. Assuming the individual processes of neutron scattering, absorption, and fission are now reasonably familiar, we see it is the cumulative effects of these processes on the distribution of neutrons in space and energy that need to be discussed. The reader should pay attention to how these processes are coupled to each other, or compete with each other in determining the neutron distribution. Besides cross sections, related quantities such as *diffusion coefficient* (*or length*), *slowing down density*, *resonance escape probability*, and *multiplication constant* will appear in the various descriptions (models) of neutron transport. These latter quantities will serve as further reminders of the underlying relation between interactions and transport.

As in all studies of particle transport phenomena, one begins with a fundamental equation describing the time-dependent distribution of the particles of interest in a medium in which the particles undergo displacements (convection or diffusion) and interactions with the medium (itself could be a system of particles, identical or distinguishable). Such an equation is typically referred to as the transport equation. It is no more than a balance relation for particles in a system where they can be *emitted* by an external source or by collision-induced reactions, and *lost* by absorption or leaking out of the system. In the case of neutron transport, the governing equation is the *neutron transport equation*. It bears close resemblance to another transport equation which is well known in gas dynamics or kinetic theory of gases and liquids, the *Boltzmann transport equation*. The neutron transport equation is linear in the neutron distribution, whereas the Boltzmann equation is quadratic in the distribution. This distinction essentially vanishes when one considers the Boltzmann equation in its linearized form, where one keeps terms to first order in an expansion about the equilibrium distribution, which is typically the Maxwellian distribution.

13.1 THE TRANSPORT EQUATION

In the study of neutron interactions and migration in any media the fundamental equation is the neutron transport equation. This equation has as its solution the time-dependent distribution function for neutrons in configuration-velocity (phase) space. Knowledge of this distribution function is sufficient to solve almost all problems of interest in the theory of nuclear reactors. However, most of the time one does not need to know the distribution function itself, the integrals over some of the phase-space coordinates such as velocity direction, energy, or position are usually sufficient.

A brief derivation of the transport equation is instructive in showing that it is just a balance relation, an accounting of gains and losses in a control volume. More elaborate derivations than what we do here can be given, including a quantum mechanical formulation [Osborn, 1967] , but we believe this is not necessary for the basic understanding of the physical content of the equation. Let us define the number density as

$$n(\underline{r}, \underline{v}, t)d^3rd^3v \equiv \text{expected no. of neutrons in } d^3r \text{ about } \underline{r} \text{ with velocities in } d^3v \text{ about } \underline{v} \text{ at time } t$$

Instead of the vector variable $\underline{v}$, it is often more convenient to use the scalar variable E and the two-dimensional unit vector $\underline{\Omega} = \underline{r}/r$, defined in Fig. 5.4, so that $\underline{v} = (v_x, v_y, v_z) \rightarrow (v, \theta, \varphi) \rightarrow (E, \theta, \varphi)$. Correspondingly, $d^3v = v^2 dv d\Omega$, with $d\Omega = \sin\theta d\theta d\phi$. Then we have

$$n(\underline{r}, \underline{v}, t)d^3rd^3v \equiv n(\underline{r}, E, \underline{\Omega}, t)d^3rdEd\Omega$$
$$= \text{expected no. of neutrons in } d^3r \text{ about } \underline{r} \text{ with energies in dE about E and going in directions in } d\Omega \text{ about } \underline{\Omega} \text{ at time } t$$

Consider a subsystem of volume V and surface S. Suppose we want to calculate the change in the number of neutrons in V with energies in dE about E and direction in $d\Omega$ about $\underline{\Omega}$ during a time interval Δt. This is given by

$$\int_V [n(\underline{r}, E, \underline{\Omega}, t+\Delta t) - n(\underline{r}, E, \underline{\Omega}, t)]d^3rdEd\Omega = gains - losses \qquad (13.1)$$

There are two contributions to the *gains*.

(1) *Fission and external source*

$$\frac{vf(E)dEd\Omega}{4\pi}\iiint_{V,\underline{\Omega}',E'} \Sigma_f(E')\phi(\underline{r}, E', \underline{\Omega}', t)dE'd\Omega'd^3r\Delta t$$
$$+\int_V S(\underline{r}, E, \underline{\Omega}, t)d^3rdEd\Omega\Delta t \qquad (13.2)$$

where $f(E)$ is the fission spectrum (see Eq. (12.17)), $\phi(\underline{r}, E, \underline{\Omega}, t) \equiv n(\underline{r}, E, \underline{\Omega}, t)v$ is the *neutron flux*, and S is the external source distribution.

(2) *Scattering*

$$\iiint_{V,E',\underline{\Omega}'} \Sigma_s(E')\phi(\underline{r}, E', \underline{\Omega}', t)d^3rdE'd\Omega'\Delta t F(E'\underline{\Omega}' \rightarrow E\underline{\Omega})dEd\Omega \qquad (13.3)$$

where $F(E'\underline{\Omega}' \rightarrow E\underline{\Omega})dEd\Omega$ = conditional probability that given a neutron scattered at (E', $\underline{\Omega}'$), its energy will be in *dE about E* and its direction will be in $d\Omega$ about $\underline{\Omega}$.

For losses there are also two terms, one for collisions and the other convective flow.

(3) *Collisions*

$$\int_V \Sigma_t(E)\phi(\underline{r}, E, \underline{\Omega}, t)d^3r dE d\Omega \Delta t$$

(4) Net flow outward

$$\int_S \underline{\Omega} \cdot \underset{\rightarrow}{\hat{n}}\, vn(\underline{r}_s, E, \underline{\Omega}, t)ds dE d\Omega \Delta t = \int_V d^3r \underline{\Omega} \cdot \underline{\nabla}\phi(\underline{r}, E, \underline{\Omega}, t)dE d\Omega \Delta t \quad (13.4)$$

where $\hat{\underline{n}}$ is the outward normal at $\underline{r}_s$, and the divergence theorem, $\int_S d\underline{s} \cdot \underline{F} = \int_V d^3r \underline{\nabla} \cdot \underline{F}$, has been applied. Putting together the gains and losses, dividing by Δt, and taking the limit of $\Delta t \to 0$, we can write the balance, Eq. (13.1), as $\int_V [\,] = 0$. Since V can be any arbitrary part of the system, the integrand [ ] must vanish identically if the integral is to vanish for any V. Thus one obtains,

$$\begin{aligned}\frac{\partial n(\underline{r}, E, \underline{\Omega}, t)}{\partial t} &= \frac{vf(E)}{4\pi} \int_{E', \underline{\Omega}'} dE' d\Omega' \Sigma_f(E')\phi(\underline{r}, E', \underline{\Omega}', t) + S(\underline{r}, E, \underline{\Omega}, t) \\ &+ \int_{E', \underline{\Omega}'} dE' d\Omega' \Sigma_s(E')\phi(\underline{r}, E', \underline{\Omega}', t)F(E'\underline{\Omega}' \to E\underline{\Omega}) \\ &- \Sigma_t(E)\phi(\underline{r}, E, \underline{\Omega}, t) - \underline{\Omega} \cdot \underline{\nabla}\phi(\underline{r}, E, \underline{\Omega}, t) \end{aligned} \quad (13.5)$$

Equation (13.5) is the neutron transport equation for a homogeneous medium. For a heterogeneous system we simply let $\Sigma(E) \to \Sigma(\underline{r}, E)$. Notice this is a *linear* equation because we have ignored the neutron-neutron interactions (the mean free path for such events is 10^8 cm or greater). Sometimes the neutron transport equation has been called the Boltzmann equation; one should be careful when using this terminology since the Boltzmann equation in kinetic theory of gases treats explicitly the collisions among the particles and is in general *nonlinear*.

The transport equation is an integro-differential equation in 7 variables. While it is much too complicated for us to attempt any kind of solution directly, either as a boundary-value or initial-value problem, all the equations in reactor theory that we will encounter can be derived from the transport equation in one approximation or another.

13.2 ENERGY MODERATION (*SLOWING DOWN*)

In neutron energy moderation one is concerned with lowering the energy of a fast neutron (~ MeV) down to the thermal energy region (0.025 eV) [Goodjohn, 1966;

Williams, 1966]. The equation that describes the slowing down process of elastic scattering can be obtained from the transport equation, Eq. (13.5), by an appropriate reduction. Another process of energy moderation, that of inelastic scattering (in the nuclear sense) also exists in special circumstances, but will not be considered here. Since we are dealing with changes in the neutron energy without regard to spatial variations, or the direction of travel, it would be simplest if we can appropriately eliminate the dependence of the flux in the transport equation on $\underline{r}$ and Ω. There are two simple ways to get rid of the spatial dependence in neutron transport. One is to integrate the transport equation over a large system and then take the system size to be infinite,

$$\phi(E) = \int \phi(\underline{r}, E) d^3r \tag{13.6}$$

$$\int \underline{\nabla} \cdot \underline{J} d^3r = \int_S (\hat{n} \cdot \underline{J}) dS \to 0 \tag{13.7}$$

This basically gets rid of the surface effects (leakage). The other is to consider a "point reactor",

$$\phi(\underline{r}, E) = \phi(E)\delta(\underline{r}) \tag{13.8}$$

and again integrate over space (along with invoking Fick's rule),

$$\int_{-\infty}^{\infty} dx \frac{\partial^2}{\partial x^2}\delta(x) = \frac{1}{2\pi}\int_{-\infty}^{\infty} dx \frac{\partial^2}{\partial x^2}\int_{-\infty}^{\infty} dk e^{ikx} = -\int_{-\infty}^{\infty} dk k^2 \delta(k) = 0 \tag{13.9}$$

In either case, one obtains

$$\Sigma_t(E)\phi(E) = S(E) + \int dE' \Sigma_s(E')\phi(E')F(E' \to E) \tag{13.10}$$

which is known as the *slowing down equation for an infinite medium*. In writing Eq. (13.10), we have combined the external source with the fission source in $S(E)$. Without specifying the scattering kernel F, an equation having the same form as Eq. (13.10) is also applicable to the thermal energy region, where up-scattering of the neutron needs to be considered. In that case the energy balance would be called the equation of *neutron thermalization* (see Sec. 13.4).

We will study Eq. (13.10) using the energy transfer kernel previously derived under the conditions of elastic scattering, target nucleus at rest, and isotropic scattering in CMCS (see Chap. 9). This kernel is well suited in the slowing down region where a neutron loses energy every time it is scattered by the nucleus (recall

Eq. (9.31)),

$$F(E' \to E) = \frac{1}{E'(1-\alpha)} \qquad \alpha E' \leq E \leq E'$$
$$= 0 \qquad \text{otherwise} \tag{13.11}$$

where $\alpha = (A-1)^2/(A+1)^2$, and $A = M/m$.

Before discussing the solution of Eq. (13.10), we make a slight digression to consider an application of Eq. (13.11) to the related problem of estimating the escape probability $P(E' \to E)$ that a neutron, scattered at E', will cross E, with no restriction on how large is the interval from E' to E and how many collisions are involved. Notice $P(E' \to E)$ differs from $F(E' \to E)$ in that the latter is for one collision and is a distribution function in the variable E. In contrast, P is not a distribution function, although it is still a probability. The exercise of estimating $P(E' \to E)$ will make clear this distinction. Since the energy range (E', E) is arbitrary, we divide this interval into a number of subintervals, ΔE_j, $j = 1, \ldots,$ such that for each subinterval we can write

$$\{p_a(\Delta E_j)\} = \{\text{avg. no. coll. req'd to cross } \Delta E_j\}$$
$$\times\{\text{prob absorp per coll}\} \tag{13.12}$$

where $p_a(\Delta E_j)$ is the probability of neutron absorption while crossing ΔE_j. The second part of Eq. (13.12), the probability of absorption each time the neturon collides, is just

$$\{\text{prob absorp per coll}\} = \Sigma_a(\Delta E_j) = \Sigma_t(\Delta E_j) \tag{13.13}$$

Equation (13.12) reveals our strategy of estimating P. To find the probability of no absorption while crossing the interval (E', E), we write it as a product of probabilities,

$$P(E' \to E) = \prod_j [1 - p_a(\Delta E_j)] \tag{13.14}$$

where $p_a(\Delta E_j)$ is to be obtained according to Eqs. (13.12) and (13.13). The advantage of doing this is that once we have Eq. (13.14), we can rewrite it as

$$\ell n P = -\sum_j p_a(\Delta E_j) \to \int_E^{E'} p_a(E'')dE'' \tag{13.15}$$

using the properties, $\ell n(ab) = \ell n(a) + \ell n(b)$, and $\ell n(1-x) \sim -x$ for $x \ll 1$. Now we see the subinterval division, ΔE_j has to be chosen small enough to make

$p_a(\Delta E_j) < 1$. From Eq. (13.15) the probability we are seeking is given by the expression,

$$P(E' \to E) = \exp\left(- \int_E^{E'} p_a(E'')dE'' \right) \tag{13.16}$$

To turn this into an explicit expression we use Eqs. (13.12) and (13.13) to find $p_a(\Delta E_j)$, which means we need to make an estimate of the average number of collisions a neutron should make in going from energy E' to energy E.

To estimate the number of collisions a neutron should make, on average, in crossing an energy interval, consider the average energy that a neutron will lose in a collision at a given energy. Suppose the neutron collision occurs at energy E' and the neutron has energy E after the collision. The probability of such an event is just given by the energy transfer kernel $F(E' \to E)$. The energy the neutron would have, on average, after a collision at E' then becomes

$$\bar{E} = \int EF(E' \to E)dE = (1+\alpha)E'/2 \tag{13.17}$$

where we have used Eq. (13.11). The energy loss per collision is $E' - \bar{E} = (1-\alpha)E'/2$. Thus the average number of collisions required to cross an interval ΔE is

$$\{\text{avg. no. coll. req'd to cross } \Delta E_j\} = \frac{2}{1-\alpha}\frac{\Delta E}{E}. \tag{13.18}$$

With this result we arrive at an explicit expression for the escape probability $P(E' \to E)$ from (13.16),

$$P(E' \to E) = \exp\left(-\frac{2}{1-\alpha} \int_E^{E'} \frac{\Sigma_a(E'')}{\Sigma_t(E'')} \frac{dE''}{E''} \right) \tag{13.19}$$

This expression contains the parameter α whose numerical value depends on mass number of the target nucleus A. In the case of a hydrogenous medium, $A = 1$ and $\alpha = 0$. For a heavy target nucleus, $A \gg 1$, then $\alpha \sim 1 - (4/A)$, and the coefficient in front of the integral in Eq. (13.19) becomes $A/2$.

Before ending this discussion, we take the opportunity to introduce another energy-like quantity, called the *lethargy* in the reactor physics literature,

$$u \equiv \ell n(E_o/E) \tag{13.20}$$

where E_o is a *reference* energy. Notice there is a one-to-one correspondence between energy and lethargy, the two variables move in opposite directions, when the energy decreases the lethargy would increase. Given that we have just calculated the average energy loss per collision, it is not surprising that the corresponding average lethargy increase per collision turns out to be also a useful quantity,

$$\langle \Delta u \rangle = \langle u - u' \rangle = \langle \ell n(E'/E) \rangle = \int_{\alpha E'}^{E'} \ell n(E'/E) F(E' \to E) dE = 1 + \frac{\alpha \ell n \alpha}{1-\alpha} \equiv \xi \tag{13.21}$$

In the two limits which are always of interest,

$$\mathrm{A} = 1, \quad \alpha = 0, \quad \xi = 1$$

$$\mathrm{A} \gg 1, \quad \alpha \sim 1, \quad \xi = 2/(A + 2/3) \sim 2/A$$

We can use Eq. (13.21) to estimate the average number of collisions to cross an interval ΔE,

$$\{\text{no. coll. to cross } \Delta E\} = \frac{\ell n(E) - \ell n(E - \Delta E)}{\xi}$$

$$= \frac{1}{\xi} \ell n \left(\frac{E}{E - \Delta E} \right) = \frac{1}{\xi} \ell n \left(1 + \frac{\Delta E}{E} \right) \sim \frac{1}{\xi} \frac{\Delta E}{E} \tag{13.22}$$

Notice that if we insert the estimate Eq. (13.22) into Eq. (13.12), we would have arrived at the estimate for P,

$$P(E' \to E) = \exp\left(- \int_{E}^{E'} \frac{\Sigma_a(E'')}{\Sigma_t(E'')} \frac{dE''}{\xi E''} \right) \tag{13.23}$$

which agrees with Eq. (13.19) in the case of heavy target nucleus, but not when $A = 1$. We conclude that when $A > 1$ one can use either Eq. (13.19) or Eq. (13.23), whereas for $A = 1$ our discussion indicates Eq. (13.19) should be preferred since it involves fewer approximations.

We now turn our attention to the solution of the slowing down equation, Eq. (13.10), for the special case of hydrogenous medium, $A = 1$, a mono-energetic source at E_o, $S(E) = S_o \delta(E - E_o)$, and furthermore we neglect absorption. Equation (13.10) then reads

$$\Sigma_s(E)\phi(E) = \frac{S_o}{E_o} + \int_{E}^{E_o} \Sigma_s(E')\phi(E') \frac{dE'}{E'} \tag{13.24}$$

This is a simple integral equation that can be solved readily. Let $G(E) = \Sigma_s(E)\phi(E)$, Eq. (13.24) is then of the form,

$$G(E) = const + \int_E^{E_o} G(E')dE'/E' \tag{13.25}$$

Differentiating both sides with respect to E gives,

$$\frac{dG(E)}{dE} = -\frac{G(E)}{E}, \quad \text{or} \quad \frac{dG}{G} = -\frac{dE}{E} \tag{13.26}$$

So,

$$G(E) = c/E \tag{13.27}$$

where c is an integration constant. To find c, take the limit as $E \rightarrow E_o$ in which case $G(E)$ should approach S_o, thus $c = S_o$.

We have shown in the energy region of neutron slowing down, the flux is given by

$$\phi(E) = \frac{S_o}{\Sigma_s(E)E} \tag{13.28}$$

The "$1/E$" behavior of the flux is very characteristic of the neutron distribution during energy moderation by elastic collisions with the target nuclei. Equation (13.28) has been obtained for the case of the hydrogenous medium. It can be extended to an arbitrary medium, $A > 1$, by writing in the same spirit as Eqs. (13.22) and (13.23),

$$\phi(E) = \frac{1}{\Sigma_s(E)\xi E}. \tag{13.29}$$

In summary, we have obtained an expression for the probability that a neutron will undergo energy moderation by elastic scattering across an arbitrary interval in the form of Eq. (13.23), which involves the quantity ξ, the average lethargy increase per collision. We have also shown the flux in the slowing down region, the region of energy below the fission spectrum and above thermal energy where neutrons can be up-scattered, is $1/E$.

13.3 DIFFUSION

To study neutron diffusion we go back to the neutron transport equation and reduce it to an equation only in the spatial variable. We first eliminate the $\underline{\Omega}$ dependence by integrating the transport equation over $\underline{\Omega}$, getting an equation with two unknowns,

$$\phi(\underline{r}, E, t) = \int d\Omega \phi(\underline{r}, E, \underline{\Omega}, t) \tag{13.30}$$

$$\underline{J}(\underline{r}, E, t) = \int d\Omega \underline{\Omega} \phi(\underline{r}, E, \underline{\Omega}, t) \tag{13.31}$$

Then we invoke Fick's law [Larmash, 1966] to eliminate $\underline{J}$, thus obtaining

$$\frac{1}{v}\frac{\partial\phi(\underline{r}, E, t)}{\partial t} = [D(E)\nabla^2 - \Sigma_t(E)]\phi(\underline{r}, E, t) + S(\underline{r}, E, t) + \nu f(E) \times \int dE' \Sigma_f(E')\phi(\underline{r}, E', t) + \int dE' \Sigma_s(E')\phi(\underline{r}, E', t) F(E' \to E) \tag{13.32}$$

To reduce further, we consider only steady-state solutions, and integrate over all energy to arrive at

$$[\overline{D}\nabla^2 + (\nu\overline{\Sigma}_f - \overline{\Sigma}_a)]\phi(\underline{r}) = -S(\underline{r}) \tag{13.33}$$

where

$$\phi(\underline{r}) = \int dE \phi(\underline{r}, E) \tag{13.34}$$

$$\overline{D} \equiv \frac{\int dE D(E)\phi(\underline{r}, E)}{\int dE \phi(\underline{r}, E)} \tag{13.35}$$

and a similar expression like Eq. (13.35) for the macroscopic cross section $\overline{\Sigma}$. The overhead bar denotes energy average weighed by the flux as indicated in Eq. (13.35). In writing Eq. (13.33) we have made use of the statement of neutron conservation,

$$\int dE F(E' \to E) = 1 \tag{13.36}$$

We need to keep in mind that in Eq. (13.33) we are also assuming the external source is time-independent, and more significantly that $\overline{D}$ is independent of position, which would be the case if $\phi(\underline{r}, E)$ were separable in $\underline{r}$ and E (this is not true in general).

Equation (13.33) is a second-order differential equation with constant coefficients. Since the *Schrödinger* equation, in the case of constant potential, is also of this form, it is worthwhile to make note of the analogy between the present problem of neutron diffusion and the problems of bound states in a potential well (Chap. 5) and scattering by a potential (Chap. 7), specifically with regard to the role of the interface boundary conditions appropriate for each problem. To keep the notations simple we drop the overhead bar on the material constants with the understanding that they are to be regarded as energy averaged quantities.

Boundary conditions

The boundary conditions to be imposed on $\phi(\underline{r})$ are quite similar to those imposed on the wave function in solving the *Schrödinger* equation. Because we are dealing with a physical quantity, the neutron distribution in space, $\phi(\underline{r})$ must be positive and finite everywhere or zero. Also, the distribution must reflect the symmetry of the problem, such as $\phi(x) = \phi(-x)$ in a slab system with $x = 0$ being at the center of the slab. Then there are the usual boundary conditions at a material interface; flux and currents must be continuous since there are no sources or sinks at such interfaces. All these conditions have counterparts in solving the wave equation.

The one boundary condition which requires some discussion is the case of neutron diffusion in a finite system where one specifies no re-entrant current across the surface (assumed to be convex) of the system. Let this surface be located at the position $x = x_o$ in a slab geometry. The physical condition is $J_-(x_o) = 0$. While the definition of J_- given in Eq. (8.10), is perfectly correct, we can evaluate J_- according to its physical meaning and by making the assumptions the scattering is *isotropic* in LCS, the medium in *non-absorbing*, and the flux is *slowly varying*. Then J_- is approximately given by the following integral (see Fig. 13.1),

$$\hat{\underline{z}} \cdot \underline{J}_- = \int_{uhs} \Sigma_s \phi(\underline{r}) \left(\frac{A \cos\theta}{4\pi r^2} \right) e^{-\Sigma_t r} d^3 r \tag{13.37}$$

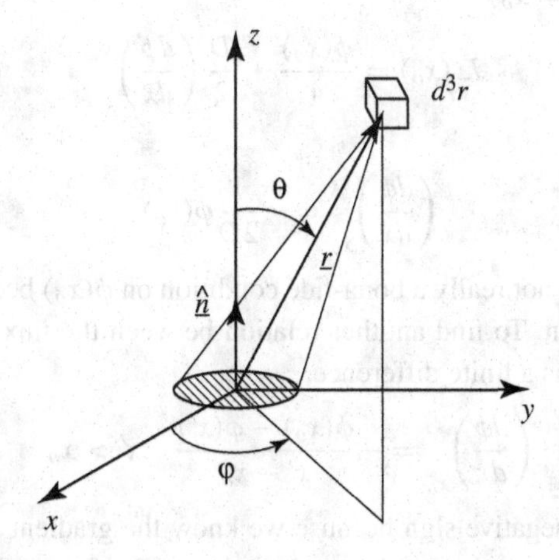

Fig. 13.1 The geometrical set up for estimating the current of neutrons crossing a unit area A at the origin (with normal $\hat{\underline{n}}$) from above after scattering isotropically in the volume element d^3r about $\underline{r}$. [Adapted from Lamarsh, 1966, p. 126.]

where the integral extends over the upper half space (*uhs*) of the medium because we are interested in all those neutrons which can cross the unit area A from above (in the direction opposite to the unit normal). The part of the integrand in the parenthesis is fraction of neutron going through the unit area A if there were no collisions along the way, with A being subtended at an angle θ from the elemental volume d^3r at $\underline{r}$. The attenuation of neutrons on the way to the unit area is taken into account by the factor $\exp(-\Sigma_t r)$. To carry out the indicated integral we need to know $\phi(\underline{r})$. Since we have assumed the flux is slowly varying we can expand about the origin (the location of the unit area) and keep only the first term in the expansion,

$$\phi(\underline{r}) \sim \phi(0) + \underline{r} \cdot \underline{\nabla}\phi|_o \tag{13.38}$$

Then,

$$J_-^z(0) = \frac{\Sigma_s}{4\pi} \int_{\varphi=0}^{2\pi} d\varphi \int_{\theta=0}^{\pi/2} \cos\theta \sin\theta d\theta \int_0^{\infty} dr e^{-\Sigma_t r}$$

$$\times [\phi_o + x(\partial\phi/\partial x)_o + y(\partial\phi/\partial y)_o + z(\partial\phi/\partial z)_o] \tag{13.39}$$

where we have taken $A = 1$. Writing out the Cartesian component x, y and z in terms of the spherical coordinates (r, θ, φ), $x = r \sin\theta \cos\varphi$, $y = r \sin\theta \sin\varphi$ and $z = r\cos\theta$, we find the φ-integration renders the terms containing x or y equal to zero. The term containing z can be easily integrated to give (after shifting the unit area from being at the origin as in Fig. 13.1 to a slab geometry with the unit area on the surface at $x = x_o$)

$$J_-(x_o) = \frac{\phi(x_o)}{4} + \frac{D}{2}\left(\frac{d\phi}{dx}\right)_{x_o} \tag{13.40}$$

or

$$\left(\frac{d\phi}{dx}\right)_{x_o} = -\frac{1}{2D}\phi(x_o) \tag{13.41}$$

Equation (13.41) is not really a bona-fide condition on $\phi(x_o)$ because the gradient $d\phi/dx$ is not known. To find another relation between the flux and gradient, we interpret the latter as a finite difference,

$$\left(\frac{d\phi}{dx}\right)_{x_o} = -\frac{\phi(x_o) - \phi(x')}{x' - x_o} \quad x' > x_o \tag{13.42}$$

where we use the negative sign because we know the gradient must be negative. Now we choose x' such that we know the value of the flux at this position. How is this possible? Suppose we choose x' to be the distance where the flux linearly extrapolates from $x = x_o$ to zero. Calling this distance $x' = x_o + \delta$ (see Fig. 13.2), we then have from Eq. (13.42)

$$\left(\frac{d\phi}{dx}\right)_{x_o} = -\frac{1}{\delta}\phi(x_o) \tag{13.43}$$

Combining this with Eq. (13.38) we obtain for the extrapolated distance $\delta = 2D$.

Conventionally instead of the physical condition of no re-entrant current, one often applies the simpler mathematical (though approximate) condition of

$$\phi(x_o + 2D) = \phi(\tilde{x}_o) = 0 \tag{13.44}$$

where $\tilde{x}_o = x_o + 2D$. Equation (13.44) is called the *extrapolated boundary condition*; because of its simplicity it is commonly adopted. One can use transport theory to do a better calculation of the extrapolated distance δ. We have seen that in simple diffusion theory this turns out to be $2D$, or $2/3\Sigma_{tr}$. The transport theory result, when there is no absorption, is $0.71/\Sigma_{tr}$. The rather small difference between diffusion theory and transport theory should not be taken to mean the flux near the surface is always accurately given by diffusion theory. As can be seen in Fig. 13.2, diffusion theory typically overestimates the flux at the surface relative to transport theory.

Diffusion kernels (Green's functions)

One can solve the neutron diffusion equation for the flux shape corresponding to various localized sources. This is tantamount to the standard problem of finding the Green's function for a point source and then integrating the result to obtain solutions for other simple source distributions. Since this kind of calculations is

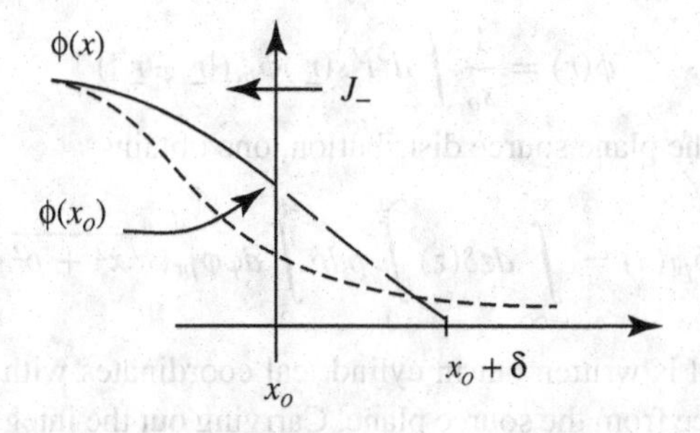

Fig. 13.2 Schematic of the extrapolated boundary condition, with δ being the distance beyond the actual boundary where a linear extrapolation of the flux at $x = x_o$ would vanish. The dotted curve shows the variation of the flux that would be obtained from transport theory. [Adapted from Lamarsh, 1966, p. 135.]

well described in the standard references [Roman, 1965], we will give only some of the results here.

Consider a plane source at $x = 0$ in an infinite medium which emits isotropically s_o *neutrons/cm*^{2}*/sec*. The diffusion equation reads

$$\left[\frac{d^2}{dx^2} - \kappa^2\right]\phi_{pl}(x) = 0 \qquad x \neq 0 \tag{13.45}$$

with $\kappa^2 = \Sigma_a/D > 0$ (κ is real). The solution for the case of a plane source is

$$\phi_{pl}(x) = \frac{s_o}{2D\kappa}e^{-\kappa|x|} \tag{13.46}$$

where we have applied the source condition,

$$[AJ(x)]_{x\to x_o} = D[d\phi(x)/dx]_{x\to x_o} = s_o/2 \tag{13.47}$$

with A again being a unit area. Equation (13.47) is simply the statement that if s_o is the number of neutrons emitted per unit area per sec from the plane source, then half of the neutrons would come out in the $+x$ direction.

Suppose now instead of a plane source we have a point source at the origin emitting s_o neutrons/sec. The diffusion equation becomes

$$(\nabla^2 - \kappa^2)\phi_{pt}(r) = 0 \qquad r \neq 0 \tag{13.48}$$

with solution

$$\phi_{pt}(r) = \frac{s_o}{4\pi r D}e^{-\kappa r} \tag{13.49}$$

Comparison of Eqs. (13.46) and (13.49) suggests that the two kernels are related, and that one can be obtained from the other. This connection is actually quite general and follows directly from the property of the Green's function. Since the diffusion equation is linear, one can superpose the contributions from different point sources to make the solution for any distributed source,

$$\phi(\underline{r}) = \frac{1}{s_o}\int d^3r' s(\underline{r}')\phi_{pt}(|\underline{r} - \underline{r}'|) \tag{13.50}$$

Applying this to the plane source distribution, one obtains

$$\phi_{pl}(x) = \int_{-\infty}^{\infty} dz\delta(z)\int_0^{\infty}\rho d\rho\int_0^{2\pi} d\varphi\phi_{pt}(\sqrt{x^2+\rho^2}) \tag{13.51}$$

where the integral is written out in cylindrical coordinates with x being the perpendicular distance from the source plane. Carrying out the integrations, one finds

$$\phi_{pl}(x) = 2\pi\int_x^{\infty} d\gamma\gamma\phi_{pt}(\gamma) \tag{13.52}$$

which one can verify is consistent with Eqs. (13.46) and (13.49). One can invert Eq. (13.52) by differentiating to give

$$\phi_{pt}(r) = -\frac{1}{2\pi}\left[\frac{d\phi_{pl}(x)}{dx}\right]_{x=r} \tag{13.53}$$

The relation Eq. (13.52) helps us to understand why the point source kernel is singular at the origin and yet the plane source kernel is regular everywhere.

13.4 THERMALIZATION

Neutron thermalization is the study of the distribution of thermal neutrons in energy, space, and time in moderating materials. It is clearly an essential topic in the analysis of nuclear systems where thermal neutrons play a dominant role. The part of neutron thermalization which concerns us in the present context is the scattering kernel that appears in the scattering gain term in the neutron transport equation (see Eq. (13.3)),

$$\int dE' d\Omega' \Sigma_s(E')\phi(\underline{r}, E', \underline{\Omega}')F(E'\underline{\Omega}' \rightarrow E, \underline{\Omega})$$

In our discussion of neutron slowing down (Sec. 13.2) we were justified in using a rather simple expression for the scattering kernel F, because in the energy range above thermal energy (~0.025 eV), neutrons lose energy upon elastic scattering and the effects of chemical binding and thermal motion of the scattering nuclei can be ignored. In the thermal energy region, neutrons can be up-scattered in elastic scattering, and chemical binding and thermal motion effects now need to be taken into account. It is also quite conventional to call the neutron scattering processes in the thermal energy region *inelastic*, in the molecular sense of energy exchange with molecular motions of the scatterer but not in the nuclear sense of excitation of nuclear levels. The latter process is clearly not relevant in the context of thermal neutron scattering, so there should be no possibility of any confusion between the two types of inelastic scattering, molecular versus nuclear. The terminology of thermal inelastic scattering therefore refers to neutron exchanging energy with the scattering nucleus, which is chemically bound and in thermal motion (therefore the nucleus is capable of giving or receiving energy). The process of energy exchange involves the interaction between the neutron and the molecular degrees of freedom, or excitation of molecular energy levels. It is also still correct to call the neutron scattering process in the thermal region elastic, provided that elastic here means nuclear interaction, that is, the neutron does not leave the scattering nucleus in an excited nuclear state. Any possible confusion would be eliminated if we use terms like "nuclear elastic" and "molecular inelastic". In both neutron slowing down and thermalization, the scattering process is "nuclear elastic", whereas in neutron slowing down the scattering process is "molecular elastic" and in neutron thermalization it is "molecular inelastic".

The purpose of this section is to present the theory of neutron "molecular inelastic" scattering. We will see that we need to extend the previous method of cross section calculation based on the phase shift analysis. While the method is sufficient to give the angular differential cross section $\sigma(\theta)$, it is not suitable for treating the new effects of thermal motion and chemical binding which dominate the behavior of thermal neutrons. In this section, our strategy is to develop the *theory of thermal neutron scattering* [Marshall, 1971, Parks, 1970] in several stages. First we introduce a different method of cross section calculation, one based on the so-called *Born approximation* which is also a very well-known approach in the general theory of scattering. Next we show that to apply the Born approximation we will need to develop a new potential for the neutron-nucleus interaction, a result that will be called the *Fermi pseudopotential*. Then we combine the Born approximation and the Fermi pseudopotential to obtain an expression for the double differential scattering cross section, the kernel that appears in the neutron transport equation, Eq. (13.5). As an application we examine the cross section to arrive at an expression for the *scattering law* $S(\alpha, \beta)$ which is part of the ENDF-B database mentioned in Sec. 3.3.

The integral equation approach to potential scattering

In Chap. 7 we studied the method of cross section calculation based on an expansion in partial waves and the determination of a phase shift for each partial wave. The phase-shift method is well suited to nuclear reactions, as demonstrated in various applications such as the optical model of nuclear reactions discussed in Chap. 8. For thermal neutron scattering, another method of calculating the scattering amplitude $f(\theta)$ turns out to be much more appropriate. This is the Born approximation where $f(\theta)$ is given by the Fourier transform of the interaction potential. It is more suitable for treating the scattering of a particle by a system of target nuclei and not just a single target nucleus. In contrast, the phase-shift method is quite powerful for potential scattering or two-body interaction, but it does not allow extension to a system of scatterers in any systematic way. Since the Born approximation is a fundamental approach in scattering theory, it is worthwhile to derive it in general and then apply it to neutron scattering.

The *Schrödinger* equation to be solved describes the scattering of an effective particle by a central potential. We will rewrite this second-order differential equation in the form of an integral equation

$$(\nabla^2 + k^2)\psi(\underline{r}) = U(r)\psi(\underline{r}) \tag{13.54}$$

where we have defined

$$U(r) \equiv \frac{2\mu}{\hbar^2} V(r) \tag{13.55}$$

We now regard the right-hand side of Eq. (13.54) as a "source" or inhomogeneous term and write down the formal solution to Eq. (13.54) as the sum of the solution to the homogeneous equation and a particular solution due to the source term. The homogeneous equation is

$$(\nabla^2 + k^2)\varphi(\underline{r}) = 0 \tag{13.56}$$

The particular solution is a superposition of the Green's function and the source term Eq. (13.55). By the Green's function we mean the solution to the equation

$$(\nabla^2 + k^2)G(\underline{r} - \underline{r}') = -\delta(\underline{r} - \underline{r}') \tag{13.57}$$

The formal solution to Eq. (13.54) is therefore

$$\psi(\underline{r}) = \varphi(\underline{r}) - \int d^3r' G(\underline{r} - \underline{r}')U(r')\psi(\underline{r}') \tag{13.58}$$

The best way to see this is indeed a general solution to Eq. (13.54) is by direct substitution. The advantage of writing the solution to the *Schrödinger* equation in this form is we can generate a series solution by assuming the second term in Eq. (13.58) is "small" in the sense to be specified below. This assumption imposes certain condition on the interaction potential $V(r)$, as we will see.

The homogeneous equation Eq. (13.56) should be familiar to us (recall Chaps. 5 and 7). We will work with the solution in the form of a plane wave,

$$\varphi(\underline{r}) = \exp(i\underline{k} \cdot \underline{r}) \equiv e^{ikz} \tag{13.59}$$

where we have taken incident wave vector to be along the z-axis. Thus Eq. (13.59) corresponds to the incident plane wave in our previous description of potential scattering in Chap. 7. The solution to the Green's function equation Eq. (13.57) is also quite well-known; it can be written in the form of a spherical outgoing wave, which is how we want to represent the scattered wave,

$$G(\underline{r} - \underline{r}') = \frac{e^{ik|\underline{r}-\underline{r}'|}}{4\pi|\underline{r} - \underline{r}'|} \tag{13.60}$$

Because Eq. (13.60) is a result one is likely to encounter in other problems, given that Eq. (13.56) is the *Helmholtz equation*. We digress a bit to show that Eq. (13.60) can be obtained using the method of Fourier transform.

Without loss of generality, we can set $\underline{r}' = 0$ in Eq. (13.57). We introduce the Fourier transform of $G(\underline{r})$ as

$$F(\underline{\kappa}) = \int d^3 r e^{-i\underline{\kappa}\cdot\underline{r}} G(\underline{r}) \tag{13.61}$$

Taking the Fourier transform of (13.57) we obtain an algebraic equation in F which gives

$$F(\underline{\kappa}) = -(k^2 - \kappa^2)^{-1} \tag{13.62}$$

To find $G(\underline{r})$ we take the inverse Fourier transform of Eq. (13.62),

$$G(\underline{r}) = (2\pi)^{-3} \int d^3\kappa e^{i\underline{\kappa}\cdot\underline{r}} (\kappa^2 - k^2)^{-1} \tag{13.63}$$

The angular integrations can be readily carried out since the only dependence is in the factor $\exp(i\kappa r \cos\theta)$,

$$G(r) = \frac{1}{2\pi^2 r} \int_0^{\infty} \kappa \sin \kappa r \frac{1}{\kappa^2 - k^2} d\kappa \tag{13.64}$$

Since $F(\underline{\kappa})$ depends only the magnitude of $\underline{\kappa}$, its Fourier transform depends only on the magnitude of $\underline{r}$. The integrand of Eq. (13.64) is manifestly an even function of κ, we can extend the integration from $-\infty$ to ∞,

$$G(r) = \frac{1}{4i\pi^2 r} \int_{-\infty}^{\infty} \kappa \frac{e^{i\kappa r}}{(\kappa + k)(\kappa - k)} d\kappa \tag{13.65}$$

So Eq. (13.65) amounts to a *contour integration*. The integrand is seen to have two simple poles on the real axis, at $\kappa = \pm k$. We will choose the path for the contour integral to close on itself in the upper half plane while enclosing the pole at $+k$ (see Fig. 13.3). This particular choice gives a contribution from the enclosed pole and nothing else. Thus,

$$G(r) = \frac{1}{4i\pi^2 r} \cdot 2\pi i \cdot k \frac{e^{ikr}}{2k} = \frac{e^{ikr}}{4\pi r} \tag{13.66}$$

which is the desired result, Eq. (13.60). This choice of the contour is motivated by the physical consideration of having $G(r)$ being in the form of a spherical outgoing wave, since then Eq. (13.58) would match the boundary condition for the scattering problem. If we had chosen the contour to enclose the other pole, $\kappa = -k$, we would have obtained a spherical incoming wave for the Green's function. What

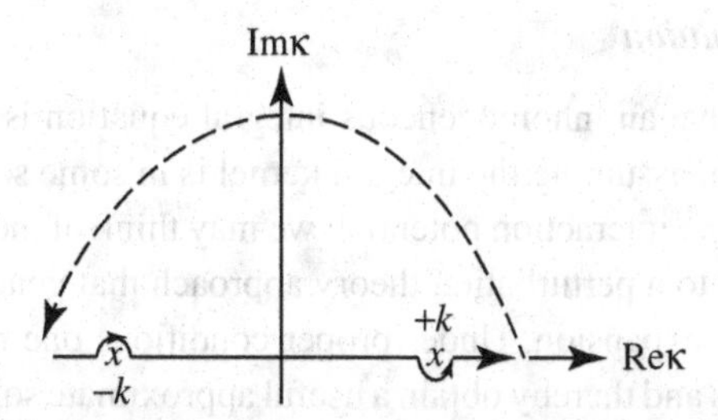

Fig. 13.3 Choice of contour for the inverse Fourier transform, Eq. (13.65), which gives a spherical outgoing wave, Eq. (13.66).

if one encloses both poles? The result would be a standing spherical wave $\sin(kr)/r$. This ends our digression.

Returning to our formal solution to the *Schrödinger* equation, we combine Eqs. (13.58) and (13.60) to find

$$\psi(\underline{r}) = e^{ikz} - \int d^3r' \frac{e^{ik|\underline{r}-\underline{r}'|}}{4\pi|\underline{r}-\underline{r}'|} U(r')\psi(\underline{r}'). \tag{13.67}$$

We will refer to Eq. (13.67) as the *integral equation of scattering*. it is entirely equivalent to the *Schrödinger* equation since no approximation has been made. The advantage of this form is that the boundary condition where one introduces the scattering amplitude has been explicitly incorporated. However, it is not yet in a form from which one can directly extract $f(\theta)$.

Because we are only interested in the solution in the asymptotic region of large r compared to the range of interaction r_o, we can simplify Eq. (13.67) by noting the presence of $U(r')$ means the integral over r' will be restricted to $r' \leq r_o$. Thus, with $r'/r \ll 1$ we can write

$$|\underline{r} - \underline{r}'| \sim r - \hat{\underline{r}} \cdot \hat{\underline{r}}' \tag{13.68}$$

in the exponent in Eq. (13.67), and just take r for $|\underline{r} - \underline{r}'|$ in the denominator. Then Eq. (13.67) becomes

$$\psi(\underline{r}) \sim e^{ikz} - \frac{e^{ikr}}{4\pi r} \int d^3r' e^{-ik\hat{\underline{r}}\cdot\underline{r}'} U(r')\psi(\underline{r}'), \quad r \gg r_o \tag{13.69}$$

We can now compare this result with the boundary condition for the scattering problem,

$$\psi(\underline{r}) \sim e^{ikz} + f(\theta)\frac{e^{ikr}}{r} \tag{13.70}$$

thereby obtaining a formal expression for the scattering amplitude. Such a result is not yet useful because it still contains the wave function ψ, which is not known. To turn Eq. (13.69) into a useful result we need to introduce an approximation for ψ.

The Born approximation

A simple way to solve an inhomogeneous integral equation is to iterate with the inhomogeneous term assuming the integral kernel is in some sense "small" (since the kernel contains the interaction potential, we may think of the potential as being "weak"). This leads to a perturbation theory approach that generates a solution in the form of a series expansion. Under proper conditions one may be justified in truncating this series and thereby obtain a useful approximate solution directly. The zeroth-order solution, which we denote by a superscript, is obtained by ignoring completely the integral term in Eq. (13.69),

$$\psi^{(o)}(\underline{r}) = e^{ikz} \tag{13.71}$$

The next approximation is to replace the wave function ψ in Eq. (13.69) by Eq. (13.71), thus giving the first-order solution. In the asymptotic region, this would read

$$\psi^{(1)}(\underline{r}) \sim e^{ikz} - \frac{e^{ikr}}{4\pi r}\int d^3r' e^{-i\underline{k}\cdot\underline{r}'} U(r') e^{i\underline{k}'\cdot\underline{r}'} \tag{13.72}$$

where we have set $\underline{k}' \equiv k\hat{\underline{z}}$. Equation (13.72) is known as the *first Born approximation*. Comparing it with Eq. (13.70) one obtains an expression for the scattering amplitude,

$$f(\theta) = -\frac{2\mu}{4\pi\hbar^2}\int d^3r e^{i\underline{\kappa}\cdot\underline{r}} V(r) \tag{13.73}$$

with

$$\underline{\kappa} \equiv \underline{k}' - \underline{k} \tag{13.74}$$

Equation (13.73) shows in the first Born approximation the scattering amplitude is given by the Fourier transform of the interaction potential. The Fourier transform variable $\underline{\kappa}$ is a wave vector, the difference between the wave vectors of the scattered and incident waves respectively, $\underline{k}'$ and $\underline{k}$. We will refer to $\underline{\kappa}$ as the *wave vector transfer* (see Fig. 13.4), and note $\hbar\underline{\kappa}$ is the momentum gained by the particle

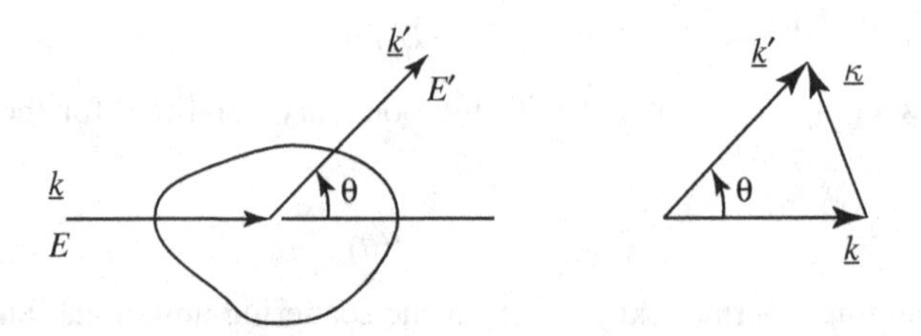

Fig. 13.4 Schematic of scattering of a particle with incoming energy E, wavevector $\underline{k}$ at an angle θ and outgoing energy E', wavevector $\underline{k}'$. The scattering sample is the encircled area, whereas the detector is located at ougoing wave vector k'. The momentum triangle shows the relation between the momentum transfer $\hbar\underline{\kappa}$ and the scattering angle, the angle between $\underline{k}$ and $\underline{k}'$.

in scattering through an angle θ. Since in potential scattering there is no energy transfer, the incident and scattered wave vectors have the same magnitude. This means the magnitude of $\underline{\kappa}$ is

$$\kappa = 2k\sin(\theta/2) \tag{13.75}$$

In the Born approximation the dependence on the scattering angle enters therefore through the wave vector transfer.

Validity of the Born approximation

The simplicity of the result of the Born approximation makes it widely useful. We find that the theory of thermal neutron scattering depends critically on the use of this approximation. Before going on to discuss the theory, it is just as important to pause and consider the condition under which this approximation can be justified. It will be seen that strictly speaking the condition for the Born Approximation is not fulfilled in the case of thermal neutron scattering. However, there is a way out of this dilemma. It involves the introduction of a *pseudopotential* in place of the actual neutron-nucleus interaction that we have studied in the neutron-proton scattering problem. (Sec. 9.1). The idea of the pseudopotential that enables the use of the Born approximation is due to *E. Fermi*.

The first Born approximation amounts to one iteration of the inhomogeneous integral equation Eq. (13.69). (We call it the first Born approximation because one can iterate further and obtain the second and higher-order Born approximation results.) The assumption is that the integral term in Eq. (13.69) is small compared to the inhomogeneous term. Since the magnitude of the latter is unity, the condition for validity can be expressed as

$$\Delta = \left| \int d^3r' \frac{e^{ikr'}}{4\pi r'} \cdot \frac{2\mu}{\hbar^2} V(r') \cdot e^{i\underline{k}\cdot\underline{r}'} \right| \ll 1 \tag{13.76}$$

where we have taken the integral term to be evaluated at $r = 0$ and require its magnitude to be small compared to unity. Since the potential is spherically symmetric, the angular integrations can be readily performed. For the radial integration we take $V(r)$ to be a spherical well with depth V_o and range r_o,

$$\begin{aligned}\Delta &= \frac{\mu V_o}{\hbar^2 k} \left| \int_o^{r_o} dr(e^{2ikr} - 1) \right| \\ &= \frac{\mu V_o}{2\hbar^2 k^2} |e^{2ikr_o} - 2ikr_o - 1| \\ &= \frac{\mu V_o}{2\hbar^2 k^2} [y^2 + 2 - 2y\sin y - 2\cos y]^{1/2} \ll 1\end{aligned} \tag{13.77}$$

where $y = 2kr_o$. The inequality Eq. (13.77) can be satisfied in two ways. The dimensionless parameter y is either small or large. For $kr_o \gg 1$, the square root quantity behaves like y. Then Eq. (13.77) gives the condition

$$\frac{V_o r_o}{\hbar v} \ll 1 \tag{13.78}$$

where $v = \hbar k/\mu$ is the incident speed of the particle. We will call Eq. (13.78) the *high-energy condition* since it is derived from $kr_o \gg 1$. For $kr_o \ll 1$, the square root in Eq. (15.30) becomes $y^2/2$ to lowest order. This then leads to the *low-energy condition*

$$\frac{\mu V_o r_o^2}{\hbar^2} \ll 1. \tag{13.79}$$

For thermal neutron scattering the appropriate condition to check is Eq. (13.79) since we are in the regime of low-energy scattering, $kr_o \ll 1$. If one were interested in electron scattering, one would find that the appropriate condition for the use of Born approximation would be the high-energy condition Eq. (13.78).

It is now a simple matter to put in appropriate numerical values for the constants in Eq. (13.79) to see whether the condition is satisfied for thermal neutron scattering. For the potential parameters we can take the values for *n-p scattering* (triplet interaction). The left-hand side of Eq. (13.79) becomes

$$\frac{1.6 \times 10^{-24} \cdot 36 \times 10^6 \times 1.6 \times 10^{-12} \cdot 4 \times 10^{-26}}{10^{-54}} = 3.7 \tag{13.80}$$

which is manifestly too large for the Born approximation to be applicable. One would like to see a magnitude typically $\sim 10^{-2}$.

The Fermi pseudopotential

It was pointed out by E. Fermi [Femi, 1936] that there is a way to formulate the thermal neutron scattering problem such that the low-energy condition Eq. (13.79) can be satisfied. One could see the reason that Eq. (13.79) is not satisfied lies in the value of V_o being too large. This is not all that surprising since nuclear forces are well-known to be strong interactions. It was observed that in addition to the condition for the Born approximation there are two other conditions in the problem which need to be preserved, the condition of low-energy scattering, $kr_o \ll 1$, and the expression for the scattering length,

$$a = -f(\theta)|_{kr_o \to 0} = \frac{m}{2\pi\hbar^2}\int d^3r V(r) \sim V_o r_o^3 \tag{13.81}$$

Although Eq. (13.81) may not be a condition in the same sense as the low-energy scattering condition and Eq. (13.79), it is nevertheless a constraint on the potential parameters. By this we mean that in formulating our theory there is freedom to choose the form of the potential *V*(*r*) so long as it still gives the correct scattering length. In other words, one may be allowed to alter *V*(*r*) in the theory, but not the magnitude of the scattering cross section. After all, the latter is a measurable quantity (a physical observable as opposed to a theoretical construct). The fact that the thermal neutron scattering theory does not involve directly the specification of a potential *V*(*r*) allowed Fermi to suggest an alteration of *V*(*r*) while still preserving the three conditions just discussed. In other words, the thermal neutron scattering theory is not a theory of the actual neutron–nucleus interaction potential *V*(*r*).

Collecting these three conditions, we see that using the values of V_o and r_o previously determined from neutron-proton scattering in the present problem of thermal neutron scattering leads us to the situation where

$$kr_o \sim 10^{-4} \tag{13.82}$$

$$\frac{\mu V_o r_o^2}{\hbar^2} \sim 3 \tag{13.83}$$

$$V_o r_o^3 = \text{a given constant} \tag{13.84}$$

which is not a happy outcome because Eq. (13.83) means we are not justified in using the Born approximation which in turn means we do not have a good method to calculate the scattering amplitude. In particular, we see in Eq. (13.78) the combination of potential parameters in the form of $V_o r_o^2$ is too large by about two orders of magnitude. What Fermi suggested was to replace the actual neutron-nucleus potential *V*(*r*) by a fictitious potential $V^*(r)$ such that all three conditions are satisfied. In other words, can we choose a $V^*(r)$ for which $kr_o^* \ll 1$, $\mu V_o^* r_o^{*2}/\hbar^2 \ll 1$, and $V_o^* r_o^{*3} = \text{constant}$ (same as Eq. (13.84))? Suppose we retain the spherical well shape, but simply scale the well depth and range so that V^* is a spherical well with depth V_o^* and range r_o^*. In particular, suppose we take

$$V_o^* = 10^{-6} V_o, \quad r_o^* = 10^2 r_o$$

How would this replacement affect the three conditions, Eqs. (13.82)–(13.84)? With the scaled (fictitious) potential the conditions become

$$kr_o^* \sim 10^{-2} \tag{13.85}$$

$$\frac{\mu V_o^* r_o^{*2}}{\hbar^2} \sim 3 \times 10^{-2} \tag{13.86}$$

$$V_o^* r_o^{*3} = V_o r_o^3 \tag{13.87}$$

which is just what we desire. Thus, replacing (V_o, r_o) by (V_o^*, r_o^*) allows one to proceed with formulating a theory of thermal neutron scattering based on the Born approximation.

The suggestion of Fermi was in effect to distort the neutron-nucleus interaction–extending the range and decreasing the depth in such a way that preserves the scattering length. Equations (13.85)–(13.87) show that by doing this one can satisfy the low-energy condition for the Born approximation. Fermi made one further suggestion. He proposed that the fictitious potential can be expressed in the form

$$V^*(r) = \frac{2\pi\hbar^2}{m} a\delta(r) \tag{13.88}$$

where a is the scattering length that ensures the calculated scattering cross section in the low-energy limit (in the sense of Eq. (7.30)) is always the correct (experimental) value. This form of the potential has come to be known as the *Fermi pseudopotential.* We will see next why this is a very important result for the calculation of the double differential scattering cross section for thermal neutrons. Physically, Eq. (13.88) describes the neutron-nucleus collision as an impulse interaction, like the collision between two hard spheres — a sudden interaction only at the moment when the spheres touch. Why should this be reasonable given we have just gone through an argument where the fictitious potential was introduced to decrease the strength of the interaction V_o in order to satisfy the condition Eq. (13.79)? The answer is that indeed for the purpose of justifying the Born approximation we should be thinking of a potential which is weaker than the $n-p$ interaction by six orders of magnitude with a range that is two orders of magnitude greater. With $r_o \sim 2 \times 10^{-13}$ cm, this would mean a pseudopotential having a range of 10^{-11} cm. What Fermi means by Eq. (13.88) is that on the scale of the wavelength of a thermal neutron (10^{-8} cm), the interaction range, regardless of whether it is 10^{-13} or 10^{-11} cm, is still so small that the interaction can be *effectively treated* as an impulse interaction occurring essentially at a point. It is in this sense that Eq. (13.88) is physically meaningful. Before closing this discussion we remark that in Eq. (13.88) the scattering length a is a parameter that is to be specified externally; it can be calculated by another theory such as the optical model (see Eq. (8.70)) or determined by experiment [Bacon, 1977]. A virtue of Eq. (13.88) is that the correct scattering cross section is already built into the potential. Recall in Chap. 7 in the low-energy limit the scattering cross section is given by the s-wave cross section, $\sigma_s = 4\pi a^2$, Eq. (7.30).

Our objective in this section is twofold, first to formulate the theory of the double differential scattering cross section $d^2\sigma/d\Omega dE$ for thermal neutrons, and then to discuss the concept of the scattering law $S(\alpha, \beta)$, the nuclear data available from standard evaluated nuclear data files. In the broader context of the physics of thermal neutron scattering, the scattering law is more conventionally denoted as $S(Q, \omega)$ and known as the *dynamic structure factor*. We prefer the latter term because of the importance of this quantity in the physical understanding of condensed matter and the field of non-equilibrium statistical mechanics.

Neutron scattering by a system of nuclei

Now that we have developed a suitable potential for use in the Born approximation, we can proceed to set up the double differential scattering cross section for quantitative calculations. Notice the present problem is more than neutron scattering by *an isolated nucleus*, it is the interaction between a neutron and *a system of scattering nuclei*. For the single scattering nucleus we could solve the problem of two-body collisions by reducing it to an effective one-body problem. This is no longer possible (or desirable) when there is more than one scattering nucleus involved. The *Schrödinger* wave equation we now need to solve describes a *many-body system*, therefore it has the complexities associated with any many-body problem. If one is to obtain a tractable solution, some key approximations must be introduced. This is why we need the simplifications provided by the Born approximation and the Fermi pseudopotential.

The many-body *Schrödinger* equation for the system (neutron plus scattering nuclei) is of the form

$$\left[\frac{p^2}{2m} + H_s + V\right]\Psi(\underline{r}, \underline{R}) = \varepsilon\Psi(\underline{r}, \underline{R}) \tag{13.89}$$

The first term is the kinetic energy operator of the neutron, H_s is the Hamiltonian of the system of scattering nuclei, and V is the neutron-nucleus interaction (for which we will take to be the Fermi pseudopotential when the time comes). The many-body wave function depends on the neutron coordinate $\underline{r}$ and the coordinates of the N scattering nuclei in the target system, denoted here by $\underline{\mathrm{R}}$, $\underline{R} = (\underline{R}_1, \underline{R}_2, \ldots, \underline{R}_N)$. In Eq. (13.89), ε is the total energy in the problem, neutron plus scattering system, which is a constant. When the neutron is far from the target there is no interaction, therefore the constant is the sum of the kinetic energy of the neutron and the energy of the scattering system (kinetic energy of the nuclei and their interaction energy). We will not need to specify what is H_s until later when we discuss explicit model calculations. For now it is sufficient to regard H_s as known in the sense that its eigenfunctions and eigenvalues exist,

$$H_s|n\rangle = \varepsilon_n|n\rangle \tag{13.90}$$

where the eigenfunction in state n is denoted by the *ket* $|n\rangle$. It is also customary to assume that the states $|n\rangle$ form a complete set, in which case $|n\rangle$ becomes an appropriate basis for an expansion,

$$\Psi(\underline{r}, \underline{R}) = \sum_n \psi_n(\underline{r})|n\rangle \tag{13.91}$$

The coefficient of expansion, $\psi_n(\underline{r})$, is the neutron wave function when the target system is in the *nth* state. It satisfies an equation obtained by inserting Eq. (13.91) into Eq. (13.89) and using the orthnormality property of the eigenstates $|n\rangle$,

$$(\nabla^2 + k_n^2)\psi_n(\underline{r}) = \frac{2m}{\hbar^2}\sum_{n'}\langle n|V|n'\rangle \Psi_{n'}(\underline{r}) \tag{13.92}$$

with

$$\frac{\hbar^2 k_n^2}{2m} = \varepsilon - \varepsilon_n \tag{13.93}$$

Equation (13.93) is simply another statement of energy conservation, the total energy which is a constant, is divided between the kinetic energy of the neutron and the energy of the target. When one changes, the other must change correspondingly. This means that when the neutron gains or loses energy that amount must come from or given to the target. Equation (13.92) is a one-body equation, an equation describing the neutron in a state that corresponds to the target being in the *nth* state. Whether this equation is difficult or easy to solve depends on the matrix elements $V_{nn'} = \langle n|V|n'\rangle$. For now we proceed without saying anything more about these matrix elements.

We observe that since Eq. (13.92) is of the same form as the wave equation, Eq. (13.54), we can treat it in the same manner as before. Thus, we can write the solution to Eq. (13.92) as the sum of a homogeneous solution and a particular solution, the latter involving the convolution of the inhomogeneous (source) term on the right hand side of Eq. (13.92) with an appropriate Green's function. In the asymptotic region ($r \gg r_o$) this equation reads

$$\psi_n(\underline{r}) \sim \delta_{nn_o} e^{i\underline{k}_i\cdot\underline{r}} + \frac{m}{2\pi\hbar^2}\frac{e^{ik_n r}}{r}\sum_{n'}\int d^3r' e^{-ik_n\hat{\underline{r}}\cdot\underline{r}'} V_{nn'}\psi_{n'}(\underline{r}') \tag{13.94}$$

In writing Eq. (13.94), we have specified the initial and final states of the target are n_o and n, respectively. That is to say, prior to the scattering the neutron approaches the target with a kinetic energy $E_i = \hbar^2 k_i^2/2m$ and the target is in the state n_o. After the scattering the neutron leaves with energy $E_f = \hbar^2 k_n^2/2m$ with the target left in the state n. Applying the first Born approximation to Eq. (13.94) and comparing the result with the boundary condition Eq. (13.70), we obtain an expression for the scattering amplitude,

$$f_{n_o n}(\theta) = -\frac{m}{2\pi\hbar^2}\int d^3r e^{i\underline{Q}_n\cdot\underline{r}} V_{nn_o}(\underline{r}) \tag{13.95}$$

where

$$\underline{Q}_n = \underline{k}_i - \underline{k}_n \tag{13.96}$$

and $\underline{k}_n = k_n\hat{\underline{r}}$. Next we bring in the Fermi pseudopotential for the neutron-nucleus interaction,

$$V(\underline{r}, \underline{R}_1, \ldots, \underline{R}_N) = \frac{2\pi\hbar^2}{m}\sum_{\ell=1}^{N} a_\ell \delta(\underline{r} - \underline{R}_\ell) \tag{13.97}$$

Notice this is now a many-body interaction potential. While the neutron interacts with all the nuclei in the target, the effect of each interaction is additive because the interaction is extremely short-ranged. If for some reason the interactions are collective in nature, then the problem of neutron scattering would become intractable given our current state of understanding. Notice also we have associated with each nucleus its own scattering length, thus allowing for the possibility that the scattering length can vary from one nucleus to the next, which would be the case if the nuclei are not all the same. Thus if two nuclei have different spin orientations, then it is to be expected that they would not scatter in the same way. Besides spin, different nuclear isotopes also can give rise to different scattering lengths. This variation of the scattering length is the reason why we need to distinguish between two kinds of neutron scattering, *coherent* and *incoherent* (see below).

Taking the matrix elements of Eq. (13.97) amounts to multiplying Eq. (13.97) by the *bra* state $\langle n|$ and the *ket* state $|n_o\rangle$, and integrating the product over the coordinates of the scattering nuclei. Without saying what these states are and actually carrying out the integration, we can interchange the matrix element integration with the r-integral indicated in Eq. (13.95),

$$f_{n_o n}(\theta) = -\sum_{\ell} a_\ell \langle n|e^{i\underline{Q}_n\cdot\underline{R}_\ell}|n_o\rangle \tag{13.98}$$

We see the matrix element operation has been shifted from the interaction potential to the the exponential factor, $\exp(i\underline{Q}_n \cdot \underline{R}_\ell)$, as a result of the impulse nature of the pseudopotential. We will come to appreciate later the significance of this step of the theoretical development.

Once we have the scattering amplitude, we can immediately write down the angular differential cross section $(d\sigma/d\Omega)_{n_o n}$ for the scattering process where the initial and final states of the target are n_o and n, respectively. Since in an actual experiment, one cannot control the state of the target either before or after scattering, the best one can do is to average over some thermal distribution of the initial states (like a Maxwellian distribution), and sum over all possible final states. Thus the angular differential cross section that one should compare with experiment is given by the expression,

$$\frac{d\sigma}{d\Omega} = \sum_{n}\sum_{n_o} P(n_o)\frac{k_n}{k_i}\frac{1}{N}|f_{n_o n}(\theta)|^2 \tag{13.99}$$

where $P(n_o)$ is the probability that the target is initially in state n_o. We will take this to be the Maxwell distribution (see Eq. (9.42)),

$$P(n) = \frac{e^{-\varepsilon_n/k_B T}}{\sum_s e^{-\varepsilon_s/k_B T}} \tag{13.100}$$

In Eq. (13.99), the factor (k_n/k_i) arises from the definition of the cross section when the incident and outgoing wave vectors no longer have the same magnitude (this is one consequence of inelastic scattering). Recall from Chap. 7 that the cross section is essentially the ratio of scattered current to incident current. Since current is density times speed, this is why the cross section is proportional to the ratio of outgoing to incoming speeds. The factor of $(1/N)$ arises because Eq. (13.99) gives the cross section per nucleus (we have N nuclei in the target). Equation (13.99) with the scattering amplitude given by Eq. (13.98) is still quite formal. But it is now the complete expression which one can actually evaluate to obtain the angular distribution of thermal neutron scattering.

The double differential scattering cross section

The double differential scattering cross section, $d^2\sigma/d\Omega dE$, as the name implies, is a differential cross section in both the solid angle of scattering and the energy of the scattered neturon. Imagine a scattering experiment such as the one schematically depicted in Fig. 13.4. What one can measure at the detector is the intensity of scattering per unit solid angle and per unit final energy. Given that we have an expression for the scattering intensity per unit solid angle, we can modify that result to ensure that the proper energy exchange occurs and thus obtain the scattering intensity per unit solid angle and final neutron energy. The appropriate modification is

$$\frac{d^2\sigma}{d\Omega dE} = \frac{1}{N}\frac{k_f}{k_i}\sum_{nn_o} P(n_o)|f_{n_o n}(\theta)|^2\delta(E_f + \varepsilon_n - E_i - \varepsilon_{n_o}) \tag{13.101}$$

The argument of the delta function is nothing but the statement of energy conservation, the sum of neutron and target energies before and after the scattering must be equal. Equation (13.101) is the basic result in the theory of thermal neutron scattering. We will make use of it in several ways from here on.

Space-time representation

It is useful to rewrite Eq. (13.101) in a representation that brings out more physical content of the double differential scattering cross section. Consider a particular representation of the delta function,

$$\delta(x) = \frac{1}{2\pi} \int_{-\infty}^{\infty} dt e^{itx} \tag{13.102}$$

Applying this to Eq. (13.101) and performing a series of rearrangements of the various exponential factors, taking care that the positions of the nuclei become time-dependent Heisenberg operators which may not commute with each other, one can obtain the following result,

$$\frac{d^2\sigma}{d\Omega dE_f} = (E_f/E_i)^{1/2} a^2 S(Q,\omega) \tag{13.103}$$

where

$$S(Q,\omega) = \frac{1}{2\pi} \int_{-\infty}^{\infty} dt e^{i\omega t} \frac{1}{N} \sum_{\ell,\ell'} \langle e^{i\underline{Q}\cdot\underline{R}_\ell(0)} e^{-i\underline{Q}\cdot\underline{R}_{\ell'}(t)} \rangle \tag{13.104}$$

Equation (13.103) shows the cross section can be separated into three parts. The first factor is merely kinematical, reflecting the fact we are dealing with the ratio of two currents. The second factor, the scattering length, represents the nuclear physics aspect of the scattering; it can vary with spin orientation and isotope distribution. The third factor, $S(Q,\omega)$, defined in Eq. (13.104) is the quantity that gives the double differential cross section the complexity and interest of a many-body problem. We will come back to this point soon. For now we note $S(Q,\omega)$ has been given the name of *dynamical structure factor*. It embodies all the information on the structure and dynamics of the nuclei in the target system that one can extract by neutron scattering. This function is the reason thermal neutron scattering has become one of the most important experimental techniques for probing the structure and dynamics of atoms and molecules in condensed matter, from physics to chemistry to biology.

Coherent and incoherent scattering

It is important to distinguish between two processes of thermal neutron scattering, coherent and incoherent. The distinction comes about because the scattering

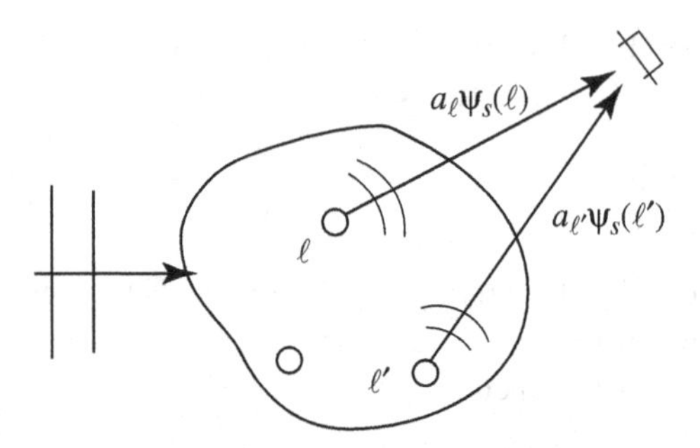

Fig. 13.5 Scattering of an incident plane wave by two nuclei, labeled ℓ and ℓ'. Each nucleus acts as an independent scattering center, emitting an outgoing spherical wave with amplitude given by the scattering length a labeled for that nucleus. The amplitude can vary from one nucleus to another depending on the spin orientation and isotope species, if any.

length of a nucleus depends on the spin orientation of the nucleus, if the nucleus has a nonzero spin, and in the case that the target is a distribution of several isotopes also on the which isotope is each nucleus. We now explain what is being taken into account in each process. Imagine the scattering of an incident plane wave by a system of target nuclei, two of which are shown in Fig. 13.5. The intensity measured at the detector is the square of the total scattered wave, the sum of individual contributions from the various scattering centers,

$$I \sim \left| \sum_{\ell=1}^{N} a_\ell \psi_s(\ell) \right|^2 = \sum_{\ell} a_\ell^2 |\psi_s(\ell)|^2 + \sum_{\ell,\ell'}{}' a_\ell a_{\ell'} \psi_s^*(\ell) \psi_s(\ell') \qquad (13.105)$$

where the prime on the second summation denotes terms for which the indices are not the same. In Eq. (13.105), we have separated the intensity into two contributions, which may be called *direct* and *interference* scattering respectively. Since the scattering length can vary with spin orientation and isotope distribution, what is actually measured at the detector is an average of Eq. (13.105) over these variations,

$$\bar{I} \sim \sum_{\ell,\ell'} \overline{a_\ell a_{\ell'}} \psi_s^*(\ell) \psi_s(\ell') \qquad (13.106)$$

where the overhead bar indicates averaging over spin and isotope distributions. Keeping in mind any quantity averaged necessarily loses its dependence on the nucleus index, we have

$$\begin{aligned} \overline{a_\ell a_{\ell'}} &= \overline{a_\ell^2} = \overline{a^2}, \qquad \text{if } \ell = \ell' \\ &= \overline{a_\ell a_{\ell'}} = \bar{a}^2, \quad \text{if } \ell \neq \ell' \end{aligned} \qquad (13.107)$$

Also we can cover both situations by writing

$$\overline{a_\ell a_{\ell'}} = (\overline{a^2} - \overline{a}^2)\delta_{\ell\ell'} + \overline{a}^2$$
$$\equiv a_{inc}^2 \delta_{\ell\ell'} + a_{coh}^2 \qquad (13.108)$$

which identifies the incoherent process to be direct scattering only, and the coherent process to be the sum of direct and interference scattering. If one does not distinguish among the scattering nuclei, one would have only coherent scattering. It is the incoherent scattering that is special in thermal neutron scattering. This is so because the scatteringy length can fluctuate with spin orientation or isotope variation.

Now that we have introduced the distinction between coherent and incoherent scattering, we go back to our double differential cross section Eq. (13.03) and perform spin and isotope averaging as indicated in Eq. (13.108). This leads to the identification of coherent and incoherent contributions to the double differential cross section,

$$\frac{d^2\sigma}{d\Omega dE_f} = (E_f/E_i)^{1/2}[a_{inc}^2 S_s(Q,\omega) + a_{coh}^2 S(Q,\omega)] \qquad (13.109)$$

where

$$S_s(Q,\omega) = \frac{1}{2\pi}\int_{-\infty}^{\infty} dt e^{i\omega t}\frac{1}{N}\sum_{\ell}\langle e^{i\underline{Q}\cdot\underline{R}_\ell(0)} e^{-i\underline{Q}\cdot\underline{R}_{\ell'}(t)}\rangle \qquad (13.110)$$

$$S(Q,\omega) = \frac{1}{2\pi}\int_{-\infty}^{\infty} dt e^{i\omega t}\frac{1}{N}\sum_{\ell,\ell'}\langle e^{i\underline{Q}\cdot\underline{R}_\ell(0)} e^{-i\underline{Q}\cdot\underline{R}_{\ell'}(t)}\rangle \qquad (13.111)$$

Notice in S_s one is concerned only with the motions of the same particle at two different times. For this reason S_s is known as the *single particle* (or *self*) *dynamic structure factor*. Although S and S_s may look quite similar, one should be aware that they describe very different properties of the scattering system. From what we have said above, it should come as no surprise that incoherent scattering gives information about the motions of particles moving individually, such as in diffusion, while coherent scattering describes the collective or cooperative motions among the target nuclei. This distinction will be brought out very clearly when one considers specific models that have been used to evaluate Eqs. (13.110) and (13.111) [Williams, 1966; Egelstaff, 1967; Parks, 1970].

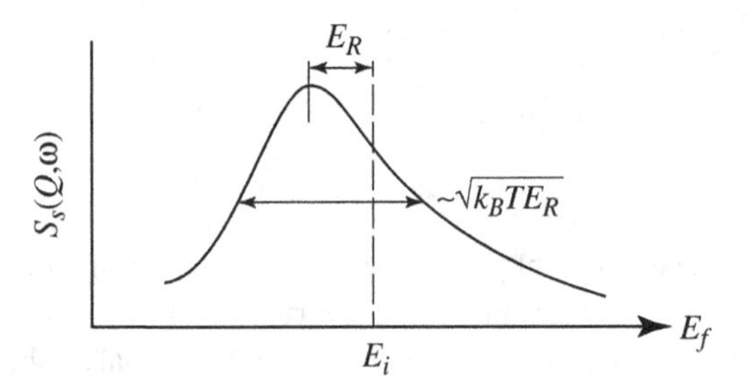

Fig. 13.6 Energy variation of the dynamic structure factor $S_s(Q, \omega)$ for an ideal-gas model. The spectrum (E_f is the energy after scattering) is the energy distribution of a thermal neutron scattered at initial energy E_i and momentum transfer $\hbar Q$. E_R is the recoil energy of the target nucleus, and T is the temperature of the ideal gas.

For a simple example of the dynamic structure factor we consider the results of a calculation of S_s using an ideal gas model. The result is illustrated in Fig. 13.6. In this case there is no distinction between coherent and incoherent scattering because there are no interference effects in an ideal gas. What one can see are the effects of thermal motion, in this case nuclei move only in straight-line trajectories following a Maxwellian distribution of velocities.

The meaning of this curve is essentially the energy distribution of the scattered neutrons which have exchanged momentum $\hbar Q$ with the scattering nuclei. Notice the peak of the distribution is not at $E_f = E_i$, as one might expect on the grounds that scattering at no energy exchange should be more likely than scattering with energy gain or loss. The peak is seen to be shifted to a lower energy by an amount E_R which is the recoil energy of the scattering nucleus, $E_R = \hbar^2 Q^2/2M$. So what we find is that the most likely process is one where the neutron loses an amount of energy to the nucleus that is equal to the recoil energy of the nucleus. The other feature to note in Fig. 13.6 is the width of the distribution is proportional the square root of the product of the thermal energy (temperature) and the recoil, the higher these two factors the larger is the range in which the scattered neutrons can have their final energies. Both features should give the reader some feeling for how neutrons with initial energy E_i can have their energies spread out after the scattering; in this spreading the neutrons can be upscattered or downscattered with an asymmetric distribution. A useful exercise for the reader is to compare the dynamic structure factor shown in Fig. 13.6 with the scattering frequency F shown in Fig. 9.3, and with the energy differential cross section shown in Fig. 9.4, and note the difference between the latter two results with regard to thermal motion effects.

13.5 CRITICALITY OF SYSTEMS WITH NEUTRON FISSION

Basic notions of neutron multiplication

Consider a homogeneous beam of neutrons having a distribution of energies. Define the distribution function $n(E)$ such that

$$n(E)dE = \text{expected number of neutrons with energies in } dE \text{ about } E \textit{ per } \text{cm}^3 \tag{13.112}$$

Suppose we ask how many neutrons in a certain energy range (and corresponding speed v) will cross an area A during a time interval Δt. Call this number

$$\phi(E)dEA\Delta t \equiv n(E)dE[Av\Delta t] \tag{13.113}$$

since all the neutrons in a certain volume $Av\ \Delta t$ will cross during Δt. In effect, we have defined in a physically intuitive way the neutron "flux"

$$\phi(E) \equiv vn(E) \tag{13.114}$$

with $n(E)$ being the number density. The interpretation of $\phi(E)$ is the expected number of neutrons with energies in *dE about E* crossing a unit area per unit time. Dimensionally, $n(E)$ is the number of neutrons per cm^3 per eV, while $\phi(E)$ is the number of neutrons per cm^2 per eV per sec. Both are therefore distributions in E, while $n(E)$ is also a spatial distribution (dimension of inverse volume).

Because we are interested in neutron reactions, we imagine the neutron beam is directed at a thin target of surface area A and thickness Δx. The rate at which neutrons in *dE about E* are incident upon the target is $\phi(E)dEA$. Now let the interaction (reaction) rate be denoted as $[\phi(E)dEA]P(\Delta x, E)$, which means we are defining $P(\Delta x, E)$ as the interaction probability (and therefore dimensionless). It will be convenient to take out the dependence on target thickness by writing $P(\Delta x, E) = \Delta x\Sigma(E)$, where

$$\Sigma(E) \equiv \text{interaction probability per unit path length (for short paths)} \tag{13.115}$$

The fact that P can be taken to be simply directly proportional to the target thickness is valid only for thin targets where one has the situation of only one interaction or none most of the time. The interaction rate then becomes

$$\Sigma(E)\phi(E)dEV$$

where

$$V = A\Delta X$$

The product $\Sigma(E)\phi(E)$, or the spatial-dependent form $\Sigma(\underline{r}, E)\phi(\underline{r}, E)$, appears frequently in reactor physics discussions; it is often called the "collision density".

We should keep in mind it is a distribution function in energy and space, and is also a rate.

We have previously introduced the macroscopic cross section $\Sigma(\underline{r}, E)$ to represent the target property that is most relevant to the consideration of neutron interaction in a material medium. Recall that $\Sigma = N\sigma$, where N is the nuclear density, the number of target nuclei per cm^3, and $\sigma(E)$ is the microscopic cross section. When the medium is not homogeneous, then nuclear density can vary with position in which case N becomes position dependent. In any event, the spatial and energy dependence of Σ is separable.

In problems of neutron multiplying systems (systems with neutron fission), we need to take into account several types of reactions. This can be expressed by writing a decomposition of the total cross section as a sum of scattering, capture, and fission cross sections. The term *absorption* will be used to denote the sum of capture and fission.

$$\sigma_t(E) = \sigma_s(E) + \sigma_a(E) \tag{13.116}$$

$$\sigma_a(E) = \sigma_c(E) + \sigma_f(E) \tag{13.117}$$

The basic reason reactor physics has its own unique flavor (and this is the reason for us to study it) is that the materials in a nuclear reactor have characteristic cross sections that make each reaction important at low or high neutron energies, but usually not both. For a thermal reactor with uranium fuel, the fission cross section of U^{235} is large (several hundred barns) at thermal energy but is only ~1 barn at ~10^4 eV, and is essentially zero at the energy where fission neutrons are emitted (~2 MeV). This behavior was previously noted in Chap. 3 (see Fig. 3.10) and also seen partially in Fig. 12.10. In contrast, the capture cross section $\sigma(n, \gamma)$ of U^{238} can be quite large in the resonance energy region (between thermal and fast), and is also non-negligible at thermal energy (see Fig. 3.11). The scattering cross section, to a first approximation, can be taken to be just a constant throughout the entire energy range of interest. Combining these variations with the fact that neutrons emitted from a fission reaction have a distribution peaked around 2 MeV, we have the situation of maintaining a chain reaction where neutrons are introduced into the system at high energy, but they need to slow down to thermal energy where the probability of their causing fission reactions to continue the chain reaction is the greatest. This is the essence of the problem of criticality in nuclear reactor physics.

Nuclear reactor criticality

A simple way of describing the criticality issue is to consider how one might estimate the critical condition to maintain a self-sustained fission chain reaction. This exercise is instructive in showing the role of different types of neutron interactions, and the interplay between materials properties and geometric factors. Suppose we separate all neutrons into either the *fast* (F) or high-energy group or the *thermal* (T) group. We begin by introducing a fast neutron into the homogenized (no spatial variation) reactor, and then follow all the contributions to the next generation of neutrons that it can possibly make (this is essentially what one would do in a Monte Carlo simulation).

Four things can happen to the neutron as it moves through the reactor. It can escape from the reactor (fast leakage) contributing nothing to the next generation, undergo a fission reaction (fast fission) contributing $\nu P_{NFL} P_{FF}$, undergo a capture reaction contributing nothing, or undergo a scattering interaction. The contribution of the fourth event is a little more complicated to estimate. We will further assume any scattering of a fast neutron will cause it to become a thermal neutron (this can be relaxed later on). Then three things can happen to this thermal neutron. It can escape from the reactor as a thermal neutron (thermal escape), undergo a capture reaction (thermal capture), or undergo a fission reaction (thermal fission). The first two events contribute nothing to the next generation while the third gives $\nu P_{NFL} P_{FS} P_{NTL} P_{TF}$.

For the various probabilities, we can estimate those pertaining to reactions using the corresponding Σ, while leaving aside the non-escape probabilities to be discussed later.

$$P_{FF} = \frac{\Sigma_{FF}}{\Sigma_{FA} + \Sigma_{FS}}, \quad P_{FS} = \frac{\Sigma_{FS}}{\Sigma_{FA} + \Sigma_{FS}}, \quad P_{TF} = \frac{\Sigma_{TF}}{\Sigma_{TA}} \tag{13.118}$$

If we add up all the contributions to the next generation, the sum k is then the multiplication constant of the reactor. That is, The neutron population of every succeeding generation changes by a factor of k. From the above scenario, we see there are two contributions to k, one from *fast fission* and the other from *thermal fission*. All other processes represent loss of neutron with no further contribution. Thus,

$$k = \nu P_{NFL}[P_{FF} + P_{NTL} P_{FS} P_{TF}] \tag{13.119}$$

In terms of k, we can now define what we mean by a critical reactor, namely, $k = 1$. For $k > 1$, the neutron population increases by a factor k for every generation, the reactor is said to be "*supercritical*". For $k < 1$, the system is "*subcritical*" and cannot maintain a self-sustained chain reaction. In the supercritical case, how quickly the reactor "runs away" (neutron flux increases too quickly for the control system to maintain the reactor in a safe operating state) is a major concern. We will come back to discuss this situation shortly.

The multiplication constant k is perhaps the most fundamental quantity in reactor physics calculations. It is conventional to recast it into a form that separates out the spatial effects of neutron escaping from the system from the materials effects of various reactions. We rewrite k as

$$k = k_\infty P_{NFL} P_{NTL} \tag{13.120}$$

$$\begin{aligned} k_\infty &= \nu \frac{P_{FF} + P_{NTL} P_{FS} P_{TF}}{P_{NTL}} \\ &= \eta f p \varepsilon \end{aligned} \tag{13.121}$$

with

$$\eta = \nu P_{TF} \tag{13.122}$$

$$p = P_{FS} \tag{13.123}$$

$$\varepsilon = 1 + \frac{P_{FF}}{P_{NTL} P_{TF} P_{FS}}. \tag{13.124}$$

The quantity k_∞ is known as the *infinite medium multiplication constant*; it is the multiplication constant when the reactor can be considered to be an infinite system (where there is no leakage). Incidentally, k is also often called the *effective* multiplication constant and denoted as k_{eff}.

Equation (13.121) shows k_∞ is the product of four factors. In reactor physics terminology, Eq. (13.121) is known as the *four-factor formula*. Here η is the average number of neutrons emitted per fission event, the thermal utilization factor f denotes the probability that the absorption of a thermal neutron takes place in the fuel rather than anywhere else (since we have only fuel in the present simple example, $f = 1$), p is the *resonance escape probability*, referring to the fact that in slowing down from fast to thermal energy the neutron must avoid any resonance absorption reaction, and ε is called the *fast fission factor* because it is the sum of fast and thermal fission contributions. We see also that the multiplication constant is just the product of k_∞ and the two non-escape probabilities.

Given that k_∞ refers to an infinite system, we may expect it to depend purely on material properties. Thus it may seem a little strange that k_∞ should involve P_{NTL} (see for example, Eq. (13.124)) which is a property that depends on the system size. A bit of reflection shows that the origin of this puzzling behavior arises from the fact that we started out in our analysis with a model of a finite system and then obtain k_∞ by extracting it from our finite-system result, Eq. (13.119). This observation indicates that there are different ways one can estimate k_∞, and the result for k_∞ can vary from one case to another. Is there a unique k_∞? We think so; it is the multiplication constant one finds when one starts at the outset with an infinite system. What about the expressions for k_∞ which seem to still involve

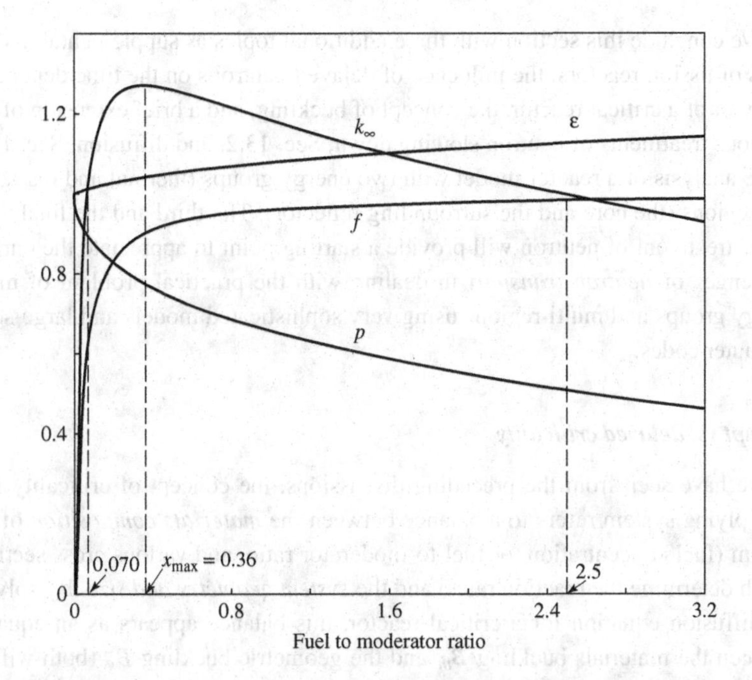

Fig. 13.7 Variation of the infinite medium multiplication constant k_∞ and three of its four components ε, f and p with the concentration ratio of fuel to moderator. The range of concentration ratio where k_∞ is greater than unity is between 0.070 and 2.5, and the maximum value k_∞ can have is about 1.27. [Adapted from Lamarsh, 1966.]

system size, such as our present estimate? We think they are all model-dependent approximations. The non-uniqueness of extracting k_∞ from finite-system models suggests that system size and material property effects, strictly speaking, are not separable in the determination of the multiplication constant.

To get a feel for the various numbers one can consult a figure like Fig. 13.7 that shows the variation of the four factors, ε, f, p and k_∞, for a homogeneous reactor. In this particular case, the reactor is a homogeneous mixture of 2% enriched uranium as fuel and H_2O as moderator at various concentration ratios (x) of fuel to moderator. Although not shown, the value η (U^{235}) is a constant at ~2.06.

The main feature to notice in Fig. 13.7 is k_∞ does not increase indefinitely with the fuel concentration. Since ε increases only slightly with x, the value of k_∞ is determined mostly by the competition between f and p. It is reasonable that with increasing x thermal utilization will initially increase rapidly but the effect then saturates. In contrast the resonance escape probability does not show any saturation and thus causes k_∞ to decrease when f starts to saturate. The peak value of k_∞ at about 1.27 sets the limit on the product of the two non-leakage probabilities if the reactor is to be critical.

We conclude this section with three additional topics as supplemental discussions of fission reactors, the influence of delayed neutrons on the time dependent behavior of a critical reactor, the concept of buckling, and a brief extension of our previous treatments of neutron slowing down, Sec. 13.2, and diffusion, Sec. 13.3, to the analysis of a reactor model with two energy groups (thermal and fast), and two regions (the core and the surrounding reflector). The third and the final topic in our treatment of neutron will provide a starting point to appreciate the current challenges of *neutron transport* in dealing with the practical problem of many energy groups and multi-regions using very sophisticated models and large-scale computer codes.

Prompt vs. delayed criticality

As we have seen from the preceding discussions, the concept of criticality in a multiplying system refers to a balance between the *materials composition* of the system (fuel concentration, or fuel-to-moderator ratio, and various cross sections which determine the reaction rates) and the *system geometry and size*. By solving the diffusion equation for a critical reactor, this balance appears as an equality between the materials buckling B_m and the geometric buckling B_g (both will be defined below). When this condition is satisfied, there is a nontrivial solution to the steady-state diffusion equation without a source, meaning that one can maintain a steady-state flux in the absence of an external source, or in other words, the neutron population ratio of the $(n+1)$th generation to the *nth* generation is *unity*.

Another way to describe critical behavior is through the time-variation of the flux as given by an appropriate kinetics equation. This is an approach we have not explored up to now. Suppose we ignore spatial and energy dependence and consider an infinite reactor in the absence of an external source. The time dependence of the flux is then described by

$$\frac{1}{v}\frac{d\phi(t)}{dt} = (\nu\Sigma_f - \Sigma_a)\phi(t) \tag{13.125}$$

with solution,

$$\phi(t) = \phi(0)\exp[(k_\infty - 1)t/\ell_p] \tag{13.126}$$

In writing Eq. (13.126), we have defined the infinite multiplication constant $k_\infty = \nu\Sigma_f/\Sigma_a$, and the prompt neutron lifetime $\ell_p = 1/v\Sigma_a$. According to this simple result, if the reactor is above critical (supercritical), $k_\infty > 1$, the flux would rise exponentially in time with a period

$$T = \ell_p/(k_\infty - 1) \tag{13.127}$$

For a rough estimate we take the lifetime ℓ_p to be typically 10^{-4} sec and $(k_\infty - 1)$ to be 10^{-3}, $T \sim 0.1$ sec. This means that in a timespan of 1 sec the flux will increase by a factor $e^{10} \sim 2.2 \times 10^4$. Such a rate of change, if true, would be uncontrollable by any practical means. (Reactors are usually controlled by the mechanical insertion or withdrawal of control rods made of materials which have high neutron absorption cross sections, for example, cadmium.)

Since reactors obviously do not actually behave in this manner (they can be operated safely), what is wrong with our estimate? The answer lies in our neglect of *delayed neutrons*. It is well known that not all the neutrons from fission come off promptly. A certain fraction of the fission neutrons, β, are delayed in that these neutrons are emitted from various decays of fission fragments, including β-decay which is a weak interaction and therefore a slow process. U^{235}, U^{238}, and Pu^{239} each has six groups of delayed neutrons, with the half life of a group varying from ~0.2 sec to ~54 sec [Lamarsh, 1966, p. 102]. The total delayed fraction β is 0.0064, 0.0148, and 0.002 respectively.

We can take into account the delayed neutrons in our estimate by revising our estimate of the period T, using instead an average neutron lifetime which is significantly longer than ℓ_p. At the same time we can refine our treatment of the multiplication effect by replacing the infinite multiplication constant k_∞ by the effective multiplication costant k. Thus, our time-dependent reactor description becomes

$$\ell \frac{d\phi(t)}{dt} = [(1-\beta)k - 1]\phi(t) + \beta\phi(0) \tag{13.128}$$

where we denote the average neutron life time as ℓ,

$$\ell = (1-\beta)\ell_p + \sum_{i=1}^{6} \ell_i \beta_i \tag{13.129}$$

The important point to note is the average life time is now of order 0.1 sec because of the delayed neutrons, the second term in Eq. (13.129). With this new value of the life time the period T then becomes $10^{-1}/10^{-3} \sim 100$ sec. Now the flux increases by a factor of e in 100 sec which makes the reactor quite controllable by moving the control rods.

We can readily solve Eq. (13.128) to obtain

$$\begin{aligned}\phi(t) &= \phi(0)\exp\{[(1-\beta)k-1]t/\ell\} \\ &\quad + \frac{\beta\phi(0)}{1-(1-\beta)k}(1-\exp\{[(1-\beta)k-1]t/\ell\})\end{aligned} \tag{13.130}$$

For $(1-\beta)k < 1$, the exponentials die out in time, the steady-state solution is

$$\phi(t) \rightarrow \frac{\beta}{1-(1-\beta)k}\phi(0) \tag{13.131}$$

So at long times, the flux increase over the initial value is

$$\phi(t)/\phi(0) \rightarrow \frac{\beta(1-\rho)}{\beta-\rho} \tag{13.132}$$

where we introduce the quantity called reactivity,

$$\rho \equiv \frac{k-1}{k} = \frac{\Delta k}{k}, \quad k = \frac{1}{1-\rho} \tag{13.133}$$

In reactor kinetics terminology, changing the multiplication constant by adding or removing fuel, or inserting or withdrawing control rods, all can be carried out by specifying the reactivity ρ.

Notice Eq. (13.132) gives an infinite jump when the amount of reactivity introduced is equal to delayed neutron fraction. When this happens, the reactor is said to be "*prompt critical*", which means that the reactivity is sufficient to make the system critical without relying on the delayed neutrons. Then we are back to a very dangerous situation where the flux increases exponentially — a bomb scenario.

The concept of buckling in nuclear reactor criticality

We turn next to the another fundamental aspect of criticality to show the interplay between materials properties of a system and its geometric attributes. To do this we use diffusion theory to estimate the non-escape probabilities that appear in the multiplication constant. It is instructive to ask what could be a measure of the reactor size besides the bare system dimensions. Recall the extrapolated boundary condition, Eq. (13.44), expresses the idea of an extrapolated distance as an incremental length beyond the actual system boundary. Thus it is not surprising that a useful geometric measure of system size should involve the extrapolated distance. How does this come about naturally in the context of boundary conditions for solving the diffusion equation? We will examine this connection through the example of a critical spherical reactor, a system in which the materials properties and the geometric size are in balance to maintain the multiplication constant at unity.

Consider a spherical reactor of radius R composed of materials for which all the cross sections, scattering, absorption, and fission, are nonzero, and there is no external source. The diffusion equation for this system is

$$(\nabla^2 + \alpha^2)\phi(\underline{r}) = 0 \quad r \leq R \tag{13.134}$$

with $\alpha^2 = (\nu\Sigma_f - \Sigma_a)/D > 0$. (Note $\alpha^2 \leq 0$ means at best $k_\infty = 1$, and any finite system must therefore be subcritical, i.e., the system cannot maintain a non-zero steady state flux in the absence of a source.). The physical solution to Eq. (13.134), after applying the condition of finite flux at the origin, is just

$$\phi(\underline{r}) = A\frac{\sin \alpha r}{r} \tag{13.135}$$

We also conclude A must be positive and αR must be $\leq \pi$. We can apply one more boundary condition, that at the reactor surface $r = R$. Since Eq. (13.135) is a homogeneous equation, we know that we will not be able to determine A. Thus the condition at R has to impose a constraint on α, the only other constant left in the description. (The analogy with energy quantization in quantum mechanics when solving the wave equation for a certain shape of the potential should be noted at this point.)

We have already seen the proper boundary condition for a material-vacuum interface is no re-entrant current, $J_-(R) = 0$. In diffusion theory it becomes approximately (see the development surrounding Eq. (13.40))

$$\frac{\phi(R)}{4} + \frac{D}{2}\underline{\hat{n}} \cdot \underline{\nabla}\phi|_{r=R} = 0, \quad \text{or} \quad \frac{R}{\phi}\frac{d\phi}{dr}\bigg|_{r=R} = -\frac{R}{2D} \tag{13.136}$$

Applying this to Eq. (13.135) gives

$$1 - \alpha R \cot \alpha R = \frac{R}{2D} \tag{13.137}$$

The solution to Eq. (13.137), $\alpha_o R$, is seen to depend on the magnitude of the ratio $R/2D$. It is close to zero if $R \ll 2D$ and close to π if $R \gg 2D$. The latter is the more physically common situation for any interesting value of D, i.e., reactor material. So we write $\alpha_o R = \pi - \varepsilon$, with ε being small. The left hand side of Eq. (13.137) then becomes

$$1 - \alpha_o R \cot \alpha_o R \sim 1 + \frac{\pi - \varepsilon}{\varepsilon} = \frac{\pi}{\varepsilon} = \frac{\pi}{\pi - \alpha_o R} \tag{13.138}$$

$$1 - \alpha_o R \cot \alpha_o R \sim \frac{\pi}{\pi - \alpha_o R} \tag{13.139}$$

Notice that

$$\alpha_o = (\pi - \varepsilon)/R = (\pi/R)(1 - \varepsilon/\pi) \sim (\pi/R)(1 + \varepsilon/\pi)^{-1} \tag{13.140}$$

From Eqs. (13.137) and (13.138) we also have

$$\frac{\pi}{\varepsilon} = \frac{R}{2D}, \quad \text{or} \quad R\varepsilon/\pi = 2D \tag{13.141}$$

Combining Eqs. (13.140) and (13.141) we obtain

$$\alpha_o \sim \pi/\tilde{R} \equiv B_g \tag{13.142}$$

with $\tilde{R}$ being the "extrapolated" radius,

$$\tilde{R} \equiv R + 2D \tag{13.143}$$

Thus we arrive at the same result as in the case of the slab reactor before. One can apply the surface boundary condition as $\phi(\tilde{R}) = 0$, rather than $J_-(R) = 0$. Equation (13.142) also serves to introduce the quantity B_g, called "*geometric buckling*" in reactor physics, presumably because it has to do with the shape ("buckling") of the flux and it depends only on the size (geometry) of the system.

The implication of Eq. (13.142) is in order for the critical spherical reactor to have a physical solution satisfying the boundary conditions in diffusion theory, the constant α has to have the value specified by Eq. (13.142). However, recall that in writing the diffusion equation Eq. (13.134), the constant α already was defined by the materials properties. We will rewrite this definition as

$$\alpha^2 = \frac{\nu\Sigma_f - \Sigma_a}{D} \equiv B_m^2 \tag{13.144}$$

thereby introducing the quantity B_m, "*materials buckling*", in analogy with the geometric buckling. Therefore, the only way to satisfy both the materials constraint, represented by Eqs. (13.134) and (13.132), and the system size constraint, represented by Eqs. (13.138) and (13.142), is to require

$$B_m^2 \equiv B_g^2 \tag{13.145}$$

We can regard this to be *the condition for system criticality*, the balance between materials properties and system size.

To see what Eq. (13.145) can lead to, we rewrite it as

$$\Sigma_a\left(\frac{\nu\Sigma_f}{\Sigma_a} - 1\right) = DB_g^2 \tag{13.146}$$

or

$$\frac{\eta f}{1 + L^2 B_g^2} = 1 \tag{13.147}$$

with $\nu\Sigma_f/\Sigma_a \equiv \eta f$ and $L^2 \equiv D/\Sigma_a$, L being called the *diffusion length*. We purposely write Eq. (13.147) in the form of a critical condition, explicitly showing the multiplication constant $k = k_\infty P_{NL}$ having the value of unity. With this identification we can pick off an expression for the non-leakage probability,

$$P_{NL} \equiv \frac{1}{1 + L^2 B_g^2} \tag{13.148}$$

Equation (13.148) is useful because it provides a quick estimate, in the context of simple diffusion theory, of the non-escape probability that appears in the multiplication constant. Going back to Eq. (13.145), we see another way to interpret the balance condition is the requirement

$$\nu\Sigma_f - \Sigma_a = DB_g^2 \tag{13.149}$$

The left hand side represents the effective cross section for "neutron gain", whereas the right hand side represents the "neutron loss", with DB_g^2 playing the role of a "leakage cross section". This observation makes it possible to compare the effects of *neutron interactions*, in the sense of scattering and reactions measured in the form of macroscopic cross sections (or the mean free path), with those of *neutron diffusion*, in the sense of diffusion and surface boundary condition in terms of D and the geometric buckling, on the same basis. An appreciation of this simple equivalence is a primary reason that we can give for studying neutron diffusion theory.

A two-group reflected reactor

In elementary reactor physics a classic problem is the *two-group two-region* reactor model. It is instructive to understand, even if only qualitatively, how one can analyze the criticality of this system, because the insight provides the basis for the extension to many practical calculations involving many energy groups and several reactor regions. The latter are surely more complicated in detail, but the basic physics is no different.

We consider fast and thermal neutrons in a reactor with an inner core and an outer reflector. We therefore distinguish cross sections, etc. and fluxes using a numerical index:

1 — fast neutrons in the core
2 — thermal neutrons in the core
3 — fast neutrons in the reflector
4 — thermal neutrons in the reflector

and write the steady-state diffusion equations as

$$\begin{aligned}
(D_1\nabla^2 - \Sigma_{a1} - \Sigma_{R1} - \nu\Sigma_{f1})\phi_1 + \nu\Sigma_{f2}\phi_2 &= 0 \\
\Sigma_{R1}\phi_1 + (D_2\nabla^2 - \Sigma_{a2})\phi_2 &= 0 \\
(D_3\nabla^2 - \Sigma_{a3} - \Sigma_{R3})\phi_3 &= 0 \\
(D_4\nabla^2 - \Sigma_{a4})\phi_4 + \Sigma_{R3}\phi_3 &= 0
\end{aligned} \tag{13.150}$$

where Σ_R denotes a "*removal*" cross section, in the sense of a neutron being transferred from the fast group to the thermal group. One can give relatively simple expressions to estimate this quantity, but we do not get into the details here because our focus is on the overall structure of the problem.

Inspection of the system of four equations, Eq. (13.150), shows the system is coupled through slowing down, diffusion, and fission. Fast neutrons in the core are produced by thermal fission in the core while they are lost by slowing down in the core and by leakage to the reflector. Fast neutrons in the reflector are produced by leakage from the core and lost by slowing down to thermal. For the thermal neutrons in the core, they are produced by slowing down and lost by leakage to the reflector. In the reflector thermal neutrons are produced by slowing down and by leakage from the core. Given this model, the mathematical problem now is to solve the coupled second-order differential equations simultaneously with appropriate conditions at the core-reflector interface and at the outer boundary of the reflector.

Although the symmetry between the fast and thermal flux in the core may not be apparent, it turns out that ϕ_1 and ϕ_2 satisfy the same set of equations. This does not mean they will be the same since each has its own boundary conditions. To bring out this symmetry, we first rewrite Eq. (13.150) in a more compact notation,

$$\begin{aligned}(\nabla^2 + \alpha_{11})\phi_1 + \alpha_{12}\phi_2 &= 0\\ \alpha_{21}\phi_1 + (\nabla^2 - \alpha_{22})\phi_2 &= 0\\ (\nabla^2 - \alpha_{33})\phi_3 &= 0\\ (\nabla^2 - \alpha_{44})\phi_4 + \alpha_{43}\phi_3 &= 0\end{aligned} \tag{13.151}$$

With

$$\begin{aligned}\alpha_{11} &= (\nu\Sigma_{f1} - \Sigma_{a1} - \Sigma_{R1})/D_1 & \alpha_{21} &= \Sigma_{R1}/D_2\\ \alpha_{12} &= \nu\Sigma_{f2}/D_1 & \alpha_{22} &= \Sigma_{a2}/D_2\\ \alpha_{33} &= (\Sigma_{a3} + \Sigma_{R3})/D_3 & \alpha_{43} &= \Sigma_{R3}/D_4\\ \alpha_{34} &= 0 & \alpha_{44} &= \Sigma_{a4}/D_4\end{aligned} \tag{13.152}$$

Notice in Eq. (13.151) ϕ_1 and ϕ_2 are explicitly coupled to each other, whereas their coupling to ϕ_3 and ϕ_4 appears only through the interface boundary conditions. Thus we can use the first two equations to obtain an equation only in ϕ_1, and another equation only in ϕ_2. The results are

$$\begin{aligned}(\nabla^2 - \alpha_{22})(\nabla^2 + \alpha_{11})\phi_1 - \alpha_{12}\alpha_{21}\phi_1 &= 0\\ (\nabla^2 + \alpha_{11})(\nabla^2 - \alpha_{22})\phi_2 - \alpha_{12}\alpha_{21}\phi_2 &= 0\end{aligned} \tag{13.153}$$

The one equation to be solved is of the form

$$(\nabla^2 + \mu^2)(\nabla^2 - \nu^2)\phi = 0, \quad \phi = \phi_1, \text{ or } \phi_2 \tag{13.154}$$

with

$$\mu^2 - \nu^2 = \alpha_{11} - \alpha_{22} \tag{13.155}$$

$$\mu^2\nu^2 = \alpha_{11}\alpha_{22} + \alpha_{12}\alpha_{21} \tag{13.156}$$

Next we consider a spherical core and write the general solutions in the form

$$\phi(r) = AZ(r) + CW(r) \tag{13.157}$$

$$(\nabla^2 + \mu^2)Z(r) = 0, \quad Z(r) \sim \frac{1}{r}\sin\mu r \tag{13.158}$$

$$(\nabla^2 - \nu^2)W(r) = 0, \quad W(r) \sim \frac{1}{r}\sinh\nu r \tag{13.159}$$

$$\begin{pmatrix}\phi_1(r)\\ \phi_2(r)\end{pmatrix} = \begin{pmatrix}A_1 & C_1\\ A_2 & C_2\end{pmatrix}\begin{pmatrix}Z(r)\\ W(r)\end{pmatrix} \tag{13.160}$$

Notice that we accept only solutions that are finite everywhere in the system.

Similarly, for the reflector region we have

$$(\nabla^2 - \alpha_{33})U(r) = 0, \quad U(r) \sim \frac{1}{r}\sinh\sqrt{\alpha_{33}}(\tilde{R} - r) \tag{13.161}$$

$$(\nabla^2 - \alpha_{44})V(r) = 0, \quad V(r) \sim \frac{1}{r}\sinh\sqrt{\alpha_{44}}(\tilde{R} - r) \tag{13.162}$$

$$\begin{pmatrix}\phi_3(r)\\ \phi_4(r)\end{pmatrix} = \begin{pmatrix}A_3 & 0\\ A_4 & C_4\end{pmatrix}\begin{pmatrix}U(r)\\ V(r)\end{pmatrix} \tag{13.163}$$

The integration constants A and C are constrained by the condition of no reentrant current at reflector surface, for which we substitute extrapolated boundary conditions, and the conditions at the core-reflector interface for which we specify continuity of flux and current,

$$\begin{aligned}
&\phi_j(\tilde{R}_2) = 0\ , \qquad\qquad j = 3, 4\\
&\phi_1(R) = \phi_3\ (R), \qquad \phi_2(R) = \phi_4(R)\\
&D_1\phi_1'(R) = D_3\ \phi_3'(R), \quad D_2\phi_2'(R) = D_4\phi_4'(R)
\end{aligned} \tag{13.164}$$

Applying these conditions one finds a set of 4 homogeneous equations for the 4 constants A_1, C_1, A_3, and C_4. For nontrivial solutions, the determinant of the coefficients of these four constants , comprised of the α's, functions Z, W, U and V

evaluated at R, and the derivatives of these functions evaluated at R, must vanish. Let the determinant be denoted as Δ and regard this as a function of the core size R. One can numerically find $\Delta(R)$ for several values of R, and then by interpolation find that value of R, say R_c, which gives $\Delta(R_c) = 0$. This is the solution of the criticality problem, namely, the size of critical core for given materials composition and concentration in the system.

This determination of the size of the critical reactor is the analogue of setting the material buckling equal to the geometric buckling in the previous case of one speed neutron diffusion equation (see Sec. 13.3). Of course one can do the calculation for a given core size and determine a material property such as the fuel to moderator ratio for the critical system.

To conclude this discussion we show in Fig. 13.8 the fast and thermal fluxes in a critical, spherical reactor fueled with U^{235} with water as moderator and reflector. It is not surprising that both fluxes have maximum at the center of the reactor, basically the same behavior as obtained when there is no reflector. What is somewhat unusual, at least at first glance, is *the peak in the thermal flux in the reflector*. This is a new characteristic feature associated with the two-group two-region system. Physically the peak signifies that fast neutrons leaking out of the core and slowing down in the reflector can more than compensate for the loss of thermal neutrons by leakage in the reflector. Mathematically, the peak in ϕ_4 arises from the sum of two terms, see Eq. (13.163), a positive contribution associated with U and a negative contribution associated with V. Regarding the latter as the new effect, we might attribute the thermal flux peaking in the reflector to

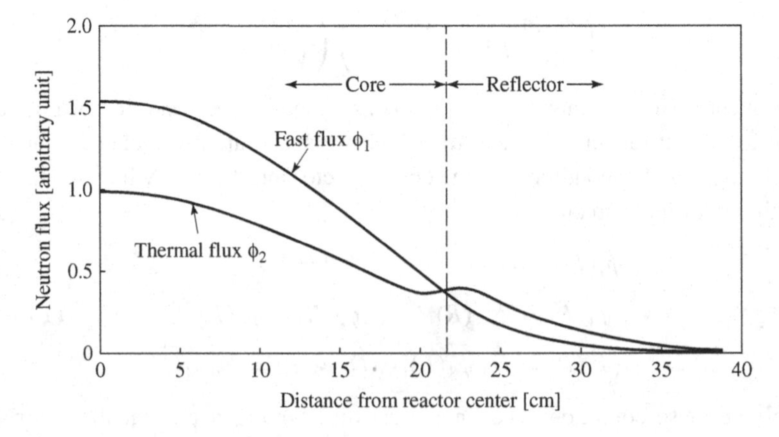

Fig. 13.8 Profiles of the fast and thermal flux in a critical spherical reactor with a core composed of a homogeneous mixture of U^{235} and water, and a water reflector [Lamarsh 1966, p. 338.]

a significant thermal absorption in the reflector, as measured by the constant α_{44} in Eq. (13.152).

REFERENCES

D. E. Bacon, *Neutron Scattering in Chemistry* (Butterworths, London, 1977).

G. I. Bell and S. Glasstone, *Nuclear Reactor Theory* (Van Nostrand Reinhold, 1970).

J.-P. Boon and S, Yip, *Molecular Hydrodynamics*, McGraw-Hill 1980 (Dover reprint 1991).

J. Byrne, *Neutrons, Nuclei and Matter* (Institute of Physics, London, 1994).

K. M. Case, F. deHoffmann, G. Placzek, *Introduction to the Theory of Neutron Diffusion*, Vol. 1, Los Alamos Scientific Laboratory, 1953.

J. J. Duderstadt and W. R. Martin, *Transport Theory* (Wiley Interscience, 1979).

P. A. Egelstaff, *An Introduction to the Liquid State* (Academic Press, London, 1967).

E. Fermi [*Ricerca Scientifica* **7**, 13 (1936), reprinted in E. Fermi, *Collected Papers* (Univercity of Chicago Press, 1952), Vol. I, p. 980.]

A. J. Goodjohn and G. C. Pomeraning, *Reactor Physics in the Resonance and Thermal Regions* (MIT Press, Cambridge, 1966), vol. I. Neutron Thermalization, vol. II. Resonance Absorption.

J. R. Lamarsh, *Introduction to Nuclear Reactor Theory* (Addison-Wesley, Reading, 1966).

W. Marshall and S. W. Lovesey, *Theory of Thermal Neutron Scattering* (Oxford, London 1971).

R. K. Osborn and S. Yip, *The Foundations ofNeutron Transport Theory* (Gordon and Breach, New York, 1967).

D. E. Parks, M. S. Nelkin, J. R. Beyster, N. F. Wikner, *Slow Neutron Scattering and Thermalization* (W. A. Benjamin, New York, 1970).

A. M. Weinberg and E. P. Wigner, *The Physical Theory of Neutron Chain Reactors* (Univercity of Chicago Press, Chicago, 1958).

M. M. R. Williams, *The Slowing Down and Thermalization of Neutrons* (North Holland, Amsterdam, 1966).

Bibliography

D. E. Bacon, *Neutron Scattering in Chemistry* (Butterworths, London, 1977).

G. I. Bell and S. Glasstone, *Nuclear Reactor Theory* (Van Nostrand Rheinhold, 1970).

H. A. Bethe and J. Ashkin, "Passage of Radiation Through Matter", in *Experimental Nuclear Physics*, E. Segrè, ed. (Wiley, New York, 1953), Vol. I, p. 166.

F. Bjorklund and S. Fernbach, *Phys. Rev.* **109**, 1295 (1958).

J. M. Blatt and V. F. Weisskopf, *Theoretical Nuclear Physics* (Wiley, New York, 1952), p. 627.

J.-P. Boon and S. Yip, *Molecular Hydrodynamics*, McGraw-Hill 1980 (Dover, Reprint 1991).

K. A. Bruecker, A. M. Lockett and M. Rotenberg, *Phys. Rev.* **121**, 255 (1961).

J. Byrne, *Neutrons, Nuclei and Matter* (Institute of Physics, Bristol, 1994).

K. M. Case, F. de Hoffmann and G. Placzek, *Introduction to the Theory of Neutron Diffusion*, Vol. 1 (Los Alamos Scientific Laboratory, 1953).

J. Chadwick, "Possible Existence of a Neutron", *Nature* **129**, 312 (1932) [Chadwick 1932a].

J. Chadwick, "The Existence of a Neutron", *Proc. Roy. Soc.* **136A**, 692 (1932) [1932b].

B. L. Cohen, *Concepts of Nuclear Physics* (McGraw-Hill, New York, 1971).

E. R. Cohen and B. N. Taylor, "The Fundamental Physical Constants", *Physics Today* (August 1995), p. BG9.

I. Curie and P. Savitch, *C. R. Acad. Sci. (Paris)* 206, 906 (1938), 1648 (1938) [Curie 1938a].

I. Curie and P. Savitch, *J. Phys. Radium* 9, 355 (1938) [Curie 1938b].

A. S. Davydov, *Quantum Mechanics* (Pergamon Press, London, 1965), pp. 306 and 578.

R. H. Dicke and J. P. Wittke, *Introduction to Quantum Mechanics* (Addison-Wesley, Reading, 1960).

H. E. Duckworth, "Evidence for Nuclear Shells from Atomic Measurements", *Nature* **170**, 158 (1952).

J. J. Duderstadt and W. R. Martin, *Transport Theory* (Wiley Interscience, 1979).

P. Domenici, *A Brighter Tomorrow* (Rowman and Littlefield, Lanham, 2004).

P. A. Egelstaff, *An Introduction to the Liquid State* (Academic Press, London, 1967).

W. S. Emmerich, in *Fast Neutron Physics*, ed. J. B. Marion and J. L. Fowler (Interscience, New York, 1963), Vol. I, p. 1072.

R. D. Evans, *The Atomic Nucleus* (McGraw Hill, New York, 1955).

E. Fermi, *Ricerca Scientifica* **7**, 13 (1936), reprinted in E. Fermi, *Collected Papers* (University of Chicago Press, 1952), Vol. I, p. 980.

E. Fermi, *Nuclear Physics*, Lecture Notes by J. Orear, A. H. Rosenfeld and R. A. Schluter (University of Chicago Press, 1949).

H. Feshbach, *Ann. Rev. Nuc. Sci.* **8**, 49 (1958).

A. Foderaro, *The Elements of Neutron Interaction Theory* (MIT Press, 1971) Chap. 4.

G. Gamow, *Z. Phys.* **51**, 204 (1928).

R. L. Garwin and G. Charpak, *Megawatts and Megatons: The Future of Nuclear Power and Nuclear Weapons* (University of Chicago Press, 2002).

A. Gersten, *Nucl. Phys. A* **96**, 288 (1967).

S. Glasstone and P. J. Dolan, *The Effects of Nuclear Weapons*, Third Edition (U.S. Government Printing Office: Washington, D.C., 1977).

H. Goldstein, C. P. Poole and J. L. Safko, *Classical Mechanics* (Addison-Wesley, Reading, 2001), Third Edition.

A. J. Goodjohn and G. C. Pomeraning, *Reactor Physics in the Resonance and Thermal Regions* (MIT Press, Cambridge, 1966), Vol. I. Neutron Thermalization, Vol. II. Resonance Absorption.

R. W. Gurney and E. U. Condon, "Wave Mechanics and Radioactive Disintegration", *Nature* **122**, 439 (1928).

O. Hahn, L. Meitner, and F. Strassmann, *Naturwissenschaften* **26**, 475 (1938).

O. Hahn and F. Strassmann, *Naturwissenschaften* **27**, 11 (1939).

O. Haxel, H. D. Jensen, H. E. Suess, "On the 'Magic Numbers' in Nuclear Structure", *Physical Review* **75**, 1766 (1949).

W. Heitler, *Quantum Theory of Radiation* (Oxford, 1955), Sec. 26.

P. E. Hodgson, *The Optical Model of Elastic Scattering* (Oxford, London, 1963).

D. J. Hughes and J. A. Harvey, Neutron Cross Sections, BNL-325 (Brookhaven National Laboratory, National Nuclear Data Center, 1955), vol. 2, Curves.

J. D. Jackson, *Classical Electrodynamics*, (Wiley, New York, 1962), Chap. 17.

S. F. Johnston, *The Neutron's Children: Nuclear Engineers and the Shaping of Identity* (Oxford University Press, 2012).

G. F. Knoll, *Radiation Detection and Measurement* (Wiley, New York, 1979).

K. S. Krane, *Introductory Nuclear Physics* (Wiley, New York, 1987).

A. M. Lane and R. G. Thomas, "R-Matrix theory of nuclear reactions", *Rev. Mod. Phys.* **30**, 257 (1958).

J. R. Lamarsh, *Nuclear Reactor Theory* (Addison-Wesley, Reading, 1966), Chap. 2.

C. M. Lederer and V. S. Shirley, eds. *Table of Isotopes* (Wiley, New York, 1978), Seventh Edition.

R. L. Liboff, *Introductory Quantum Mechanics* (Holden Day, New York, 1980).

D. L. Livesey, *Atomic and Nuclear Physics* (Blaisdell, Waltham, 1966).

J. E. Lynn, *The Theory of Neutron Resonance Reactions* (Oxford, London, 1968).

G. H. Marcus, *Nuclear Firsts: Milestones in the Road to Nuclear Power Development* (American Nuclear Society, La Grange Park, 2010).

P. Marmier and E. Sheldon, *Physics of Nuclei and Particles* (Academic Press, New York, 1969), Vols. I and II.

W. Marshall and S. W. Lovesey, *Theory of Thermal Neutron Scattering* (Oxford, London 1971).

MIT Department of Nuclear Science and Engineering, "*Science. Systems. Society. A New Strategy for the Department of Nuclear Science and Engineering*," (2010) http://web.mit.edu/nse/pdf/spotlights/2011/NSE_StrategicPlan_overview.pdf

M. G. Mayer, "On Closed Shells in Nuclei", *Physical Review* **74**, 235 (1948).

M. G. Mayer and H. F. Jensen, *Elementary Theory of Nuclear Shell Structure* (Wiley, New York, 1955).

V. McLane, C. L. Dunford, P. F. Rose, Neutron Cross Sections, 4th ed. (Academic Press, New York, 1988), vol. 2 Neutron Cross Section Curves.

L. Meitner and O. R. Frisch, "Disintegration of uranium by neutrons: A new type of nuclear reaction", *Nature* **143**, 239 (1939).

W. E. Meyerhof, *Elements of Nuclear Physics* (McGraw Hill, New York, 1967).

P. M. Morse and H. Feshbach, *Methods of Theoretical Physics* (McGraw-Hill, New York, 1953).

N. F. Mott and H. S. W. Massey, *The Theory of Atomic Collisions* (Oxford University Press, London, 1949).

R. L. Murray, *Nuclear Energy* (Butterworth-Heinemann, Amsterdam, 2009), Chap. 26.

R. A. Nelson, *Physics Today* (August 1995), p. BG15.

I. Noddack, *Angew. Chem.* **47**, 653 (1934).

R. K. Osborn and S. Yip, *The Foundations of Neutron Transport Theory* (Gordon and Breach, New York, 1967).

D. E. Parks, J. R. Beyster, M. S. Nelkin and N. F. Wikner, *Slow Neutron Scattering and Themalization* (W. A. Benjamin, New York, 1970).

S. Pearlstein, "Evaluated Nuclear Data Files", in *Advances in Nuclear Science and Technology* (Academic Press, New York, 1975), Vol. 8.

A. Penzias, *Rev. Mod. Phys.* **51**, 425 (1979).

F. Perey and B. Buck, *Nucl. Phys.* **32**, 353 (1962).

M. A. Preston, *Physics of the Nucleus* (Addison-Wesley, Reading, 1962).

F. Reines and C. L. Cowan, "Detection of the Free Neutrino", *Physical Review* **92** , 830 (1953).

R. Rhodes, *The Making of the Atomic Bomb* (Simon & Schuster, New York, 1986).

M. E. Rose, *Elementary Theory of Angular Momentum* (Wiley, New York, 1957).

P. Roman, *Advanced Quantum Theory* (Addison-Wesley, Reading, 1965), Chap. 3.

L. Sartori, "Effects of Nuclear Weapons," *Physics Today,* March 1983.

G. R. Satchler, *Introduction to Nuclear Reactions* (Halsted, New York, 1980).

L. I. Schiff, *Quantum Mechanics*, McGraw-Hill, New York, 1955).

P. Schofield, *The neutron and its applications, 1982: plenary and invited papers from the conference to mark the 50th anniversary of the discovery of the neutron held at Cambridge, 13–17 September 1982* (Institute of Physics, Bristol, 1983), Institute of Physics conference series, no. 64.

E. Segrè, *Nuclei and Particles* (W. A. Benjamin, New York, 1965).

E. Segrè, "The Discovery of Nuclear Fission", *Physics Today* **42**, 38 (1989).

Bal Raj Sehgal, ed. *Nuclear Safety in Light Water Reactors* (Elsevier, Amsterdam, 2012).

B. E. Watts, *Phys. Rev.* **87**, 1037 (1952).

A. M. Weinberg and Eugene P. Wigner, *The Physical Theory of Neutron Chain Reactors* (University of Chicago Press, 1958).

E. J. Williams, *Rev. Mod. Phys.* **17**, 217 (1945).

M. M. R. Williams, *Slowing Down and Thermalization of Neutrons* (North-Holland, Amsterdam, 1966).

D. Wilmore and P. E. Hodgson, *Nucl. Phys.* **55**, 673 (1964).

Problems

Chapter 1 Context and Perspective

Problem 1

Write a brief statement of purpose for this book based on your understanding of what was said in the Preface. What can you make out from the Table of Contents, and your reading of Chap. 1?

Problem 2

Compare you answer to Problem 1 with the actual syllabus of the class you are presently taking (if applicable). Discuss how you can best use the book under your own circumstances.

Problem 3

Study the one-page letter to the editor of J. Chadwick (1932) and in your own words (you may paraphrase Chadwick's sentences but do not copy whole paragraphs) write an abstract of the results he was reporting. The aim is to state in not more than five sentences (the more concise the better), what was new. You are not being asked to say he was right or wrong, but to just condense what was said in his one-page letter. Your abstract therefore is an even more brief statement of what are the new results and their implications (if any). You will find this exercise is not as simple as it may look. You should also find that unless you have a good idea of what Chadwick was trying to say, you would not be able to properly condense his report into an abstract. Notice Chadwick did not have to write an abstract of his letter, presumably because the letter was meant to be a short communication. In present days we tend to see abstracts even for short communications. Chadwick's letter was remarkably effective and a good example of technical writing. You can find it readily on the internet.

***Problem* 4**

In what sense is the discovery of the neutron an unexpected event?

***Problem* 5**

Read Segre's account of the discovery of fission in *Physics Today*. In what sense can one say this discovery was an unexpected event?

***Problem* 6**

Discuss in your own words the lessons one can learn from the three nuclear incidents described in Chap 1.

***Problem* 7**

Following up on the previous problem, suppose one argues that the best thing our society can do to minimize the probability and impact of "future" nuclear incidents is to *invest in more science and technology* (to level the playing field in our struggles against *Mother Nature* so to speak), discuss what would be the logic in this point of view.

***Problem* 8**

How would you explain the meaning of *Science-System-Society* to students of NSE?

Chapter 2 Organization

***Problem* 1**

Based on the table of contents, Fig. 2.1, and the description given in Chap. 2, which chapters of this book can be considered to form the core of NSE? Briefly discuss in what way they are the most fundamental.

***Problem* 2**

What is the reason that the basic references for the study of NRI are books written in 1960's through 1980's?

***Problem* 3**

Locate the 5 key references discussed in Sec. 2.2 either in the library or on the internet. Get to know each one by browsing through them to get an idea of the style of the book and the contents. Make some notes on these references so that in the future you know where to look for additional reading should you run into questions that are not covered in the present book. If possible, treat these references as an extension of the coverage of standard text materials in nuclear physics.

***Problem* 4**

Spend a few minutes browsing through this book in the same way as you did in Problem 3. Get to know the layout of this book and the way the chapters are grouped and placed in a certain flow. Keep in mind the benefits of going back and forth among the chapters much as having access to different parts of a building all the time rather than using one room after another only in a sequential manner.

***Problem* 5**

What are the two types of problems given at the end of each chapter? Discuss generally the circumstances in which one type of problem is more appropriate than the other. In your opinion, which type is more helpful in learning the basic concepts?

Chapter 3 Nuclear Properties and Data

***Problem* 1**

What are the nuclear radiations (particles) that will be studied in this book? In what way are they most relevant to the discipline of NSE? Give some examples of their interactions that are of particular interest to you.

***Problem* 2**

Compute the wavelengths of the following nuclear radiation particles:

neutrons at thermal energy, 1 eV, 1 keV, 1 MeV
photons at 1 eV, 1 keV, 1.02 MeV
electrons at 1 eV, 1 keV, 0.511 MeV

Compare these wavelengths and discuss briefly any significance of the energies specified in connection with the interactions that you think could be of interest in this book. In other words, give some reasons why these particles at these energies could be relevant to NSE.

Problem 3

Besides energy and linear momentum, what other physical properties do neutrons and protons have that could influence their interactions?

Problem 4

Sketch the nuclear density distribution. Discuss what relation this distribution could have regarding the nature of nuclear interaction energy or nuclear forces (force is derivative of the interaction energy).

Problem 5

On the internet, browse the *Nuclide Chart* and the *Table of Isotopes*. Make some notes summarizing the essential information that each one provides. Discuss briefly how the data in these two tabulations complement each other.

Problem 6

Sketch the neutron cross section $\sigma(E)$ for water, graphite, U^{235} and U^{238} over the energy range from thermal to MeV.

Chapter 4 Nuclear Binding and Stability

Problem 1

Use the semiempirical mass formula to calculate the mass of O^{16} and Pb^{208}. Tabulate the individual percentage contribution to the average binding energy per nucleon from the volume, surface, Coulomb, asymmetry, and pairing terms. Compare your mass and B/A results with experimental values from the Nuclide Chart, and also with the contributions shown in Fig. 4.5.

Problem 2

Use the semi-empirical mass formula to determine which isobars with $A = 136$ should be stable. Indicate the various modes of decay and the predicted stable isotope. Compare your results with experimental data. Discuss what one learns from this exercise.

***Problem* 3**

Using the semiempirical mass formula and the Table of Nuclides, calculate a few (enough) points to construct an overall B/A curve. Sketch this curve along with the contributions from the various terms in the mass formula. Compare your results with the trends indicated in Fig. 4.5.

***Problem* 4**

The reaction, $^3\text{H} + {}^1\text{H} \rightarrow {}^3_2\text{He} + n$, has a Q-value of -0.764 Mev. Tritium ^3H also undergoes β-decay with end-point energy of 0.0185 Mev. Find the difference between the neutron and hydrogen mass in MeV. Draw an energy level diagram showing the levels involved in the reaction and the β^-decay, then indicate in your diagram the proton separation energy S_p, the Q-value, and the end-point energy T_{max}.

***Problem* 5**

Suppose you do not know the gamma ray energy that is given off when the proton absorbs a neutron, but you have the Chart of Nuclides. How do you go about determining the energy of the gamma (state the steps but do not do the math)? Can you also find out the cross section value for this reaction?

***Problem* 6**

Since the binding energy per nucleon (B/A) of $_2\text{He}^4$, at 7.07 Mev, is higher than that of $_3\text{Li}^6$, at 5.33 Mev, it may appear that the former is more stable than the latter and therefore Li^6 should spontaneously undergo α-decay,

$$_3\text{Li}^6 \rightarrow {}_2\text{He}^4 + {}_1\text{H}^2$$

where the B/A of $_1\text{H}^2$ is 1.11 Mev. Determine whether this is indeed the case.

***Problem* 7**

(a) Explain the usefulness of the Mass Parabola.
(b) Suppose you are asked to analyze the Mass Parabola using the empirical binding energy formula, show which is more important between the Coulomb term

and the asymmetry term. (The coefficients of the two terms are $a_c = 0.72$ Mev and $a_a = 23.5$ Mev.)

**Problem* 8

The binding energy per nucleon of ${}^6_3$Li is about 5.3 Mev while that of ${}^4_2$He is 7.1 Mev. Does this mean that the former is unstable against α-decay? Explain. (*Note*: The binding energy of the deuteron ^{2}H is 2.25 Mev.)

**Problem* 9

Give a short and concise answer to each of the following questions.

(a) Derive using a sketch the asymmetry term in the empirical mass formula.
(b) Explain what is β^+-decay, then state and justify the energy condition for this process.

**Problem* 10

You are told that all nuclides follow a B/A curve of the form

$$\begin{aligned} B/A &= E_B \qquad A_1 \le A \le A_2 \\ &= 0.1 E_B \quad \text{otherwise} \end{aligned}$$

with $A_2 > A_1$. What can you conclude qualitatively about the stability of these nuclides? Is fission possible? How about fusion?

Chapter 5 Nuclear Energy Levels and Models

Problem 1

(a) Starting with the time-dependent *Schrödinger* wave equation, derive an equation for the probability density $P(\underline{r}, t) = \Psi^*(\underline{r}, t)\Psi(\underline{r}, t)$ in the standard form of the continuity equation, and in this way obtain the expression for the particle current $\underline{J}(\underline{r}, t)$.
(b) Apply your expression for $\underline{J}(\underline{r}, t)$ to a plane wave, $\exp[i\underline{k} \cdot \underline{r}]$, where $\underline{k}$ is the wave vector, then give a physical interpretation of your result. How does the current differ from the flux which is defined as the particle number density times the particle speed (as in a beam of particles of a given flux)?

***Problem* 2**

Solve the 1D time-independent *Schrödinger* wave equation for a particle in a square well by applying the BC for an infinite well. Sketch and discuss the first three eigenvalues and eigenfunctions, and then indicate in your sketch how the solutions are affected if the well were instead *very steep but not infinite.*

***Problem* 3**

(a) Write down the time-independent *Schrödinger* wave equation of a system of two particles, mass m_1 and m_2, interacting through a central potential *V(r)*, where *r* is the separation distance between the two particles, $r = \left|\underline{r}_1 - \underline{r}_2\right|$. Introduce the center-of-mass $\underline{R} = [m_1\underline{r}_1/(m_1+m_2) + m_2\underline{r}_2/(m_1+m_2)]$ and relative $\underline{r} = \underline{r}_1 - \underline{r}_2$ coordinates and show that the wave equation for two particles naturally separates into two equations, one for the center-of-mass coordinate and another for the relative coordinate. Discuss what is interesting about each equation. What do your results say about the reduction of a 2-body collision problem into an effective one-body problem of a particle moving in a potential field?

(b) Carry out the same reduction for the Newton's equation.of motion for the system of two colliding particles.

***Problem* 4**

Solve the 1D time-independent *Schrödinger* wave equation for the bound states of a particle in a square well, choosing the even-parity solutions for the interior wave function. Discuss what determines the number of bound states that one can have. Compare your results with the determination of bound states using the odd-parity solutions. Sketch and discuss qualitatively the first two eigenvalues and eigenfunctions that one would obtain under the approximation the potential well is very deep.

***Problem* 5**

Solve the 1D time-independent *Schrödinger* wave equation for three lowest bound states of a particle of mass m in a square well of depth V_o and width L. Sketch the energy levels and their corresponding wavefunctions, and indicate the parity of each state. Is it physically reasonable that any given potential should have at least one bound state (comment on whether your solutions are consistent with your expectations)? Given that you also know the qualitative nature of the bound states in 3D, what can you say about the existence of a threshold for a spherical well potential?

***Problem* 6**

Solve the 3D problem of a particle of mass m fully enclosed in a cubical box with side length L by determining the energy levels and corresponding normalized wavefunctions. Discuss what is the connection between this and the problem of a 3D cubical well.

***Problem* 7**

Derive the $\ell = 0$ (s-wave) radial wave equation from the time-independent *Schrödinger* wave equation for a particle in a spherical well of width r_o and depth V_o. Solve this equation for the ground state wave function and the corresponding bound-state energy. Sketch and discuss briefly your results and then indicate what changes would occur if the well were to become *very steep and very narrow.*

***Problem* 8**

In classical mechanics a particle incident upon a potential with range r_o at an impact parameter b would be scattered if $b < r_o$, but if b were greater than r_o then there would be no interaction. Use this simple picture to show that in the scattering of a neutron at low energy, by which we mean $kr_o \ll 1$, with $E = \hbar^2 k^2/2m$, *only the s-wave interaction is important*. Take $r_o = 1.5$ F, what is the range of neutron energy where this approximation is valid?

***Problem* 9**

Consider the one-dimensional problem of a particle of mass m and energy E incident upon a potential barrier of height V_o and width L ($V_o > E$) going from left to right. Derive the following expression for the transmission coefficient

$$T = \left[1 + \frac{V_o^2}{4E(V_o - E)} \sinh^2 \kappa L\right]^{-1}$$

where $\kappa = \sqrt{2m(V_o - E)}/\hbar$. Sketch the variation of T with κL. What physical interpretation can you give for the dimensionless quantity κL? For nuclear physics problems which is the more realistic limit between thin barrier and thick barrier (why)?

***Problem* 10**

Consider Figs. 5.10, 5.11, and 5.12. Explain in your own words the information given in each figure. Is the information consistent from one figure to another? What is the overall picture of nuclear stability when the figures are taken together?

***Problem* 11**

Explain the addition of spin and orbital angular momentum operators to form the total angular momentum operator, $\underline{j} = \underline{S} + \underline{L}$, in terms of the eigenfunctions and eigenvalues of the these operators. In what way is addition of angular momentum operators different from the geometric addition of two classical vectors?

***Problem* 12**

Explain the operation of angular momentum addition in general, where one adds two angular momentum operators, $\underline{L}$ and $\underline{S}$, to form a total angular momentum operator $\underline{j}$, $\underline{j} = \underline{L} + \underline{S}$. Apply your explanation to the particular case where $\underline{L}$ and $\underline{S}$ are the orbital and spin angular momentum operators of a nucleon respectively.

***Problem* 13**

Explain how one arrives at the representation (eigenfunction or ket) that diagonalizes the Hamiltonian with spin-orbit coupling. Discuss what is the rationale for choosing the representation that also diagonalizes L^2 and S^2, where $\underline{L}$ is the orbital angular momentum and $\underline{S}$ is the spin angular momentum.

***Problem* 14**

(a) Sketch the energy levels of nucleons as given by a central force potential, a parabolic well (see Fig. 5.13). Explain the spectroscopic notation used in labeling each level, and how is degeneracy of each level determined. What is the significance of these results concerning the stability of nuclei?
(b) Sketch the energy levels of nucleons given by a shell-model potential with spin-orbit coupling (see Fig. 5.16), and explain the notation. What new features do you get with this model compared to the model in part (a)?

***Problem* 15**

On the basis of the single-particle shell model with spin-orbit coupling (Fig. 5.16), predict the ground-state spin and parity of the following nuclides:

$$Li^6, N^{14}, Mn^{55}, Nb^{93}, Xe^{131}, Au^{197}, Pb^{207}, Bi^{209}$$

Compare your results with experimental data (e.g., Nuclide Chart). In the case of any discrepancy between your predictions and experiments, give an explanation.

***Problem* 16**

Write a brief essay (not more than two pages) summarizing the essential arguments that lead to the explanation of the magic numbers in terms of the nuclear shell model.

****Problem* 17**

(a) Among the energy levels of a central force potential is a level labeled 1d. Suppose we now add a spin-orbit interaction term to the Hamiltonian such as in the shell model. Using the spectroscopic notation, label the new levels that evolve from this 1d level. Specify how many nucleons can go into each of the new levels, and explicitly write out the quantum numbers specifying the wave function of each nucleon.
(b) In an odd-odd nucleus the last neutron and proton go into a $1d_{3/2}$ and a $1g_{9/2}$ level respectively. Use the shell model to predict the spin and parity of this nucleus.
(c) A beta decay occurs between initial state (3^-) and final state (3^+), while a gamma decay occurs between (2^-) and (4^+). What is the dominant mode of decay in each case?
(d) The binding energy per nucleon of ${}^6_3$Li is about 5.3 Mev while that of ${}^4_2$He is 7.1 Mev. Does this mean that the former is unstable against α-decay? Explain. (Note: The binding energy of the deuteron ^{2}H is 2.25 Mev.)

****Problem* 18**

In a one-dimensional system with a square well potential, depth V_o and range r_o, is it possible to have at least one bound state no matter what the values of V_o and r_o? What happens in three dimensions with a spherical well potential, depth V_o and range r_o? In each case, explain your answer with a sketch of the wave function. [*Note*: you should answer this question without going through any derivation.]

***Problem* 19**

A particle of mass m is just barely bound by a one-dimensional potential well of width L. Find the value of the depth V_o.

***Problem* 20**

In the following sketches, say briefly what you are trying to show.

(a) The wave function of a free particle with energy E ($E > 0$) in a 1D system is written as $\psi = A \sin kx$, $E = \hbar^2 k^2/2m$, and A is some constant. Now introduce a potential barrier of height V_o (with $V_o < E$) and range $2L$, located at the origin. Sketch the wave function everywhere in the system to show the effects of the potential.
(b) Repeat (a) for a potential well instead of a barrier.
(c) Consider a particle barely bound by the potential well (b). Sketch the wave function everywhere in the system.

***Problem* 21**

In solving the 3D wave equation for the bound states for a spherically symmetric well, what new features appear (relative to the 1D problem) in labeling the states and in determining the degeneracy? Sketch what you expect would be the radial wave functions for the ground state and first excited state for a spherical well.

***Problem* 22**

The radial part of the 3D wave equation can be expressed in the form

$$[D_r^2 + (A^2/r^2) + V(r)]R_\ell(r) = ER_\ell(r)$$

where $D_r^2 = \frac{1}{r^2}\frac{d}{dr}\left(r^2\frac{d}{dr}\right)$ and A is a constant. It is customary to simplify this equation by making the transformation, $R_\ell(r) = u_\ell(r)/r$. Carry out this operation to obtain the equation for $u_\ell(r)$. Discuss the reason for making this transformation. Is this a purely mathematical step, or is there some physical motivation involved?

***Problem* 23**

Consider a 1D potential barrier $V(x) = V_o$, $0 < x < L$, and 0 everywhere else, with $V_o > 0$. Suppose a particle is incident on this barrier from the left, with $0 < E < V_o$. Think of the configuration space divided into three regions, (1) $x < 0$, (2) $0 < x < L$, and (3) $x > L$.

(a) Write down the form of the wave function $\psi(x)$ that is physically acceptable in each region (do not apply interface boundary conditions at $x = 0$ and L). Sketch $|\psi(x)|^2$ across the three regions
(b) Repeat question (a) for a potential well, with $V(x)$ now given by $-|V_o|$, $0 < x < L$, and 0 everywhere else.
(c) Can one have $E < 0$ solutions in parts (a) and (b)? Explain briefly.

**Problem* 24

Explain briefly what is the reason for introducing the total momentum, $\underline{j} = \underline{L} + \underline{S}$, in the Nuclear Shell model. How does this affect the labeling of the states?

**Problem* 25

Summarize briefly how the shell model is able to predict the magic numbers. State only the most important steps (commensurate with how much time you have). You can quote anything discussed in the chapter without further explanation, such as the coupling of two angular momenta, their eigenfunctions and eigenvalues, and the labeling of energy levels (so long as you use the same expressions and notations).

Chapter 6 Nuclear Disintegrations and Decays

Problem 1

Suppose one claims the most fundamental statement regarding radioactive disintegration is summarized in Eq. (6.1). Explain what one could mean by this. In your discussion consider the significance of the *decay constant* as the fundamental quantity that connects observations (measurements) and theoretical models (understanding).

Problem 2

In the spirit of the previous problem, consider how one may connect the *decay constant* to a *collision cross section* (Sec. 3.4), and in this way gain a unified view of the two phenomena involving nuclear transitions being treated in this book, spontaneous disintegration and collision-induced reactions.

Problem 3

(a) You are given the series disintegration $A \rightarrow B \rightarrow C \rightarrow D$ (stable), with decay constants $\lambda_A, \lambda_B, \lambda_C$ respectively. Find the number of atoms of B, C, and D collected in a time t if the activity of A is held constant. (This is the approximation for a very long-lived source.)

(b) Tellurium is bombarded by deuterons in a cyclotron, the reaction being

$$\mathrm{Te}^{130}(d, p)\mathrm{Te}^{131} \rightarrow \mathrm{I}^{131} \rightarrow \mathrm{Xe}^{131}\text{(stable)}$$

with both decays being β^-, and the half lives are 30 hours and 8 days respectively. The bombardment conditions are such that the (d, p) reaction is equivalent to a source strength of 2 mCi. Find

(i) the initial rate of I^{131} production, in μCi per hour,

(ii) the number of Xe^{131} atoms produced during a single 6-hour bombardment,

(iii) the total number of Xe^{131} atoms obtained eventually if the target were allowed to stand undisturbed for several months following a 6-hour bombardment.

Problem 4

Consider the series disintegration $A \rightarrow B \rightarrow C \rightarrow D$ (stable), with decay constants $\lambda_A, \lambda_B, \lambda_C$ respectively. Set up the governing equations and solve for the number of atoms of B, C, and D at time t given initial values of A_o, B_o, C_o, and D_o at $t = 0$. Sketch you results for the case where the parent is very long-lived, then repeat for the case where the grand daughter is long lived. Discuss and contrast the behavior in these two situations.

Problem 5

Write a short essay (two pages or less) describing the genealogy of α-particle emitting heavy nuclides (see Evans, 1955, pp. 517–523). Explain briefly what is meant by the Thorium, Neptunium, Uranium, and Actinium series. What happens to the α- and β-particles emitted by a decay series?

****Problem 6***

The basic idea underlying the C^{14} dating method is that radioactive C^{14} is produced continuously in the atmosphere by cosmic rays and finds its way into living plants and animals by carbon exchange. When the plant or animal dies the exchange process stops. How would you use this idea to determine the age of an antique wooden chair (when the tree was cut), that is, what activity would you measure and how would you interpret the data? Give a sketch of your method. (Note: The equilibrium concentration of C^{14} in the air is a known constant. C^{14} undergoes β-decay to N^{14} which is stable.)

***Problem** 7

Radioisotope B decays to C (decay constant λ_B) which can undergo two decay processes, one going to a nucleus D with constant $\lambda_C^{'}$ and a competing process going to another nucleus E with constant $\lambda_C^{''}$. Both D and E are stable. Write down the equations describing the four nuclides $N_B(t), \ldots, N_E(t)$ in the decay chain. Solve the equations directly using the initial condition $N_B(t = 0) = N_B^o$, $N_C(0) = N_D(0) = N_E(0) = 0$, and then demonstrate that your solutions are indeed correct.

***Problem** 8

At time $t = 0$ you are given an atom that can decay through either of two channels, a and b, with known decay constants λ_a and λ_b. Find the probability that it will decay by channel a during the time interval between t_1 and t_2, with t_1 and t_2 arbitrary. Interpret your result.

***Problem** 9

Nuclide A can decay to B with decay constant $\lambda_1$and to C with decay constant λ_2. B is stable but C decays to E with decay constant λ_4. E is stable. There is another way nuclide C can be produced. Nuclide D also decays to C with decay constant λ_3. Find the concentration of $C(t)$ given that at time $t = 0$ the concentrations of all the nuclides are known (A_o, B_o, C_o, D_o, E_o). Interpret your result to show it is physically reasonable. What nuclides are conserved in time?

***Problem** 10

Consider the reflection of a particle with mass m and energy E incident from the left upon a 1D potential barrier, $V(x) = V_o, x > 0$, and $V(x) = 0, x < 0$. Find the reflection coefficient R for $E > V_o$. Investigate the limit of $E \to V_o$.

***Problem** 11

Consider a beam of particles, each of mass m and having kinetic energy E, incident from the left upon a square barrier of height V_o and width L, with $E < V_o$. Inside the barrier the wave function is of the form

$$\psi(x) = ae^{\kappa x} + be^{-\kappa x}$$

where you are given the result

$$\frac{a}{b} = \frac{1 + ik/\kappa}{1 - ik/\kappa} e^{-2\kappa L}$$

with $\hbar^2 k^2/2m = E$ and $\hbar^2 \kappa^2/2m = V_o - E$.

(a) Using the information given on $\psi(x)$ inside the barrier, derive an approximate expression for the transmission coefficient T for the case of thick barriers, $\kappa L \gg 1$.

(b) Sketch qualitatively the absolute square of the wave function $|\psi(x)|^2$ everywhere and indicate the spatial dependence of $|\psi(x)|^2$ wherever it is known. Does $|\psi(x)|^2$ vanish at any point?

***Problem* 12**

Suppose you are given the result for the transmission coefficient T for the barrier penetration problem, one-dimensional barrier of height V_o extending from $x = 0$ to $x = L$,

$$T = \left[1 + \frac{V_o^2}{4E(V_o - E)} \sinh^2 KL\right]^{-1}$$

where $K^2 = 2m(V_o - E)/\hbar^2$ is positive ($E < V_o$).

(a) From the expression given deduce T for the case $E > V_o$ without solving the wave equation again.

(b) Deduce T for the case of a square well potential from the result for a square barrier.

***Problem* 13**

(a) You are told the reaction $C(d, p)^{13}C^{14}$ has a resonance at a deuteron energy E_d (LCS), and following this, C^{14} undergoes β-decay to N^{14}. Draw the energy level diagram for this situation in which you show explicitly how the following energies can be calculated in terms of known masses and E_d: (1) kinetic energy available for reaction T_o, (2) Q value for the reaction, (3) deuteron separation energy, (4) proton separation energy, and (5) Q_β.

(b) On the basis of (a), predict whether or not the reaction $B(\alpha, n)^{11}N^{14}$ will have a resonance, and if so, at what energy of the α particle this will occur. (Since you are not given numerical values, you should leave your answer in terms of defined quantities such as masses and various energies.)

***Problem* 14**

Give a brief and concise answer to each of the following.

(a) What is the physical picture of the model used to estimate the decay constant in alpha decay (give sketch)? Why does the model give an upper limit for the decay constant?
(b) What is *electron capture* and with what process does it compete?
(c) What is *internal conversion* and with what process does it compete?
(d) Give a sketch of the variation of the neutron cross section of C in the energy region below 0.1 MeV and explain the features.

***Problem* 15**

Radioactive Na^{24} undergoes β^-decay to Mg^{24} (see energy level diagram below). In a measurement one finds a distribution of electrons having an end-point energy of 1.39 Mev, but in addition two groups of positrons with distinct end-point energies are also observed. There were no Ne^{24} found.

(a) What could be the process giving rise to the positrons?
(b) What are the expected end-point energies of the two positron groups?
(c) What are the decay modes for the indicated transitions?

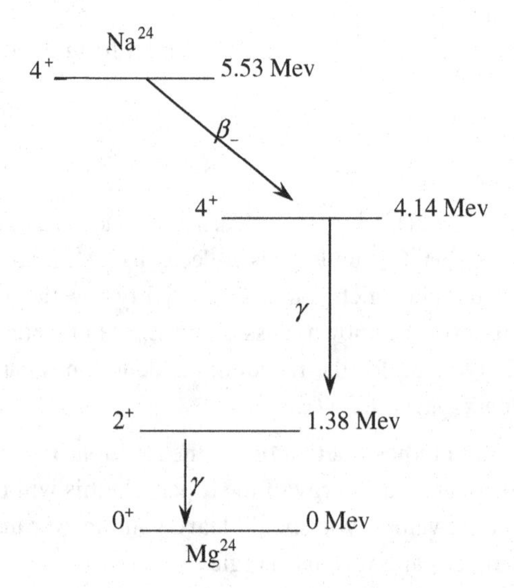

***Problem* 16**

For the energy-level diagram shown below, identify all the transitions that can take place. For each transition determine the decay mode and the decay energy (Q-value).

$_{Z}X^{A}$

3^{+} ——— 2.42 Mev

4^{-} ——— 1.57 Mev

$_{Z-1}Y^{A}$

4^{+} ——— 0 Mev

***Problem* 17**

The decay scheme of ^{80}Br is shown below. Classify the various decay modes and estimate all the decay constants that you can.

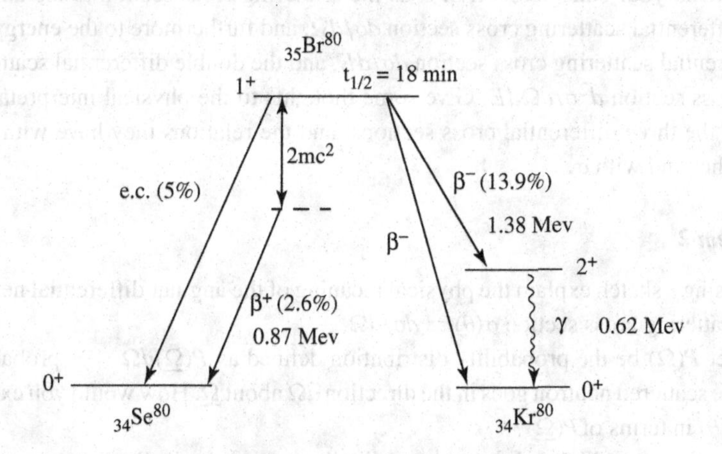

***Problem* 20**

The table below shows initial and final nuclear states. For each case, state whether natural radioactive decay is possible, and if so, give the dominant decay mode and specify any energy requirements, and briefly explain your answers.

	Initial				Final			
Case	A	Z	I	Parity	A	Z	I	Parity
1	A	Z	3/2	+	A	Z	7/2	+
2	A	Z	0	−	A	Z	0	−
3	A	Z	3/2	+	A	$Z-1$	1/2	−
4	A	Z	0	+	A	$Z+1$	3/2	+

Chapter 7 Collision Cross Sections

Problem 1

(a) It is well known the cross section σ is a measure of the probability of having a reaction, yet σ has the dimension of an area. Give a physical definition of σ that shows clearly its connection to the probability of a reaction and at the same time explains why σ has the dimension that it does. Give a simple interpretation of the ratio: {reaction probability}/σ. Discuss whether your definition of σ allows σ to be calculated.

(b) Extend your consideration of σ as the scattering cross section to the angular differential scattering cross section $d\sigma/d\Omega$, and furthermore to the energy differential scattering cross section $d\sigma/dE$, and the double differential scattering cross section $d^2\sigma/d\Omega dE$. Give some thoughts to the physical interpretations of the three differential cross sections, and the relations they have with each other and with σ.

Problem 2

(a) Using a sketch explain the physical meaning of the angular differential neutron scattering cross section $\sigma(\theta) \equiv d\sigma/d\Omega$.

(b) Let $P(\underline{\Omega})$ be the probability distribution defined as $P(\underline{\Omega})d\Omega$ = probability the scattered neutron goes in the direction $d\Omega$ about $\underline{\Omega}$. How would you express $\sigma(\theta)$ in terms of $P(\underline{\Omega})$?

(c) If you were told that the scattering distribution is spherically symmetric, what would you write for $\sigma(\theta)$?

Problem 3

Derive the expression for the angular differential scattering cross section $\sigma(\theta)$ for s-wave scattering, then obtain the expression for the scattering cross section σ.

Problem 4

Calculate the neutron scattering cross section of C^{12} for thermal neutrons. Assume a potential well with depth $V_o = 36$ MeV and range $r_o = 1.4 \times A^{1/3} F$ and consider only the s-wave contribution. Compare your result with the experimental value ($\sigma = 5$ barns) and discuss any significance.

Problem 5

Derive the general expression for the angular differential scattering cross section $\sigma(\theta)$ and incident particles of *arbitrary energy* E, then obtain the expression for the scattering cross section σ.

Problem 6

In the case of neutron-proton scattering, it can be shown (see Sec. 9.1) that the scattering cross section of thermal neturons in hydrogen obtained using the method of phase shift gives a value of 2.3 barns, whereas the experimental value is 20 barns.

(a) Demonstrate whether the assumption of s-wave scattering is justified.
(b) What other reasons can one give for the discrepancy between calculation and experiment (to answer this part, you should look ahead and read Sec. 9.1)?

Problem 7

Verify the direct derivation (showing all the intermediate steps) of the following results.

(a) Equation (7.3)
(b) Equation (7.8)
(c) Equation (7.18)
(d) Equation (7.24)
(e) Equation (7.25).

Problem 8

Calculate the neutron scattering cross section of C^{12} for thermal neutrons. Assume a potential well with depth $V_o = 36$ MeV and range $r_o = 1.4 \times A^{1/3} F$ and consider only the s-wave contribution. Compare your result with the experimental value and discuss any significant discrepancy. How does your result compare with the cross section of H^1? Explain the difference between neutron scattering by C^{12} and by H^1 in terms of the physics of neutron-nucleus interactions.

Problem 9

Calculate the neutron scattering cross section of O^{16} for thermal neutrons. Assume a potential well with depth $V_o = 36$ Mev and range $r_o = 1.4 \times A^{1/3}$ F and consider only the s-wave contribution. Compare your result with the experimental value and discuss any significant discrepancy. How does your result compare with the cross section of H^1? Explain the difference between neutron scattering by O^{16} and by H^1 in terms of the physics of neutron-nucleus interactions

**Problem* 10

(a) Calculate the phase shift δ_o for s-wave scattering of a particle of mass m and incident energy E, with $E < V_o$, by a potential barrier $V(r) = V_o, r < r_o$, and $V(r) = 0, r > r_o$.
(b) Simplify your result by going to the limit of low-energy scattering. Examine the total scattering cross section in this limit. Sketch σ as a function of $k_o r_o$, where $k_o^2 = 2mV_o/\hbar^2$ and indicate the value of σ in the infinite barrier limit, $k_o r_o \to \infty$.

**Problem* 11

Consider the scattering of a particle of mass m and incident kinetic energy E by a spherical well potential, depth V_o and range r_o. You are given the following information. The s-wave scattering cross section is $\sigma_o = (4\pi/k^2)\sin^2\delta_o(k)$, where $\delta_o(k)$ is an energy-dependent phase shift. In the case of low-energy scattering, i.e., $kr_o \ll 1$, $\delta_o(k)$ is given by

$$\tan\delta_o(k) \sim (k/K \cot Kr_o)(1 - Kr_o \cot Kr_o) \qquad (*)$$

where $\hbar^2 K^2/2m = V_o$, and $\hbar^2 k^2/2m = E$.

(a) Define the scattering length a and express σ_o in terms of a.
(b) Find a from Eq. (*) [Hint: you can take $\delta_o(k)$ to be small]. Discuss (and give a sketch) the behavior of σ_o as a function of Kr_o. Do you see a connection between the behavior of σ_o and the calculation of bound states in a deep spherical well potential?

(c) What changes do you expect if the scattering is by a spherical barrier of height V_o with the same range (give a sketch)?

***Problem* 12**

The boundary condition $\psi(r) \sim \exp(i\underline{k} \cdot \underline{r}) + f(\theta)e^{ikr}/r$, $r \gg r_o$, is applied to calculate the angular differential scattering cross section. Using the fact that s-wave scattering is spherically symmetric and writing the exterior wave function as $\psi = A\sin(kr + \delta)$, derive the relation between the scattering amplitude $f(\theta)$ and the phase shift δ_o for s-wave.

***Problem* 13**

(a) Calculate the phase shift δ_o for s-wave scattering of a particle of mass m and incident kinetic energy E by a potential barrier of height $V_o(V_o > E)$ and range r_o.
(b) Simplify your result by taking the low-energy limit, where $kr_o \ll 1$ and $\delta_o \ll 1$. Find the scattering cross section σ. Sketch σ as a function of $k_o r_o$, with $k_o^2 = 2mV_o/\hbar^2$, and show the limiting values of σ for small and large $k_o r_o$.
(c) Consider applying your result to describe neutron-proton scattering, with $V_o =$ 36 MeV and $r_o = 2 \times 10^{-13}$ cm. Give an estimate of the value of σ from (a) and compare it to the experimental value. If there is a discrepancy, what is the implication? Discuss further whether it is appropriate (reasonable) to apply the present model.

***Problem* 14**

Consider the scattering of low-energy neutrons by a nucleus which acts like an impenetrable sphere of radius R.

(a) Solve the radial wave equation to obtain the phase shift δ_o.
(b) Given that the angular differentil scattering cross section for s-waves is $(d\sigma/d\Omega)_o = (1/k^2)\sin^2\delta_o(k)$, use your result from (a) to find the total scattering cross section σ_o. Suppose we apply this calculation to n-p scattering and use for R the radius of the deuteron, $R = \hbar\sqrt{m_n E_B}$, where m_n is the neutron mass and E_B is the ground state energy of the deuteron. Find R in unit of F ($1\ F = 10^{-13}$ cm), and σ_o in barns.
(c) Does your result agree with the experimental value of neutron scattering cross section of hydrogen? If not, explain the reason for the discrepancy.

***Problem** 15

Consider a beam of collimated, monoenergetic neutrons (energy E) incident upon a thin target (density N atoms per cc) of area A and thickness Δx at a rate of I neutrons/sec. Assume the cross sectional area of the beam is greater than A. An energy sensitive detector subtended at an angle θ with respect to the incident beam direction is set up to measure the number of neutrons per second scattered into a small solid angle $d\Omega$ about the direction $\underline{\Omega}$ <u>and</u> into a small energy interval dE' about E'. Let this number be denoted by Π.

(a) Define the double (energy and angular) differential scattering cross section $d^2\sigma/d\Omega dE'$ in terms of the physical situation described above such that you can relate this cross section to the scattering rate Π and any other quantity in the problem. (You may find it helpful to draw a diagram of the specified arrangement.)
(b) How is $d^2\sigma/d\Omega dE'$ related to the angular and energy differential cross sections, $d\sigma/d\Omega$ and $d\sigma/dE'$, respectively (no need to define the latter, assume they are known)?

Chapter 8 Nuclear Reaction Fundamentals

Problem 1

Interpret Fig. 8.4, in particular the meaning of the family of curves (curves for ϑ above and below 90°). Explain the significance of $(E_1)_{90°}$.

Problem 2

Derive Eq. (8.15) from Eq. (8.4), then show how Eq. (8.17) is obtained.

Problem 3

Discuss what parts of Fig. 8.5 can be related to the Q-equation, Eq. (8.4). What features of the cross sections in Fig. 8.5 can be related to any of the cross sections shown in Figs. 3.6 through 3.11?

Problem 4

Give a comparative description of Eqs. (7.40) and (8.56). Explain the origin of these two results and discuss their similarities and differences.

Problem **5**

Discuss in what sense one may regard the optical model of nuclear reactions to be a theory.

Problem **6**

Summarize what the results of Table 8.1 tell us about the optical model approach to cross section calculations. What improvements do you expect from further refinements?

Problem **7**

On the basis of the comparison between predictions and measurements shown in Fig. 8.6, what general conclusions can you draw?

Problem **8**

As a follow up to Problem 6, what are the implications of Fig. 8.9 regarding the effects of the nonlocal optical model formulation?

Problem **9**

Explain briefly the basic idea of the R-matrix formalism. Interpret Eq. (8.79) in terms of parameters that have physical meanings of resonance energies and line widths.

Problem **10**

Explain Eq. (8.84). In your discussion consider the answers one would give to Problem 4.

Problem **11**

Discuss the similarities and differences between Eqs. (7.40) and (8.84).

Chapter 9 Neutron Scattering

Problem* **1

(a) Calculate the thermal neutron scattering cross section of hydrogen using the phase shift method and the approximation that the scattering potential is an impenetrable sphere of radius R, where R is the nuclear radius of the deuteron (estimate by whatever reasonable means you like). Solve the s-wave radial wave equation to determine the phase shift, the corresponding angular differential cross section and the total cross sectionσ_o. Evaluate the numerical value of σ_o.

(b) Compare you numerical value with the value obtained in this chapter assuming a square-well potential, and with the experimental value. Explain the implications of any agreement or discrepancy in this comparison.

(c) Can you imagine a way to modify your calculation (still assuming the impenetrable sphere model for the interaction) to treat the neutron-proton scattering as spin dependent? Comment on whether the concept of the scattering length is still applicable.

*** *Problem* 2**

Consider the preceding problem but now take the interaction potential to be a spherical barrier of height V_o and range $r_o = R$ (the deuteron nuclear radius), where $V_o \gg$ thermal energy.

(a) Solve the s-wave radial wave equation to obtain the dispersion relation that involves the wavenumbers K, k and the phase shift $\delta_o(k)$, with $\hbar^2 K^2/2m = V_o$ and $\hbar^2 k^2/2m = E$. Denote this relation as Eq. $(*)$.

(b) Define the scattering length a and find a from Eq. $(*)$ [Hint: you can take $\delta_o(k)$ to be small].

(c) Give a sketch of σ_o as a function of Kr_o. Discuss briefly and comment on whether your results in Problems 1 and 2 are consistent?

***Problem* 3**

Consider neutron elastic scattering where the target nucleus can be taken to be at rest.

(a) Derive the Q-equation, Eq. (8.4), and show that Eq. (9.16) is a solution.

(b) Derive Eqs. (9.23) and (9.24) using Fig. 9.2(c).

***Problem* 4**

(a) Show that the conservation of kinetic energy and linear momentum during an elastic collision requires that in the CMCS the speed of each particle is the same before and after the collision.

(b) Does the relative speed of particle 1 with respect to particle 2 change during an elastic collision, (i) in the CMCS, and (ii) in LCS?

***Problem* 5**

Consider the problem of neutron elastic scattering in the notation of this chapter.

(a) Find an expression for E_4 involving E_1 and angle γ (the angle between the recoil direction of M_4 and the x-axis). Is there a one-to-one relation between E_4 and γ?
(b) Eliminate γ in favor of θ_c in your result in (a), then use the fact that the scattering is isotropic in CMCS to find the distribution $P(E_4)$, the probability per unit energy that the recoil nucleus will have energy E_4.
(c) How would you obtain $P(E_4)$ directly from the result for $F(E \to E')$ derived in Eq. (9.31). How general is your relation between P and F?

***Problem* 6**

Derive the relation between the angular differential cross sections for neutron elastic scattering in LCS and CMCS, $\sigma(\theta)$ and $\sigma(\theta_c)$, then use this relation to discuss the behavior of $\bar{\mu}$ and $\bar{\mu}_c$, where $\mu = \cos\theta$.

***Problem* 7**

Answer each part separately and concisely.

(a) In solving $H\psi = E\psi$ as an effective one-body problem (after making the transformation from a two-body problem), how are the mass and momentum operator in H, and the energy E related to the mass, velocity (or momentum) of the two interacting particles?
(b) Give a physical argument showing that for low energy scattering, $kr_o \ll 1$, only the s-wave contribution is important.
(c) Understanding the neutron scattering cross section involves the concept of singlet and triplet scattering lengths. Explain where these scattering lengths come from, give their values, and explain the meaning of the sign of each length.
(1) Discuss briefly the significance of the neutron absorption cross sections of H_2O and D_2O in the historical development of nuclear power.

***Problem 8**

The solution to the Q-equation is given in Eqs. (8.5)–(8.7) $\sqrt{E_3} = s \pm \sqrt{s^2 + t}$, where

$$s = \frac{\sqrt{M_1 M_3 E_1}\cos\theta}{M_3 + M_4} \quad \text{and} \quad t = \frac{M_4 Q + (M_4 - M_1)E_1}{M_3 + M_4}$$

For an endothermic reaction with $M_4 > M_1$, find the threshold energy E_{th}.

***Problem 9**

Consider the reaction

$$a + b \rightarrow c + d$$

with particle b stationary and Q arbirary.

(a) Calculate the velocities of the CM before and after the collision. What is the effect of Q on the relation between them?
(b) Calculate the total momentum in CMCS after the collision. Does this result depend on the value of Q? Does it depend on particle b being stationary?

***Problem 10**

In neutron elastic scattering by hydrogen where the target nucleus is assumed to be at rest, the ratio of final to initial neutron energy is $E'/E = (1/2)(1+\cos\theta_c)$, where θ_c is the scattering angle in CMCS. Suppose you are told the angular distribution of the scattered neutrons is proportional to $\cos\theta_c$ for $0 \le \theta_c \le \pi/2$ and is zero for all other values of θ_c. Find the corresponding energy distribution $F(E \rightarrow E')$. Sketch your result and discuss how it is different from the case of isotropic angular distribution.

***Problem 11**

(a) Angular differential scattering cross sections (in CMCS, *unit is millibarn per steradian*) for C^{12} and U^{238} for two incident neutron energies, 0.5 MeV and 14 MeV, are shown in the sketch (adapted from Lamarsh, 1966). Interpret (explain) all the features, including the difference between *carbon* and *uranium*.

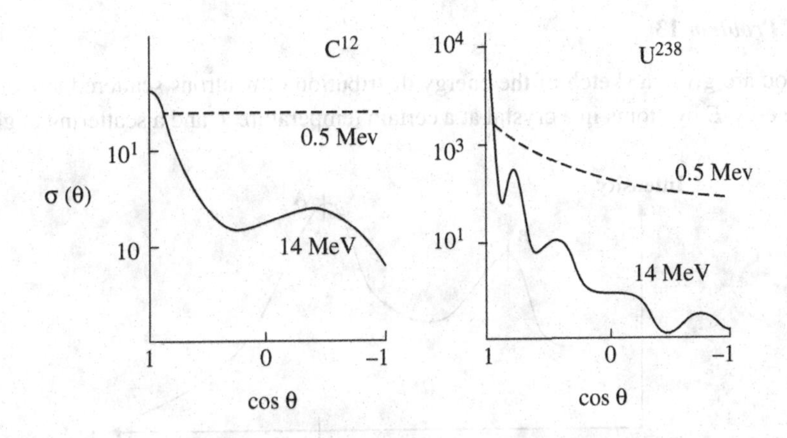

(b) Write down the relation between the neutron scattering cross section $\sigma_{meas}(v)$ measured in the laboratory (which depends on the temperature T of the target) and the theoretical cross section for s-wave scattering at low energy calculated using the method of phase shift. Define all the quantities. Sketch the behavior of $\sigma_{meas}(v)$ in the energy range from thermal up to the energy where one sees free-atom scattering. Briefly discuss the significance of your sketch.

Problem 12

In neutron radiation damage studies an incident neutron with energy E collides with an atom in the crystal lattice, called the primary knock-on atom (*PKA*), causing it to recoil and displace a number of other atoms in the solid.

(a) Calculate the average recoil energy of the *PKA* (mass A). State all assumptions involved in your calculation.
(b) Let the number of atoms displaced by a *PKA* with energy T be denoted by $\nu(T)$ which behaves as follows

$$\begin{aligned} \nu(T) &= 0 && \text{for } 0 < T < E_d \\ &= T/2E_d && E_d < T < L_c \end{aligned}$$

where E_d is the energy to displace an atom, and L_c is the energy above which all the energy loss is by electronic excitation. Find the average number of atoms displaced as a function of E in the range, $0 < E < L_c$. Call this quantity N_E. (*Note*: Neither ν nor N_E are distributions.)

***Problem* 13**

You are given a sketch of the energy distribution of neutrons scattered at thermal energy E by atoms in a crystal at a certain temperature T and a scattering angle θ.

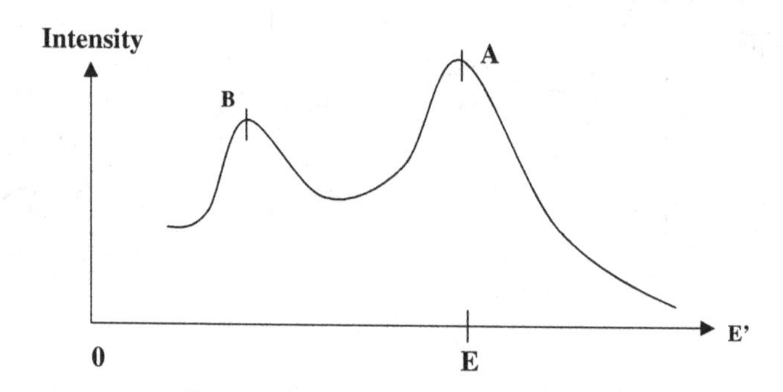

(a) What is the underlying process giving rise to the peak marked A? Peak B? For each peak, how do you expect the intensity to vary with T and with θ (explain briefly)?

(b) Suppose now you are told that the energy distribution of photons scattered at arbitrary energy E by free electrons at a fixed scattering angle is also given qualitatively by the sketch. What processes are responsible for the two peaks in this case? What do you expect when the photon E is large or small compared to the electron rest mass energy? Would you expect the position of peak B to vary with the scattering angle? Explain.

Chapter 10 Gamma Attenuation

***Problem* 1**

Consider Compton scattering of a photon. Derive the following four expressions: (i) the Compton shift, (ii) energy of the scattered photon, (iii) energy of the recoiling electron, and (iv) the relation between the angles of the scattered photon and the recoiling electron.

***Problem* 2**

Define concisely what is Compton scattering. Derive the relation between incident gamma energy $\hbar\omega$ and scattered gamma energy $\hbar\omega'$ for Compton scattering which also involves the scattering angle θ. What is the similarity (and difference) between this relation and the corresponding relation involving incident and scattered energies in neutron elastic scattering?

Problem 3

Calculate the energy distribution of Compton electrons for an incident photon of energy 1.20 MeV. Give your results as the energy differential cross section, $d\sigma/dT$, in units of 10^{-25} cm^2/MeV, at electron energies 0.25, 0.5, 0.75, 0.9 and 0.99 MeV. Make a sketch of $d\sigma/dT$ and compare with the corresponding results in Fig. 10.5.

*Problem 4

(a) State what is Thomson scattering. Sketch the Klein-Nishina cross section $d\sigma_C/d\Omega$ in general, then point out the characteristic features of Thomson scattering.

(b) Show how one can use $d\sigma_C/d\Omega$ to find the energy distribution of the Compton electron.

*Problem 5

(1) You are given the Klein-Nishina cross section for Compton scattering.

$$dc/d\Omega = \frac{r_e^2}{2}\left(\frac{\omega'}{\omega}\right)\left[\frac{\omega}{\omega'} + \frac{\omega'}{\omega} - 2\sin^2\theta\cos^2\phi\right]$$

Find the cross section for Thompson scattering for unpolarized radiation. Sketch qualitatively your result and comment on any interesting feature.

(b) Indicate (do not carry out any complicated manipulations) how you would find the energy distribution of the Thompson scattered electrons using the result from (a).

(c) Sketch the energy distributions for Compton scattering at two energies, say 0.5 MeV and 2.5 MeV, and comment on all the characteristic features. Do you expect the Compton and Thompson electron distributions to be similar or different (give some explanation)?

*Problem 6

When a beam of photons with wavelength 0.7 A is scattered by carbon at various scattering angles, the observed spectra behave in a way as sketched below schematically. What is the process responsible for the peak at 0.7 A? What is the process responsible for the other peak in the spectrum? Comment on the variation with scattering angle of the two peaks.

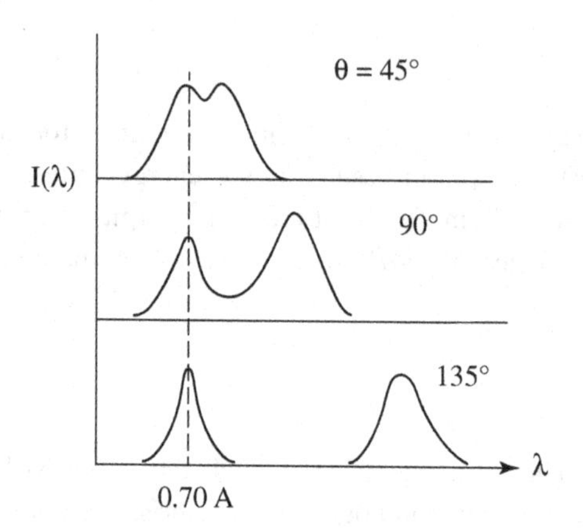

Problem 7

A sodium iodide detector in the shape of a 7 cm cube is bombarded by a beam of 2.8 Mev gamma radiation normal to one face of the cube.

(a) What fraction of the gamma radiation is detected?
(b) What fraction of the detected gamma appears in the photo peak, the Compton distribution, and the pair peaks, assuming no re-absorption of Compton gamma or annihilation quanta?
(c) Make a rough estimate of the relative fraction of pair events that appear in the full-energy (photo) peak, in the one-escape peak, and in the two-escape peak. Compare your result with Fig. 10.18.
(*Note*: The attenuation coefficients that you will need can be found in Evans, Chap. 25, Sec. 1. For 0.51 Mev photons, $\mu = 0.33\,\text{cm}^{-1}$. For 2.8 Mev photons, $\mu = 0.135\,\text{cm}^{-1}$, $\mu_\tau = 2.5 \times 10^{-3}\,\text{cm}^{-1}$, $\mu_C = 0.113\,\text{cm}^{-1}$, $\mu_\kappa = 0.020\,\text{cm}^{-1}$.)

****Problem 8***

Consider the measurement of monoenergetic gammas (energy $\hbar\omega$) in a scintillation detector whose size is small compared to the mean free path of the secondary gammas produced by interactions of the incident (primary) gammas in the detector.

(a) Sketch the pulse-height spectrum of low-energy gammas, say $\hbar\omega < 500$ kev. Explain briefly the important characteristics of this spectrum in terms of the different interactions that can take place.
(b) Repeat (a) for higher-energy gammas, $\hbar\omega > 2$ Mev.
(c) What other peaks can appear in the pulse-height spectrum if the detector were not small? Give a sketch and explain briefly.

**Problem 9*

Sketch the peaks that one would observe in the pulse-height spectra of a large detector with good energy resolution in the presence of a 2-Mev gamma ray source, including any radiation from the background. For each peak identify the radiation interaction process that gives rise to it and indicate the energy at which this peak would appear.

Chapter 11 Charged Particles Stopping

Problem 1

Give a derivation of the Rutherford scattering cross section

$$\frac{d\sigma}{d\Omega} = \frac{1}{4}\left(\frac{zZe^2}{mv^2}\right)\frac{1}{\sin^4(\theta/2)}$$

Define all the quantities. State the physical meaning of this formula and its significance regarding charged particle interactions with matter. (Note: You are expected to consult a reference book for this problem. You should cite your reference and give enough details in your derivation to demonstrate that you have worked through all the steps.)

Problem 2

The range-energy relationship for protons in air at 1 atm. and 15°C is given in Fig. 11.3. From this curve deduce the energy loss (in MeV cm^2/gm) curve for the same energy range. Compare your result with Fig. 11.1 and with your own calculations using the Bethe formula.

Problem 3

Estimate the contributions to the stopping power due to ionization and to radiation for the passage of electrons with energy E in aluminum. Consider $E = 0.1, 0.5, 1, 2,$ and 4 MeV. Express your results in both MeV/mm and MeV cm^2/gm, and compare your values with those given in Fig. 11.7. Discuss the significance of your results and the comparison.

Problem 4

Show that alpha particles and protons having the same initial speed (high speed but nonrelativistic) have approximately the same range in any stopping material. Examine Fig. 11.5 to see which particle should have a slightly longer range. Justify your answer.

**Problem* 5

In analyzing the energy loss per unit path length of a charged particle (ze) moving with speed v, an atomic electron in the medium located at an impact parameter b will gain an amount of kinetic energy equal to

$$T = \frac{2(ze)^2}{mv^2b^2}$$

from collision with the charged particle, where m is the electron mass. Suppose you are told to ignore the binding energy of the atomic electrons, so that all the electron in a collision cylinder of radius b, thickness db, and length Δx are ejected. Find the number of electrons per unit path length ejected with kinetic energy in dT about T (Hint: Think of this number as a distribution.)

**Problem* 6

Give a short and concise answer to each of the following questions.

(a) Derive using a sketch the asymmetry term in the empirical mass formula.
(b) Explain what is β^+-decay, then state and justify the energy condition for this process.
(c) Sketch schematically the curve for the stopping power of a heavy charged particle in a high-Z medium in the energy range, zero to three times its rest mass energy. Label all the characteristic energies that you know, and explain what physical processes are represented in the curve.
(d) Show that the range of an α-particle and a proton, both having the same initial speed, will be approximately the same.
(e) Is *bremsstrahlung* an elastic or inelastic process (explain)? Why is this process more important for electron than for proton in problems of interest in this book?

***Problem* 7**

Explain briefly under what conditions energy loss by radiation is important. Give a sketch of the mass absorption coefficients for electrons in the range (0, 5 MeV), comparing two absorbers, Pb and Al. Justify the relative magnitudes you have sketched.

***Problem* 8**

Sketch the energy variations of the stopping power (energy loss per unit path) of both electrons and protons in lead (in the same figure). Discuss all the features of these two curves that you know.

***Problem* 9**

On the basis of the Bethe-Bloch formula, the stopping power of a material for incident electrons (with kinetic energy T_e) can be related to that for incident alpha particles (with kinetic energy T_α). Denoting the two by $-(dT/dx)_e$ and $-(dT/dx)_\alpha$ respectively, sketch the two curves on the same graph to show how knowing one allows you to find the other.

****Problem* 10**

Discuss briefly the significance of each of the following. Give a definition whenever it is appropriate. (If you use the same notations as in this book, you may assume the symbols are already defined.)

(a) The asymmetry term in the empirical mass formula.
(b) Mass parabolas for isobars for even A (give a sketch).
(c) Mass or energy requirements for electron capture.
(d) Secular equilibrium in radioactive decay.
(e) Bragg curve for charged particles (give sketch).
(f) Bethe formula for stopping power and its relativistic corrections.
(g) Charge and mass dependence of *bremsstrahlung* intensity.
(h) Mass absorption coefficient for charged particles.
(f) Sketch schematically the curve for the stopping power of a heavy charged particle in a high-Z medium in the energy range zero to three times its rest mass energy. Label all the characteristic energies that you know, and explain what physical processes are represented in the curve.
(g) Show that the range of an α-particle and a proton, both having the same initial speed, will be approximately the same.

(h) Is *bremsstrahlung* an elastic or inelastic process (explain)? Why is this process more important for electron than for proton in problems of interest to the class.

***Problem* 11**

(a) Sketch the stopping power of electrons in a high Z absorber such as lead over an energy range from zero to 10 times the rest mass energy. Explain all characteristic features as to their physical origin and give simple formulas describing that portion of the curve where possible.
(b) You are given the range-energy relation as a function $R(T)$, where R is the range and T is the energy of the charged particle. How would you deduce from this the stopping power $-(dT/dx)$?

***Problem* 12**

(a) Explain the idea of expressing energy losses in terms of mass absorption, $-dT/dw$.
(b) Sketch on one graph the mass absorption energy losses for electrons in Al and Pb showing ionization and radiation losses in the energy range 0–5 MeV. Explain why the curves for Al and Pb are actually somewhat different.

Chapter 12 Neutron Reactions

***Problem* 1**

Explain in your own words the significance of Fig. 12.1. Draw the corresponding diagram for an exothermic reaction. Compare the two reactions (diagrams) with regard to the essential features of energy-level diagrams for a general reaction considered in Fig. 8.1, and described by Eq. (8.4).

***Problem* 2**

Discuss the connection between the concept of compound nucleus formation and decay as depicted in Fig. 12.2 and the two neutron resonance cross sections expressed in Eqs. (12.14) and (12.15) by considering all the terms in the two expressions.

***Problem* 3**

Sketch the energy variation of an observed resonance in (a) neutron elastic scattering (resonance scattering in the presence of potential scattering), and (b) neutron inelastic scattering. Comment on the characteristic features in the cross sections, especially the low-energy behavior below the resonance. What is the connection between the energy at which the observed cross sections show a peak and the energy of the nuclear level associated with the resonance? (You may assume it is the same level in both cases.)

***Problem* 4**

Show in what sense can one say the neutron resonant cross section for radiative capture, Eq. (12.14), can account for the characteristic features of Fig. 12.4.

***Problem* 5**

Discuss any connection between the neutron elastic scattering as described by Eq. (12.15) and the cross section behavior shown in Figs. 8.5 and 9.5. Recall Eq. (9.43) and discuss what connection it may have with the quantities just mentioned.

***Problem* 6**

Consider the compound nucleus reaction of inelastic scattering of neutrons at energy T_1 (LCS) by a nucleus ${}^{A}_{Z}X$.

(a) Draw the energy level diagram showing the different energies that one can use to describe this reaction (including the Q value).
(b) Write down the corresponding Breit-Wigner cross section in terms of some of the energies shown in (a). Define all the parameters appearing in your expression.

***Problem* 7**

Consider the reaction $a + b \rightarrow c + d$, where Q is nonzero and particle b is stationary. What can you say about the magnitude and direction of the velocity of the center-of-mass before and after the reaction?

***Problem* 8**

Consider neutron inelastic scattering at incident energy T_1 (LCS) where the reaction can be written as

$$n + {}_{Z}X^{A} \rightarrow ({}_{Z}X^{A+1}) \rightarrow n' + ({}_{Z}X^{A*})$$

Is the Q-value positive or negative? Draw the energy-level diagram depicting the kinetic and rest-mass energies involved in the reaction, and show how the kinetic energies are related to T_1 and any other energies in the problem. You may assume target nucleus is at rest.

***Problem* 9**

Explain what one is trying to depict in Fig. 12.9. Compare this figure with Fig. 6.11 and discuss similarities and differences.

***Problem* 10**

Explain Eq. (12.16) along with the results given in Table 12.2. Discuss whether or not this formulation is applicable to thermal fission in U^{235} and U^{238}.

***Problem* 11**

Describe the behavior of the neutron cross sections shown in Fig. 12.10. Compare this figure with Fig. 3.11 (for U^{238}), and discuss the differences in neutron reactions between these two uranium isotopes. Give some qualitative considerations to how one can explain the differences in the fission cross sections.

***Problem* 12**

The evolution of the fission event is depicted in Fig. 12.11. Describe in your own words how the event occurs in space and time. Based on this information, what can you conclude about the spatial distribution of the heat generation in a nuclear reactor? Do you know of any special considerations regarding heat generation in time?

Chapter 13 Neutron Transport

***Problem* 1**

Write down the Neutron Transport Equation, Eq. (13.5), defining all the terms in the equation. Derive the streaming term in the Transport equation. Explain the physical meaning of each term as a contribution to a "balance relation".

***Problem* 2**

Looking at Eq. (13.5) as a balance relation, one can interpret each term as a contribution associated with either streaming (spatial) or collision (energy) effects. Notice these terms enter in an additive fashion (not multiplied together). Discuss what this means in terms of treating spatial distribution effects separately from energy distribution effects mathematically and physically. Briefly indicate whether you expect this to be a serious limitation of the neutron transport equation.

***Problem* 3**

The neutron transport equation is too complicated for us to discuss its spatial and energy behavior together. The equation is much more tractable when it is reduced to an equation only in space or in energy. The former reduction is called the diffusion approximation (Sec. 13.2), whereas the latter leads to two problems, one is known as the problem of neutron slowing down (Sec. 13.3) and the other is the problem of neutron thermalization (Sec. 13.4). As a follow up to the previous problem, discuss which of the terms in the balance relation are involved in spatial and energy reductions. What assumptions, if any, are involved in each case?

***Problem* 2**

Explain what is Fick's rule of diffusion and show how it is used in reducing the neutron transport equation to the neutron diffusion equation. What assumption is being made. Physically under what conditions do you expect this assumption to start to breakdown?

***Problem* 4**

Derive the escape probability $P(E' \to E)$. Discuss how you would calculate this quantity using a Monte Carlo code such as MCNP. Discuss the conditions under which you would or would not expect the MCNP calculation to agree with a numerical evaluation of your analytical expression.

***Problem* 5**

Consider one-dimensional neutron transport in a purely scattering medium where the cross section $\sigma(x)$ is a known function of position. Apply the same kind of argument used for Problem 4 to derive an expression for the probability $P(x' \to x)$ that a neutron will go the distance from x' to x without scattering, where $x > x'$ but otherwise both are arbitrary. Discuss your result as compared to $P(E' \to E)$.

***Problem* 6**

Define the variable lethargy u and show that the average increase in lethargy per collision is given by

$$\xi = 1 + \frac{\alpha \ell n \alpha}{1-\alpha} \sim 2/A \quad \text{for } A \gg 1$$

where $\alpha = \lfloor (A-1)^2/(A+1)^2 \rfloor$. Discuss how ξ can be used to estimate the collision density in a medium where $A > 1$.

***Problem* 7**

Derive Eq. (13.40) and explain in what way is this result useful.

***Problem* 8**

By following the derivation of Eq. (13.40) obtain the corresponding expression for $J_+(x_o)$. Using the results for $J_\pm(x_o)$ discuss under what condition is Fick's rule valid.

***Problem* 9**

Derive the diffusion kernels for a point source and a plane source, and verify your results are consistent with Eqs. (13.52) and (13.53). Explain why the point source kernel is called a Green's function. (In other words, what is the property of a *Green's function*?)

***Problem* 10**

Explain the concept of extrapolated boundary in solving neutron diffusion problems. For what situations is this boundary condition valid? When is it not valid?

***Problem* 11**

Explain what is the physical meaning of the geometric buckling in solving the neutron diffusion equation for a critical spherical reactor, and derive an expression for it.

***Problem* 12**

Show how the neutron multiplication constant k can be written in the form of Eq. (13.120) and define all the quantities that appear in this expression in terms of cross sections and any other similar basic nuclear data. Give typical values for each of the four components of the infinite medium multiplication constant. Which do you think is larger between the fast and thermal nonleakage probabilities?

***Problem* 13**

Derive the Born approximation in neutron scattering, starting with the integral equation approach to potential scattering.

***Problem* 14**

Investigate the validity of the Born approximation for thermal neutron scattering to show whether or not the approximation is valid.

***Problem* 15**

Give a detailed discussion of what is the Fermi psuedopotential. Why is it useful in thermal neutron scattering?

***Problem* 16**

Show the double differential neutron scattering cross section can be expressed in terms of the dynamic structure factor, $S(Q, \omega)$. Discuss briefly why this quantity is useful in the study of structure and dynamics of physical systems.

***Problem* 17**

Analyze Eq. (13.137) to show how one can arrive at Eq. (13.142). What is the point of this analysis? How can one extract a "leakage cross section" from Eq. (13.148)?

***Problem* 18**

Construct a flow chart that shows the sequence of events leading to the expression of the multuiplication constant k in the form of Eq. (13.120).

***Problem* 19**

Give a simple derivation of the infinite medium multiplication constant k_{∞}. Discuss typical values for each of the factors. You may use the information shown in Fig. 13.7 (be sure to state fully what kind of reactor system you are considering).

Problem **20**

Explain the significance of Fig. 13.7 in terms of the four-factor formula for the infinite-medium multiplication constant, Eq. (13.121). Discuss the physical meaning of each of the factors and interpret its variation with fuel/moderator ratio.

Problem **21**

Discuss the importance of delayed neutrons in the understanding of reactor kinetics. Using Eqs. (13.126) and (13.129) explain physically the difference between prompt and delayed critical, and their manifestations in practice.

Problem **22**

What is the significance of Eq. (13.145) which states the equality of geometric and material bucklings? Generally speaking, does this condition have anything to do with the interface boundary condition that we have previously applied in solving the Schrödinger wave equation in Chaps. 5, 7 and 8? Be as specific as you can. Discuss why you think there should be a connection between calculations of bound and scattering states and solving the neutron diffusion equation.

Problem **23**

Discuss the special behavior of the thermal flux in the reflector region, ϕ_4, seen in Fig. 13.8. Explain physically the reason why it can show a peak in contrast to the other three fluxes.

Index